LONDON MATHEMATICAL SOCIETY LECTURE NOTE SERIES

Managing Editor: Professor N. J. Hitchin, Mathematical Institute,
University of Oxford, 24-29 St Giles, Oxford OX1 3LB, United Kingdom

The titles below are available from booksellers, or from Cambridge University Press at www.cambridge.org/mathematics

London Mathematical Society Lecture Note Series. 338

Surveys in Geometry and Number Theory: Reports on contemporary Russian mathematics

Edited by

NICHOLAS YOUNG
University of Leeds

CAMBRIDGE
UNIVERSITY PRESS

CAMBRIDGE
UNIVERSITY PRESS

University Printing House, Cambridge CB2 8BS, United Kingdom

One Liberty Plaza, 20th Floor, New York, NY 10006, USA

477 Williamstown Road, Port Melbourne, VIC 3207, Australia

314-321, 3rd Floor, Plot 3, Splendor Forum, Jasola District Centre, New Delhi - 110025, India

103 Penang Road, #05-06/07, Visioncrest Commercial, Singapore 238467

Cambridge University Press is part of the University of Cambridge.

It furthers the University's mission by disseminating knowledge in the pursuit of education, learning and research at the highest international levels of excellence.

www.cambridge.org
Information on this title: www.cambridge.org/9780521691826

© Cambridge University Press 2007

First published 2007

A catalogue record for this publication is available from the British Library

ISBN 978-0-521-69182-6 Paperback

Contents

Preface

This volume is a showcase for the continuing vitality of Russian mathematics in fields related to algebraic geometry. The Eastern European scientific diaspora may have weakened the magnificent edifice of Russian mathematics, but the Russian school had both strength and depth, and there remains a great deal of important scientific activity in the country. Universities continue to attract some of the most able students into mathematics, and their graduates have the knowledge and enthusiasm to be effective participants in the global mathematical endeavour.

There are great difficulties facing new Russian 'Candidates of Science' in mathematics. It is rare for a young person to gain a living wage as a lecturer and researcher alone. It requires dedication, self-sacrifice and a willingness to look for other sources of income for a mathematician to become established while remaining in Russia. It is not surprising that many talented mathematicians seek and find employment abroad. Despite these handicaps there are strong research groups that continue to foster new talents.

In the fields of algebraic geometry and algebraic number theory there are healthy groups, particularly those centred around the Steklov Institute in Moscow. To give some examples, in the birational geometry of 3-folds there is a group of four well-established experts who support about 10 research students and postdoctoral fellows. Another group of specialists pioneered the idea of the derived category of coherent sheaves on a variety (up to equivalence) as a geometric invariant of the variety, analogous to K-theory or cohomology theories, and continues to work in this fruitful area. Another strong strand of research is in algebraic and complex versions of quantisation relating to special geometries such as special Lagrangian fibrations in mirror symmetry.

The London Mathematical Society set up the 'Young Russian Mathematicians' scheme to help these mathematicians to visit the UK and to provide them with some financial support. Visitors give lectures in this country and write a survey article on the work of their research groups, for which they receive payment. This is the first volume of such survey articles to be published by the Society.

Nicholas Young
Department of Pure Mathematics
Leeds University.

Affine embeddings of homogeneous spaces

Ivan V. Arzhantsev

Introduction

Throughout the paper G denotes a connected reductive algebraic group, unless otherwise specified, and H an algebraic subgroup of G. All groups and algebraic varieties considered are over an algebraically closed field $\mathbb{K}$ of characteristic zero, unless otherwise specified. Let $\mathbb{K}[X]$ be the algebra of regular functions on an algebraic variety X and $\mathbb{K}(X)$ the field of rational functions on X provided X is irreducible. Our general references are [30] for algebraic groups and [56, 37, 29] for algebraic transformation groups and invariant theory.

Affine embeddings: definitions. Let us recall that an irreducible algebraic G-variety X is said to be an *embedding* of the homogeneous space G/H if X contains an open G-orbit isomorphic to G/H. We shall denote this relationship by $G/H \hookrightarrow X$. Let us say that an embedding $G/H \hookrightarrow X$ is *affine* if the variety X is affine. In many problems of invariant theory, representation theory and other branches of mathematics, only affine embeddings of homogeneous spaces arise. This is why it is reasonable to study specific properties of affine embeddings in the framework of a well-developed general embedding theory.

Which homogeneous spaces admit an affine embedding? It is easy to show that a homogeneous space G/H admits an affine embedding if and only if G/H is quasi-affine (as an algebraic variety). In this situation, the subgroup H is said to be *observable* in G. A closed subgroup H of G is observable if and only if there exist a rational finite-dimensional G-module V and a vector $v \in V$ such that the stabilizer G_v coincides with H. (This follows from the fact that any affine G-variety may be realized as a closed invariant subvariety in a finite-dimensional G-module [56, Th.1.5].) There is a nice group-theoretic description of

1

observable subgroups due to A. Sukhanov: a subgroup H is observable in G if and only if there exists a quasi-parabolic subgroup $Q \subset G$ such that $H \subset Q$ and the unipotent radical H^u is contained in the unipotent radical Q^u, see [63], [29, Th.7.3]. (Let us recall that a subgroup Q is said to be *quasi-parabolic* if Q is the stabilizer of a highest weight vector in some G-module V.)

It follows from Chevalley's theorem that any subgroup H without non-trivial characters (in particular, any unipotent subgroup) is observable. By Matsushima's criterion, a homogeneous space G/H is affine if and only if H is reductive. (For a simple proof, see [42] or [4]; a characteristic-free proof can be found in [57].) In particular, any reductive subgroup is observable. A description of affine homogeneous spaces G/H for non-reductive G is still an open problem.

Complexity of reductive group actions. Now we define the notion of complexity, which we shall encounter many times in the text. Let us fix the notation. By $B = TU$ denote a Borel subgroup of G with a maximal torus T and the unipotent radical U. By definition, the *complexity* $c(X)$ of a G-variety X is the codimension of a B-orbit of general position in X for the restricted action $B : X$. This notion firstly appeared in [45] and [70]. Now it plays a central role in embedding theory. By Rosenlicht's theorem, $c(X)$ is equal to the transcendence degree of the field $\mathbb{K}(X)^B$ of rational B-invariant functions on X. A normal G-variety X is called *spherical* if $c(X) = 0$ or, equivalently, $\mathbb{K}(X)^B = \mathbb{K}$. A homogeneous space G/H and a subgroup $H \subseteq G$ are said to be *spherical* if G/H is a spherical G-variety.

Rational representations, the isotypic decomposition and G-algebras. A linear action of G in vector space W is said to be *rational* if for any vector $w \in W$ the linear span $\langle Gw \rangle$ is finite-dimensional and the action $G : \langle Gw \rangle$ defines a representation of an algebraic group. Since any finite-dimensional representation of G is completely reducible, it is easy to prove that W is a direct sum of finite-dimensional simple G-modules.

Let $\Xi_+(G)$ be the semigroup of dominant weights of G. For any $\lambda \in \Xi_+(G)$, denote by W_λ the sum of all simple submodules in W of highest weight λ. The subspace W_λ is called an *isotypic component* of W of weight λ, and the decomposition

$$W = \oplus_{\lambda \in \Xi_+(G)} W_\lambda$$

is called *the isotypic decomposition* of W.

If G acts on an affine variety X, the linear action $G : \mathbb{K}[X]$, $(gf)(x) := f(g^{-1}x)$, is rational [56, Lemma 1.4]. (Note that for irreducible X the action on rational functions $G : \mathbb{K}(X)$ defined by the same formula is not rational.) The isotypic decomposition

$$\mathbb{K}[X] = \oplus_{\lambda \in \Xi_+(G)} \mathbb{K}[X]_\lambda$$

and its interaction with the multiplicative structure on $\mathbb{K}[X]$ give important technical tools for the study of affine embeddings.

An affine G-variety X is spherical if and only if $\mathbb{K}[X]_\lambda$ is either zero or a simple G-module for any $\lambda \in \Xi_+(G)$ [32].

Suppose that $\mathfrak{A}$ is a commutative associative algebra with unit over $\mathbb{K}$. If G acts on $\mathfrak{A}$ by automorphisms and the action $G : \mathfrak{A}$ is rational, we say that $\mathfrak{A}$ is a G-*algebra*. The algebra $\mathbb{K}[X]$ is a G-algebra for any affine G-variety X. Moreover, any finitely generated G-algebra without nilpotents arises in this way.

We conclude the introduction with a review of the contents of this survey.

One of the pioneering works in embedding theory was a classification of normal affine $SL(2)$-embeddings due to V. L. Popov, see [52, 37]. In the same period (early seventies) the theory of toric varieties was developed. A toric variety may be considered as an equivariant embedding of an algebraic torus T. Such embeddings are described in terms of convex fans. Any cone in the fan of a toric variety X represents an affine toric variety. This reflects the fact that X has a covering by T-invariant affine charts. In 1972, V. L. Popov and E. B. Vinberg [55] described affine embeddings of quasi-affine homogeneous spaces G/H, where H contains a maximal unipotent subgroup of G. In Section 1 we discuss briefly these results together with a more recent one: a remarkable classification of algebraic monoids with a reductive group G as the group of invertible elements (E. B. Vinberg [71]). This is precisely the classification of affine embeddings of the space $(G \times G)/\Delta(G)$, where $\Delta(G)$ is the diagonal subgroup.

In Section 2 we consider connections of the theory of affine embeddings with Hilbert's 14th problem. Let H be an observable subgroup of G. By the Grosshans theorem, the following conditions are equivalent:

1) the algebra of invariants $\mathbb{K}[V]^H$ is finitely generated for any G-module V;

2) the algebra of regular functions $\mathbb{K}[G/H]$ is finitely generated;

3) there exists a (normal) affine embedding $G/H \hookrightarrow X$ such that

$$\mathrm{codim}_X(X \setminus (G/H)) \geq 2$$

(such an embedding is called *the canonical embedding* of G/H).

It was proved by F. Knop that if $c(G/H) \leq 1$ then the algebra $\mathbb{K}[G/H]$ is finitely generated. This result provides a large class of subgroups with a positive solution of Hilbert's 14th problem. In particular, Knop's theorem together with Grosshans' theorem on the unipotent radical P^u of a parabolic subgroup $P \subset G$ includes almost all known results on Popov-Pommerening's conjecture (see 2.2). We study the canonical embedding of G/P^u from a geometric view-point. Finally, we mention counterexamples to Hilbert's 14th problem due to M. Nagata, P. Roberts, and R. Steinberg.

In Section 3 we introduce the notion of *an affinely closed space*, i.e. an affine homogeneous space admitting no non-trivial affine embeddings, and discuss the result of D. Luna related to this notion. (We say that an affine embedding $G/H \hookrightarrow X$ is *trivial* if $X = G/H$.) Affinely closed spaces of an arbitrary affine algebraic group are characterized and some elementary properties of affine embeddings are formulated.

Section 4 is devoted to affine embeddings with a finite number of orbits. We give a characterization of affine homogeneous spaces G/H such that any affine embedding of G/H contains a finite number of orbits. More generally, we compute the maximal number of parameters in a continuous family of G-orbits over all affine embeddings of a given affine homogeneous space G/H. The group of equivariant automorphisms of an affine embedding is also studied here.

Some applications of the theory of affine embeddings to functional analysis are given in Section 5. Let $M = K/L$ be a homogeneous space of a connected compact Lie group K, and $C(M)$ the commutative Banach algebra of all complex-valued continuous functions on M. The K-action on $C(M)$ is defined by the formula $(kf)(x) = f(k^{-1}x)$, $k \in K$, $x \in M$. We shall say that A is an *invariant algebra* on M if A is a K-invariant uniformly closed subalgebra with unit in $C(M)$. Denote by G and H the complexifications of K and L respectively. Then G is a reductive algebraic group with reductive subgroup H. There exists a correspondence between finitely generated invariant algebras on M and affine embeddings of G/F with some additional data, where F is an observable subgroup of G containing H. This correspondence was introduced by V. M. Gichev [25], I. A. Latypov [38, 39] and, in a more algebraic way, by E. B.

Vinberg. We give a precise formulation of this correspondence and reformulate some facts on affine embeddings in terms of invariant algebras. Some results of this section are new and not published elsewhere.

The last section is devoted to G-algebras. It is easy to prove that any subalgebra in the polynomial algebra $\mathbb{K}[x]$ is finitely generated. On the other hand, one can construct many non-finitely generated subalgebras in $\mathbb{K}[x_1,\ldots,x_n]$ for $n \geq 2$. More generally, every subalgebra in an associative commutative finitely generated integral domain $\mathfrak{A}$ with unit is finitely generated if and only if $\mathrm{Kdim}\,\mathfrak{A} \leq 1$, where $\mathrm{Kdim}\,\mathfrak{A}$ is the Krull dimension of $\mathfrak{A}$ (Proposition 6.5). In Section 6 we obtain an equivariant version of this result. The problem was motivated by the study of invariant algebras in the previous section. The proof of the main result (Theorem 6.3) is based on a geometric method for constructing a non-finitely generated subalgebra in a finitely generated G-algebra and on properties of affine embeddings obtained above. In particular, the notion of an affinely closed space is crucial for the classification of G-algebras with finitely generated invariant subalgebras. The arguments used in this text are slightly different from the original ones [9]. A characterization of G-algebras with finitely generated invariant subalgebras for non-reductive G is also given in this section.

Acknowledgements. These notes were initiated by my visit to Manchester University in March, 2003. I am grateful to this institution for hospitality, to Alexander Premet for invitation and organization of this visit, and to the London Mathematical Society for financial support. The work was continued during my stay at Institut Fourier (Grenoble) in April-July, 2003. I would like to express my gratitude to this institution and especially to Michel Brion for the invitation, and for numerous remarks and suggestions. Special thanks are due to F. D. Grosshans and D. A. Timashev for useful comments.

1 Remarkable classes of affine embeddings

1.1 Affine toric varieties

We begin with some notation. Let T be an algebraic torus and $\Xi(T)$ the lattice of its characters. A T-action on an affine variety X defines a $\Xi(T)$-grading on the algebra $\mathbb{K}[X] = \oplus_{\chi \in \Xi(T)} \mathbb{K}[X]_\chi$, where $\mathbb{K}[X]_\chi = \{f \mid tf = \chi(t)f$ for any $t \in T\}$. (This grading is just the isotypic decomposition, see the introduction.) If X is irreducible, then the set $L(X) = \{\chi \mid \mathbb{K}[X]_\chi \neq 0\}$ is a submonoid in $\Xi(T)$.

Definition 1.1. An *affine toric variety* X is a normal affine T-variety with an open T-orbit isomorphic to T.

Below we list some basic properties of T-actions:

- An action $T : X$ has an open orbit if and only if $\dim \mathbb{K}[X]_\chi = 1$ for any $\chi \in L(X)$. In this situation $\mathbb{K}[X]$ is T-isomorphic to the semigroup algebra $\mathbb{K}L(X)$.
- An action $T : X$ is effective if and only if the subgroup in $\Xi(T)$ generated by $L(X)$ coincides with $\Xi(T)$.
- Suppose that $T : X$ is an effective action with an open orbit. Then the following conditions are equivalent:

  1) X is normal;
  2) the semigroup algebra $\mathbb{K}L(X)$ is integrally closed;
  3) if $\chi \in \Xi(T)$ and there exists $n \in \mathbb{N}$, $n > 0$, such that $n\chi \in L(X)$, then $\chi \in L(X)$ (the saturation condition);
  4) there exists a solid convex polyhedral cone K in $\Xi(T) \otimes_{\mathbb{Z}} \mathbb{Q}$ such that $L(X) = K \cap \Xi(T)$.

In this situation, any T-invariant radical ideal of $\mathbb{K}[X]$ corresponds to the subsemigroup $L(X) \setminus M$ for a fixed face M of the cone K. This correspondence defines a bijection between T-invariant radical ideals of $\mathbb{K}[X]$ and faces of K.

The proof of these properties can be found, for example, in [23]. Summarizing all the results, we obtain

Theorem 1.2. *Affine toric varieties are in one-to-one correspondence with solid convex polyhedral cones in the space $\Xi(T) \otimes_{\mathbb{Z}} \mathbb{Q}$; and T-orbits on a toric variety are in one-to-one correspondence with faces of the cone.*

The classification of affine toric varieties will serve us as a sample for studying more complicated classes of affine embeddings. Generalizations of a combinatorial description of toric varieties were obtained for spherical varieties [45, 33, 18], and for embeddings of complexity one [68]. In this more general context, the idea that normal G-varieties may be described by some convex cones becomes rigorous through the method of U-invariants developed by D. Luna and Th. Vust. The essence of this method is contained in the following theorem (see [72, 37, 54, 29]).

Theorem 1.3. *Let $\mathfrak{A}$ be a G-algebra and U a maximal unipotent subgroup of G. Consider the following properties of an algebra:*

- *it is finitely generated;*
- *it has no nilpotent elements;*
- *it has no zero divisors;*
- *it is integrally closed.*

If (P) is any of these properties, then the algebra $\mathfrak{A}$ *has property (P) if and only if the algebra* $\mathfrak{A}^U$ *has property (P).*

We try to demonstrate briefly some applications of the method of U-invariants in the following subsections.

1.2 Normal affine $SL(2)$-embeddings

Suppose that the group $SL(2)$ acts on a normal affine variety X and there is a point $x \in X$ such that the stabilizer of x is trivial and the orbit $SL(2)x$ is open in X. We say in this case that X is a *normal $SL(2)$-embedding*.

Let U_m be a finite extension of the standard maximal unipotent subgroup in $SL(2)$:

$$U_m = \left\{ \begin{pmatrix} \epsilon & a \\ 0 & \epsilon^{-1} \end{pmatrix} \mid \epsilon^m = 1,\ a \in \mathbb{K} \right\}.$$

Theorem 1.4 ([52]). *Normal non-trivial $SL(2)$-embeddings are in one-to-one correspondence with rational numbers $h \in (0, 1]$. Furthermore,*

- $h = 1$ *corresponds to a (unique) smooth $SL(2)$-embedding with two orbits:* $X = SL(2) \cup SL(2)/T$;
- *if* $h = \frac{p}{q} < 1$ *and* $(p, q) = 1$, *then* $X = SL(2) \cup SL(2)/U_{p+q} \cup \{pt\}$, *and* $\{pt\}$ *is an isolated singular point in X.*

The proof of Theorem 1.4 can be found in [52], [37, Ch. 3]. Here we give only some examples and explain what the number h (which is called the *height* of X) means in terms of the algebra $\mathbb{K}[X]$.

Example 1.5. 1) The group $SL(2)$ acts tautologically on the space $\mathbb{K}^2$ and by conjugation on the space $\mathrm{Mat}(2 \times 2)$. Consider the point

$$x = \left\{ \begin{pmatrix} 1 & 0 \\ 0 & -1 \end{pmatrix}, \begin{pmatrix} 1 \\ 0 \end{pmatrix} \right\} \in \mathrm{Mat}(2 \times 2) \times \mathbb{K}^2$$

and its orbit

$$SL(2)x = \{(A, v) \mid \det A = -1, \mathrm{tr}\, A = 0, Av = v, v \neq 0\}.$$

It is easy to see that the closure

$$X = \overline{SL(2)x} = \{(A, v) \mid \det A = -1, \operatorname{tr} A = 0, Av = v\}$$

is a smooth $SL(2)$-embedding with two orbits, hence X corresponds to $h = 1$.

2) Let $V_d = \langle x^d, x^{d-1}y, \ldots, y^d \rangle$ be the $SL(2)$-module of binary forms of degree d. It is possible to check that

$$X = \overline{SL(2)(x, x^2 y)} \subset V_1 \oplus V_3$$

is a normal $SL(2)$-embedding with the orbit decomposition $X = SL(2) \cup SL(2)/U_3 \cup \{pt\}$, hence X corresponds to $h = \frac{1}{2}$.

An embedding $SL(2) \hookrightarrow X$, $g \to gx$ determines the injective homomorphism $\mathfrak{A} = \mathbb{K}[X] \to \mathbb{K}[SL(2)]$ with $Q\mathfrak{A} = Q\mathbb{K}[SL(2)]$, where $Q\mathfrak{A}$ is the quotient field of $\mathfrak{A}$. Let U^- be the unipotent subgroup of $SL(2)$ opposite to U. Then

$$\mathbb{K}[SL(2)]^{U^-} = \{f \in \mathbb{K}[SL(2)] \mid f(ug) = f(g),\ g \in SL(2), u \in U^-\}$$
$$= \mathbb{K}[A, B],$$

where $A \begin{pmatrix} a & b \\ c & d \end{pmatrix} = a$ and $B \begin{pmatrix} a & b \\ c & d \end{pmatrix} = b$.

Below we list some facts ([37, Ch. 3]) that allow us to introduce the height of an $SL(2)$-embedding X.

- If $\mathfrak{C}$ is an integral F-domain, where F is a unipotent group, then $Q(\mathfrak{C}^F) = (Q\mathfrak{C})^F$. In particular, if $\mathfrak{C} \subseteq \mathfrak{A}$ and $Q\mathfrak{A} = Q\mathfrak{C}$, then $Q(\mathfrak{A}^{U^-}) = Q(\mathfrak{C}^{U^-})$.

- Suppose that $\lim_{t \to 0} \begin{pmatrix} t & 0 \\ 0 & t^{-1} \end{pmatrix} x$ exists. Then $A \in \mathbb{K}[SL(2)] \subset \mathbb{K}(X)$ is regular on X.

- Let $\mathfrak{D} \subset \mathbb{K}[x, y]$ be a homogeneous integrally closed subalgebra in the polynomial algebra such that $Q\mathfrak{D} = \mathbb{K}(x, y)$ and $x \in \mathfrak{D}$. Then $\mathfrak{D}$ is generated by monomials.

  In our situation, the algebra $\mathfrak{D} = \mathfrak{A}^{U^-} \subset \mathbb{K}[A, B]$ is homogeneous because it is T-stable (since T normalizes U^-).

- There exists rational $h \in (0, 1]$ such that

$$\mathfrak{A}^{U^-} = \mathfrak{A}(h) = \langle A^i B^j \mid \frac{j}{i} \leq h \rangle.$$

Moreover, for any rational $h \in (0, 1]$ the subspace $\langle SL(2)\mathfrak{A}(h) \rangle \subset \mathbb{K}[SL(2)]$ is a subalgebra.

Remark . While normal $SL(2)$-embeddings are parametrized by a discrete parameter h, there are families of non-isomorphic non-normal $SL(2)$-embeddings over a base of arbitrary dimension [13].

Remark . A classification of $SL(2)$-actions on normal three-dimensional affine varieties without open orbit can be found in [6, 5].

1.3 HV-varieties and S-varieties

In this subsection we discuss the results of V. L. Popov and E. B. Vinberg [55]. Throughout G denotes a connected and simply connected semisimple group.

Definition 1.6. An *HV-variety* X is the closure of the orbit of a highest weight vector in a simple G-module.

Let $V(\lambda)$ be the simple G-module with highest weight λ and v_λ a highest weight vector in $V(\lambda)$. Denote by λ^* the highest weight of the dual G-module $V(\lambda)^*$.

- $X(\lambda) = \overline{Gv_{\lambda^*}}$ is a normal affine variety consisting of two orbits: $X(\lambda) = Gv_{\lambda^*} \cup \{0\}$.
- $\mathbb{K}[X(\lambda)] = \mathbb{K}[Gv_{\lambda^*}] = \oplus_{m \geq 0} \mathbb{K}[X(\lambda)]_{m\lambda}$, any isotypic component $\mathbb{K}[X(\lambda)]_{m\lambda}$ is a simple G-module, and

$$\mathbb{K}[X(\lambda)]_{m_1\lambda}\mathbb{K}[X(\lambda)]_{m_2\lambda} = \mathbb{K}[X(\lambda)]_{(m_1+m_2)\lambda}.$$

- The algebra $\mathbb{K}[X(\lambda)]$ is a unique factorization domain if and only if λ is a fundamental weight of G.

Example 1.7. 1) The quadratic cone $KQ_n = \{x \in \mathbb{K}^n \mid x_1^2 + \cdots + x_n^2 = 0\}$ $(n \geq 3)$ is an *HV*-variety for the tautological representation $SO(n) : \mathbb{K}^n$. (In fact, the group $SO(n)$ is not simply connected and we consider the corresponding module as a $\text{Spin}(n)$-module.) It follows that KQ_n is normal and it is factorial if and only if $n \geq 5$.

2) The Grassmannian cone $KG_{n,m}$ $(n \geq 2, 1 \leq m \leq n-1)$ (i.e. the cone over the projective variety of m-subspaces in $\mathbb{K}^n$) is an *HV*-variety associated with the fundamental $SL(n)$-representation in the space $\bigwedge^m \mathbb{K}^n$, hence it is factorial.

Definition 1.8. An irreducible affine variety X with an action of a connected reductive group G is said to be an *S-variety* if X has an open G-orbit and the stabilizer of a point in this orbit contains a maximal unipotent subgroup of G.

Any S-variety may be realized as $X = \overline{Gv}$, where $v = v_{\lambda_1^*} + \cdots + v_{\lambda_k^*}$ is a sum of highest weight vectors $v_{\lambda_i^*}$ in some G-module V. We have the isotypic decomposition

$$\mathbb{K}[X] = \oplus_{\lambda \in L(X)} \mathbb{K}[X]_\lambda,$$

where $L(X)$ is the semigroup generated by $\lambda_1, \ldots, \lambda_k$, any $\mathbb{K}[X]_\lambda$ is a simple G-module, and $\mathbb{K}[X]_\lambda \mathbb{K}[X]_\mu = \mathbb{K}[X]_{\lambda+\mu}$. The last condition determines uniquely (up to G-isomorphism) the multiplicative structure on the G-module $\mathbb{K}[X]$. This shows that there is a bijection between S-varieties and finitely generated submonoids in $\Xi_+(G)$.

Consider the cone $K = \mathbb{Q}_+ L(X)$. As in the toric case, normality of X is equivalent to the saturation condition for the semigroup $L(X)$, and G-orbits on X are in one-to-one correspondence with faces of K. On the other hand, there are phenomena which are specific for S-varieties. For example, the complement to the open orbit in X has codimension ≥ 2 if and only if $\mathbb{Z}L(X) \cap \Xi_+(G) \subseteq \mathbb{Q}_+ L(X)$ (this is never the case for non-trivial toric varieties). Also, an S-variety X is factorial if and only if $L(X)$ is generated by fundamental weights.

Finally, we mention one more result on this subject. Say that an action $G : X$ on an affine variety X is *special* (or *horospherical*) if there is an open dense subset $W \subset X$ such that the stabilizer of any point of W contains a maximal unipotent subgroup of G.

Theorem 1.9 ([54]). *The following conditions are equivalent:*

- *the action $G : X$ is special;*
- *the stabilizer of any point on X contains a maximal unipotent subgroup;*
- $\mathbb{K}[X]_\lambda \mathbb{K}[X]_\mu \subseteq \mathbb{K}[X]_{\lambda+\mu}$ *for any $\lambda, \mu \in \Xi_+(G)$.*

1.4 Algebraic monoids

The general theory of algebraic semigroups was developed by M. S. Putcha, L. Renner and E. B. Vinberg. In this subsection we recall briefly the classification results following [71].

Definition 1.10. An *(affine) algebraic semigroup* is an (affine) algebraic variety S with an associative multiplication

$$\mu : \ S \times S \to S,$$

which is a morphism of algebraic varieties. An algebraic semigroup S is *normal* if S is a normal algebraic variety.

Any algebraic group is an algebraic semigroup. Another example is the semigroup $\mathrm{End}(V)$ of endomorphisms of a finite-dimensional vector space V.

Lemma 1.11. *An affine algebraic semigroup S is isomorphic to a closed subsemigroup of $\mathrm{End}(V)$ for a suitable V. If S has a unit, one may assume that it corresponds to the identity map of V.*

Proof. The morphism $\mu : S \times S \to S$ induces the homomorphism

$$\mu^* : \mathbb{K}[S] \to \mathbb{K}[S] \otimes \mathbb{K}[S] \quad , \quad f(s) \mapsto F(s_1, s_2) := f(s_1 s_2).$$

Hence $f(s_1 s_2) = \sum_{i=1}^n f_i(s_1) h_i(s_2)$. Consider the linear action $S : \mathbb{K}[S]$ defined by $(s * f)(x) = f(xs)$. One has $\langle Sf \rangle \subseteq \langle f_1, \ldots, f_n \rangle$, i.e. the linear span of any 'S-orbit' in $\mathbb{K}[S]$ is finite-dimensional and the linear action $S : \langle Sf \rangle$ defines an algebraic representation of S. Take as V any finite-dimensional S-invariant subspace of $\mathbb{K}[S]$ containing a system of generators of $\mathbb{K}[S]$.

Suppose that S is a monoid, i.e. a semigroup with unit. We claim that the action $S : V$ defines a closed embedding $\phi : S \to \mathrm{End}(V)$. Indeed, there are $\alpha_{ij} \in \mathbb{K}[S]$ such that $s * f_i = \sum_j \alpha_{ij}(s) f_j$. The equalities $f_i(s) = (s * f_i)(e) = \sum_j \alpha_{ij}(s) f_j(e)$ show that the homomorphism $\phi^* : \mathbb{K}[\mathrm{End}(V)] \to \mathbb{K}[S]$ is surjective.

The general case can be reduced to the previous one as follows: to any semigroup S one may add an element e with relations $e^2 = e$ and $es = se = s$ for any $s \in S$. Then $\tilde{S} = S \sqcup \{e\}$ is an algebraic monoid. $\square$

If $S \subseteq \mathrm{End}(V)$ is a monoid, then any invertible element of S corresponds to an element of $GL(V)$. Conversely, if the image of s is invertible in $\mathrm{End}(V)$, then it is invertible in S. Indeed, the sequence of closed subsets $S \supseteq sS \supseteq s^2 S \supseteq s^3 S \supseteq \ldots$ stabilizes, and $s^k S = s^{k+1} S$ implies $S = sS$. Hence the group $G(S)$ of invertible elements is open in S and is an algebraic group. Suppose that $G(S)$ is dense in S. Then S may be considered as an affine embedding of $G(S)/\{e\}$ (with respect to left multiplication).

Proposition 1.12. *Let G be an algebraic group. An affine embedding $G/\{e\} \hookrightarrow S$ has a structure of an algebraic monoid with G as the group of invertible elements if and only if the G-equivariant G-action on the open orbit by right multiplication can be extended to S, or, equivalently, S is an affine embedding of $(G \times G)/\Delta(G)$, where $\Delta(G)$ is the diagonal in $G \times G$.*

Proof. If S is an algebraic monoid with $G(S) = G$ and $G(S)$ is dense in S, then $G \times G$ acts on S by $((g_1, g_2), s) \mapsto g_1 s g_2^{-1}$ and the dense open $G \times G$-orbit in S is isomorphic to $(G \times G)/\Delta(G)$.

For the converse, we give two independent proofs in their historical order.

Proof One (the reductive case). (E. B. Vinberg [71]) An algebraic monoid S is *reductive* if the group $G(S)$ is reductive and dense in S. The multiplication $\mu : G \times G \to G$ corresponds to the comultiplication $\mu^* : \mathbb{K}[G] \to \mathbb{K}[G] \otimes \mathbb{K}[G]$. Any $(G \times G)$-isotypic component in $\mathbb{K}[G]$ is a simple $(G \times G)$-module isomorphic to $V(\lambda)^* \otimes V(\lambda)$ for $\lambda \in \Xi_+(G)$ [37]. It coincides with the linear span of the matrix entries of the G-module $V(\lambda)$. This shows that μ^* maps an isotypic component to its tensor square, and for any $(G \times G)$-invariant subspace $W \subset \mathbb{K}[G]$ one has $\mu^*(W) \subset W \otimes W$. Thus the spectrum S of any $(G \times G)$-invariant finitely generated subalgebra in $\mathbb{K}[G]$ carries the structure of an algebraic semigroup. If the open $(G \times G)$-orbit in S is isomorphic to $(G \times G)/\Delta(G)$, then $G(S) = G$. Indeed, G is dense in S and for any $s \in G(S)$ the intersection $sG \cap G \neq \emptyset$, hence $s \in G$.

Proof Two (the general case). (A. Rittatore [59]) If the multiplication $\mu : G \times G \to G$ extends to a morphism $\mu : S \times S \to S$, then μ is a multiplication because μ is associative on $G \times G$. It is clear that $1 \in G$ satisfies $1s = s1 = s$ for all $s \in S$. Consider the right and left actions of G given by

$$G \times S \to S, \quad gs = (g, 1)s,$$

$$S \times G \to S, \quad sg = (1, g^{-1}s).$$

These actions define coactions $\mathbb{K}[S] \to \mathbb{K}[G] \otimes \mathbb{K}[S]$ and $\mathbb{K}[S] \to \mathbb{K}[S] \otimes \mathbb{K}[G]$, which are the restrictions to $\mathbb{K}[S]$ of the comultiplication $\mathbb{K}[G] \to \mathbb{K}[G] \otimes \mathbb{K}[G]$. Hence the image of $\mathbb{K}[S]$ lies in

$$(\mathbb{K}[G] \otimes \mathbb{K}[S]) \cap (\mathbb{K}[S] \otimes \mathbb{K}[G]) = \mathbb{K}[S] \otimes \mathbb{K}[S],$$

and we have a multiplication on S. The equality $G(S) = G$ may be proved as above. $\square$

For the rest of this section we assume that G is reductive. For $\lambda_1, \lambda_2 \in \Xi_+(G)$, we denote by $\Xi(\lambda_1, \lambda_2)$ the set of $\lambda \in \Xi_+(G)$ such that the G-module $V(\lambda_1) \otimes V(\lambda_2)$ contains a submodule isomorphic to $V(\lambda)$. Since any $(G \times G)$-isotypic component $\mathbb{K}[G]_{(\lambda^*, \lambda)}$ in $\mathbb{K}[G]$ is the linear span of the matrix entries corresponding to the representation of G in $V(\lambda)$, the product $\mathbb{K}[G]_{(\lambda_1^*, \lambda_1)} \mathbb{K}[G]_{(\lambda_2^*, \lambda_2)}$ is the linear span of the matrix entries corresponding to $V(\lambda_1) \otimes V(\lambda_2)$. This shows that

$$\mathbb{K}[G]_{(\lambda_1^*, \lambda_1)} \mathbb{K}[G]_{(\lambda_2^*, \lambda_2)} = \oplus_{\lambda \in \Xi(\lambda_1, \lambda_2)} \mathbb{K}[G]_{(\lambda^*, \lambda)}.$$

Since every $(G \times G)$-isotypic component in $\mathbb{K}[G]$ is simple, any $(G \times G)$-invariant subalgebra in $\mathbb{K}[G]$ is determined by the semigroup of dominant weights that appear in its isotypic decomposition, and it is natural to classify reductive algebraic monoids S with $G(S) = G$ in terms of the semigroup that determines $\mathbb{K}[S]$ in $\mathbb{K}[G]$.

Definition 1.13. A subsemigroup $L \subset \Xi_+(G)$ is *perfect* if it contains zero and $\lambda_1, \lambda_2 \in L$ implies $\Xi(\lambda_1, \lambda_2) \subset L$.

Let $\mathbb{Z}\Xi_+(G)$ be the group generated by the semigroup $\Xi_+(G)$. This group may be realized as the group of characters $\Xi(T)$ of a maximal torus of G.

Theorem 1.14 ([71]). *A subset $L \subset \Xi_+(G)$ defines an affine algebraic monoid S with $G(S) = G$ if and only if L is a perfect finitely generated subsemigroup generating the group $\mathbb{Z}\Xi_+(G)$.*

The classification of normal affine reductive monoids is more constructive. We fix some notation. The group $G = ZG'$ is an almost direct product of its center Z and the derived subgroup G'. Fix a Borel subgroup B_0 and a maximal torus $T_0 \subset B_0$ in G'. Then $B = ZB_0$ (resp. $T = ZT_0$) is a Borel subgroup (resp. a maximal torus) in G. Let N (resp. N_0, N_1) denote the $\mathbb{Q}$-vector space $\Xi(T) \otimes_\mathbb{Z} \mathbb{Q}$ (resp. $\Xi(T_0) \otimes_\mathbb{Z} \mathbb{Q}$, $\Xi(Z) \otimes_\mathbb{Z} \mathbb{Q}$). Then $N = N_1 \oplus N_0$. The semigroup of dominant weights $\Xi_+(G)$ (with respect to B) is a subsemigroup in $\Xi(T) \subset N$. By $\alpha_1, \dots, \alpha_k \in N_1$ denote the simple roots of G with respect to B, and by $C \subset N$ (resp. $C_0 \subset N_0$) the positive Weyl chamber for the group G (resp. G') with respect to $\alpha_1, \dots, \alpha_k$.

Theorem 1.15 ([71]). *A subset $L \subset \Xi_+(G)$ defines a normal affine algebraic monoid S with $G(S) = G$ if and only if $L = \Xi_+(G) \cap K$, where K is a closed convex polyhedral cone in N satisfying the conditions:*

- $-\alpha_1, \dots, -\alpha_k \in K$;
- *the cone $K \cap C$ generates N.*

The monoid S has a zero if and only if, in addition, the cone $K \cap N_1$ is pointed and $K \cap C_0 = \{0\}$.

A characteristic-free approach to the classification of reductive algebraic monoids via the theory of spherical varieties was developed in [59]. Another interesting result of [59] is that any reductive algebraic monoid is affine. Recently A. Rittatore announced a proof of the fact that any algebraic monoid with an affine algebraic group of invertible elements is affine.

2 Connections with Hilbert's 14th Problem

2.1 *Grosshans subgroups and the canonical embedding*

Let H be a closed subgroup of $GL(V)$. Hilbert's 14th problem (in its modern version) may be formulated as follows: characterize subgroups H such that the algebra of polynomial invariants $\mathbb{K}[V]^H$ is finitely generated. It is a classical result that for reductive H the algebra $\mathbb{K}[V]^H$ is finitely generated. For non-reductive linear groups this problem seems to be very far from a complete solution.

Remark . Hilbert's original statement of the problem was the following:

For a field $\mathbb{K}$, let $\mathbb{K}[x_1, \ldots, x_n]$ denote the polynomial ring in n variables over $\mathbb{K}$, and let $\mathbb{K}(x_1, \ldots, x_n)$ denote its field of fractions. If K is a subfield of $\mathbb{K}(x_1, \ldots, x_n)$ containing $\mathbb{K}$, is $K \cap \mathbb{K}[x_1, \ldots, x_n]$ finitely generated over $\mathbb{K}$?

Since $\mathbb{K}[V]^H = \mathbb{K}[V] \cap \mathbb{K}(V)^H$, our situation may be regarded as a particular case of the general one.

Let us assume that H is a subgroup of a bigger reductive group G acting on V. (For example, one may take $G = GL(V)$.) The intersection of a family of observable subgroups in G is an observable subgroup. Define *the observable hull $\hat{H}$* of H as the minimal observable subgroup of G containing H. The stabilizer of any H-fixed vector in a rational G-module contains $\hat{H}$. Therefore $\mathbb{K}[V]^H = \mathbb{K}[V]^{\hat{H}}$ for any G-module V, and it is natural to solve Hilbert's 14th problem for observable subgroups.

The following famous theorem proved by F. D. Grosshans establishes a close connection between Hilbert's 14th problem and the theory of affine embeddings.

Theorem 2.1 ([27, 29]). *Let H be an observable subgroup of a reductive group G. The following conditions are equivalent:*

1) *for any G-module V the algebra $\mathbb{K}[V]^H$ is finitely generated;*

2) *the algebra* $\mathbb{K}[G/H]$ *is finitely generated;*

3) *there exists an affine embedding* $G/H \hookrightarrow X$ *such that*

$$\operatorname{codim}_X(X \setminus (G/H)) \geq 2.$$

Definition 2.2. 1) An observable subgroup H in G is said to be a *Grosshans subgroup* if $\mathbb{K}[G/H]$ is finitely generated.

2) If H is a Grosshans subgroup of G, then $G/H \hookrightarrow X = \operatorname{Spec} \mathbb{K}[G/H]$ is called *the canonical embedding* of G/H, and X is denoted by $CE(G/H)$.

Note that any normal affine embedding $G/H \hookrightarrow X$ with $\operatorname{codim}_X(X \setminus (G/H)) \geq 2$ is G-isomorphic to the canonical embedding [29]. A homogeneous space G/H admits such an embedding if and only if H is a Grosshans subgroup.

By Matsushima's criterion, H is reductive if and only if $CE(G/H) = G/H$. For non-reductive subgroups, $CE(G/H)$ is an interesting object canonically associated with the pair (G, H). It allows us to reformulate algebraic problems concerning the algebra $\mathbb{K}[G/H]$ in geometric terms.

2.2 Popov-Pommerening's conjecture and Knop's theorem

Theorem 2.3 ([28, 19],[29, Th. 16.4]). *Let* P^u *be the unipotent radical of a parabolic subgroup* P *of* G. *Then* P^u *is a Grosshans subgroup of* G.

Proof. Let $P = LP^u$ be a Levi decomposition and U_1 a maximal unipotent subgroup of L. Then $U = U_1 P^u$ is a maximal unipotent subgroup of G, and $\mathbb{K}[G]^U = (\mathbb{K}[G]^{P^u})^{U_1}$. We know that $\mathbb{K}[G]^U$ is finitely generated (Theorem 1.3). On the other hand, Theorem 1.3 implies that the L-algebra $\mathbb{K}[G]^{P^u}$ is finitely generated if and only if $(\mathbb{K}[G]^{P^u})^{U_1}$ is, hence $\mathbb{K}[G]^{P^u}$ is finitely generated. (Another proof, using an explicit codimension 2 embedding, is given in [28].) $\qquad\square$

Let us say that a subgroup of a reductive group G is *regular* if it is normalized by a maximal torus in G. Generalizing Theorem 2.3, V. L. Popov and K. Pommerening conjectured that any observable regular subgroup is a Grosshans subgroup. At the moment a positive answer is known for groups G of small rank [64, 66, 65], and for some special classes of regular subgroups (for example, for unipotent radicals of parabolic subgroups of Levi subgroups of G [29]). Lin Tan [65] constructed explicitly canonical embeddings for regular unipotent subgroups in $SL(n)$, $n \leq 5$. A strong argument in favour of Popov-Pommerening's conjecture is given in [14, Th. 4.3] in terms of finite generation of induced modules, see also [29, § 23].

Another powerful method for checking that the algebra $\mathbb{K}[G/H]$ is finitely generated is provided by the following theorem of F. Knop.

Theorem 2.4 ([35, 29]). *Suppose that G acts on an irreducible normal unirational variety X. If $c(X) \leq 1$, then the algebra $\mathbb{K}[X]$ is finitely generated.*

Corollary . *If H is observable in G and $c(G/H) \leq 1$, then H is a Grosshans subgroup.*

2.3 The canonical embedding of G/P^u

Since the unipotent radical P^u of a parabolic subgroup P is a Grosshans subgroup of G, there exists a canonical embedding $G/P^u \hookrightarrow CE(G/P^u)$. Such embeddings provide an interesting class of affine factorial G-varieties, which was studied in [12]. Let us note that the Levi subgroup $L \subset P$ normalizes P^u, hence acts G-equivariantly on G/P^u and on $CE(G/P^u)$. By $V_L(\lambda)$ denote a simple L-module with the highest weight λ. Our approach is based on the analysis of the $(G \times L)$-module decomposition of the algebra $\mathbb{K}[G/P^u]$ given by

$$\mathbb{K}[G/P^u] = \bigoplus_{\lambda \in \Xi_+(G)} \mathbb{K}[G/P^u]_\lambda,$$

where $\mathbb{K}[G/P^u]_\lambda \cong V(\lambda)^* \otimes V_L(\lambda)$ is the linear span of the matrix entries of the linear maps $V(\lambda)^{P^u} \to V(\lambda)$ induced by $g \in G$, considered as regular functions on G/P^u. (In fact, our method works for any affine embedding $G/P^u \hookrightarrow X$, where L acts G-equivariantly.) The multiplication structure looks like

$$\mathbb{K}[G/P^u]_\lambda \cdot \mathbb{K}[G/P^u]_\mu = \mathbb{K}[G/P^u]_{\lambda+\mu} \oplus \bigoplus_i \mathbb{K}[G/P^u]_{\lambda+\mu-\beta_i},$$

where $\lambda + \mu - \beta_i$ runs over the highest weights of all "lower" irreducible components in the L-module decomposition

$$V_L(\lambda) \otimes V_L(\mu) = V_L(\lambda + \mu) \oplus \dots.$$

Here we list the results from [12].

- Affine $(G \times L)$-embeddings $G/P^u \hookrightarrow X$ are classified by finitely generated subsemigroups S of $\Xi_+(G)$ having the property that all highest weights of the tensor product of simple L-modules with highest weights in S belong to S, too. Furthermore, every choice of the generators $\lambda_1, \dots, \lambda_m \in S$ gives rise to a natural G-equivariant embedding $X \hookrightarrow \mathrm{Hom}(V^{P^u}, V)$, where V is the sum of simple G-modules of highest weights $\lambda_1, \dots, \lambda_m$. The convex cone Σ^+ spanned by S is precisely the dominant part of the cone Σ spanned by the weight polytope of V^{P^u}. In the case $X = CE(G/P^u)$, the semigroup S coincides with $\Xi_+(G)$ and Σ is the span of the dominant Weyl chamber by the Weyl group of L. In particular, if G is simply connected and semisimple then there is a natural inclusion

$$CE(G/P^u) \subset \bigoplus_{i=1}^{l} \mathrm{Hom}(V(\omega_i)^{P^u}, V(\omega_i)),$$

where $\omega_1, \dots, \omega_l$ are the fundamental weights of G.

- The $(G \times L)$-orbits in X are in bijection with the faces of Σ whose interiors contain dominant weights, the orbit representatives being given by the projectors onto the subspaces of V^{P^u} spanned by eigenvectors of eigenweights in a given face. For the canonical embedding, the $(G \times L)$-orbits correspond to the subdiagrams in the Dynkin diagram of G such that no connected component of such a subdiagram is contained in the Dynkin diagram of L. We also compute the stabilizers of points in $G \times L$ and in G, and the modality of the action $G : X$.

- We classify smooth affine $(G \times L)$-embeddings $G/P^u \hookrightarrow X$. In particular, the only non-trivial smooth canonical embedding corresponds to $G = SL(n)$, P is the stabilizer of a hyperplane in $\mathbb{K}^n$, and

$$CE(G/P^u) = \mathrm{Mat}(n, n-1)$$

with the G-action by left multiplication.

- The techniques used in the description of affine $(G \times L)$-embeddings of G/P^u are parallel to those developed in [69] for the study of equivariant compactifications of reductive groups. An analogy with monoids becomes more transparent in view of the bijection between our affine embeddings $G/P^u \hookrightarrow X$ and a class of algebraic monoids M with the group of invertibles L, given by $X = \mathrm{Spec}\,\mathbb{K}[G \times^P M]$.

- Finally, we describe the G-module structure of the tangent space of $CE(G/P^u)$ at the G-fixed point, assuming that G is simply connected and simple. This space is obtained from $\bigoplus_i \mathrm{Hom}(V(\omega_i)^{P^u}, V(\omega_i))$ by removing certain summands according to an explicit algorithm. The tangent space at the fixed point is at the same time the minimal ambient G-module for $CE(G/P^u)$.

2.4 Counterexamples

The famous counterexample of Nagata to Hilbert's 14th problem [49]
yields a 13-dimensional unipotent subgroup H in $SL(32)$ acting natu-
rally in $V = \mathbb{K}^{32}$ such that the algebra of invariants $\mathbb{K}[V]^H$ is not finitely
generated. This shows that the algebra $\mathbb{K}[SL(32)/H]$ is not finitely generated, or, equivalently, the complement of the open orbit in any affine
embedding $SL(32)/H \hookrightarrow X$ contains a divisor.

Nagata's construction was simplified by R. Steinberg. He proved that
$\mathbb{K}[V]^H$ is not finitely generated for the following 6-dimensional commutative unipotent linear group:

$$
H = \left\{ \begin{pmatrix} 1 & c_1 & & & & \\ 0 & 1 & & & 0 & \\ & & \ddots & \ddots & & \\ & & & \ddots & \ddots & \\ & 0 & & & 1 & c_9 \\ & & & & 0 & 1 \end{pmatrix}, \ \sum_{j=1}^{9} a_{ij} c_j = 0, \ i = 1, 2, 3 \right\},
$$

where the nine points $P_j = (a_{1j} : a_{2j} : a_{3j})$ are nonsingular points on
an irreducible cubic curve in the projective plane, their sum has infinite
order in the group of the curve, and $V = \mathbb{K}^{18}$ (see [62] for details).

Another method of obtaining counterexamples was proposed [60] by
P. Roberts. Consider the polynomial algebra $R = \mathbb{K}[x, y, z, s, t, u, v]$ in
7 variables over a not necessarily algebraically closed field $\mathbb{K}$ of characteristic zero with the grading $R = \oplus_{n \geq 0} R_n$ determined by assigning the
degree 0 to x, y, z and the degree 1 to s, t, u, v. The elements s, t, u, v
generate a free R_0-submodule in R considered as R_0-module. Choosing
a natural number $m \geq 2$, Roberts defines an R_0-module homomorphism
on this submodule

$$ f : R_0 s \oplus R_0 t \oplus R_0 u \oplus R_0 v \to R_0 $$

given by $f(s) = x^{m+1}$, $f(t) = y^{m+1}$, $f(u) = z^{m+1}$, $f(v) = (xyz)^m$. The
submodule $\mathrm{Ker}\, f$ generates a subalgebra of R, which is denoted by A. It
is proved in [60] that the $\mathbb{K}$-algebra $B = R \cap QA$ is not finitely generated.
(Roberts shows how to construct an element in B of any given degree
which is not in the subalgebra generated by elements of lower degree.)
A linear action of a 12-dimensional commutative unipotent group on
19-dimensional vector space with the algebra of invariants isomorphic to
the polynomial algebra in one variable over B is constructed in [1].

For a recent development in this direction, see [21, 22].

3 Some properties of affine embeddings

3.1 Affinely closed spaces and Luna's theorem

Definition 3.1. An affine homogeneous space G/H of an affine algebraic group G is called *affinely closed* if it admits only the trivial affine embedding $X = G/H$.

Assume that G is reductive; then H is reductive if G/H is affinely closed. By $N_G(H)$, $C_G(H)$ denote the normalizer and centralizer respectively of H in G, and by $W(H)$ denote the quotient $N_G(H)/H$. It is known that $N_G(H)^0 = H^0 C_G(H)^0$ and both $N_G(H)$ and $C_G(H)$ are reductive [44, Lemma 1.1].

Theorem 3.2 ([43])**.** *Let H be a reductive subgroup of a reductive group G. The homogeneous space G/H is affinely closed if and only if the group $W(H)$ is finite. Moreover, if G acts on an affine variety X and the stabilizer of a point $x \in X$ contains a reductive subgroup H such that $W(H)$ is finite, then the orbit Gx is closed in X.*

Remark . The last statement may be reformulated: if H is reductive, the group $W(H)$ is finite, and $H \subset H' \subset G$, where H' is observable, then H' is reductive and G/H' is affinely closed.

Remark . Let H be a Grosshans subgroup of G. The following conditions are equivalent:

- H is reductive and $W(H)$ is finite;
- H is reductive and for any one-parameter subgroup $\mu : \mathbb{K}^* \to C_G(H)$ one has $\mu(\mathbb{K}^*) \subseteq H$;
- the algebra $\mathbb{K}[G/H]$ does not have non-trivial G-invariant ideals and does not admit non-trivial G-invariant $\mathbb{Z}$-gradings;
- the algebra $\mathbb{K}[G/H]$ does not have non-trivial G-invariant ideals and the group of G-equivariant automorphisms of $\mathbb{K}[G/H]$ is finite.
- no invariant subalgebra in $\mathbb{K}[G/H]$ admits a non-trivial G-invariant ideal.

Example 3.3. 1) Let $\rho : H \to SL(V)$ be an irreducible representation of a reductive group H. Then the space $SL(V)/\rho(H)$ is affinely closed ($W(\rho(H))$ is finite by the Schur Lemma).

2) If T is a maximal torus of G, then $W(T)$ is the Weyl group and G/T is affinely closed.

Proposition 3.4. *Let G be an affine algebraic group. The following conditions are equivalent:*

- *any monoid S with $G(S) = G$ and $\overline{G(S)} = S$ coincides with G;*
- *the group G/G^u is semisimple.*

Proof. Let G be reductive. The space $(G \times G)/\Delta(G)$ is affinely closed if and only if the group $N_{G \times G}(\Delta(G))/\Delta(G)$ is finite. But this is exactly the case when the center of G is finite. The same arguments work for any G (Theorem 3.9). $\qquad\qquad\qquad\qquad\qquad\qquad\qquad\qquad\qquad\qquad\qquad$ □

We now give a proof of Theorem 3.2 in terms of so-called adapted (or optimal) one-parameter subgroups following G. Kempf [31, Cor.4.5].

We have to prove that if G/H' is a quasi-affine homogeneous space that is not affinely closed and $H \subset H'$ is a reductive subgroup, then there exists a one-parameter subgroup $\nu : \mathbb{K}^* \to C_G(H)$ such that $\nu(\mathbb{K}^*)$ is not contained in H. There is an affine embedding $G/H' \hookrightarrow X$ with a G-fixed point o, see Section 3.5. Denote by x the image of eH' in the open orbit on X. By the Hilbert-Mumford criterion, there exists a one-parameter subgroup $\gamma : \mathbb{K}^* \to G$ such that $\lim_{t \to 0} \gamma(t)x = o$. Moreover, there is a subgroup γ that moves x 'most rapidly' toward o. Such a γ is called *adapted to x*; for the precise definition see [31, 56]. For adapted γ, consider the parabolic subgroup

$$P(\gamma) = \{ g \in G \mid \lim_{t \to 0} \gamma(t) g \gamma(t)^{-1} \text{ exists in } G \}.$$

Then $P(\gamma) = L(\gamma)U(\gamma)$, where $L(\gamma)$ is the Levi subgroup that is the centralizer of $\gamma(\mathbb{K}^*)$ in G, and $U(\gamma)$ is the unipotent radical of $P(\gamma)$. By [31], [56, Th. 5.5], the stabilizer $G_x = H'$ is contained in $P(\gamma)$. Hence there is an element $u \in U(\gamma)$ such that $uHu^{-1} \subset L(\gamma)$.

We claim that $\gamma(\mathbb{K}^*)$ is not contained in uHu^{-1}. In fact, γ is adapted to the element ux, too [31, Th. 3.4], hence $\gamma(\mathbb{K}^*)$ is not contained in the stabilizer of ux. Thus $u^{-1}\gamma u$ is the desired subgroup ν.

Conversely, suppose that there exists $\nu : \mathbb{K}^* \to C_G(H)$ and $\nu(\mathbb{K}^*)$ is not contained in H. Consider the subgroup $H_1 = \nu(\mathbb{K}^*)H$. The homogeneous fibre space $G *_{H_1} \mathbb{K}$, where H acts on $\mathbb{K}$ trivially and H_1/H acts on $\mathbb{K}$ by dilation, is a two-orbit embedding of G/H. $\qquad$ □

3.2 Affinely closed spaces in arbitrary characteristic

In this subsection we assume that $\mathbb{K}$ is an arbitrary algebraically closed field. Suppose that G acts on an affine variety X. In positive characteristic, the structure of an algebraic variety on the orbit Gx of a point $x \in X$ is not determined (up to G-isomorphism) by the stabilizer

$H = G_x$, and it is natural to consider the isotropy subscheme $\tilde{H}$ at x, with H as the reduced part, identifying Gx and $G/\tilde{H}$. There is a natural bijective purely inseparable and finite morphism $\pi : G/H \to G/\tilde{H}$ [30, 4.3, 4.6]. The following technical proposition shows that this difficulty does not play an essential role for affinely closed spaces.

Proposition 3.5 ([9, Prop. 8]). *The homogeneous space G/H is affinely closed if and only if $G/\tilde{H}$ is affinely closed.*

Definition 3.6. We say that an affinely closed homogeneous space G/H is *strongly affinely closed* if for any affine G-variety X and any point $x \in X$ fixed by H the orbit Gx is closed in X.

By Theorem 3.2, in characteristic zero any affinely closed space is strongly affinely closed.

The following notion was introduced by J.-P. Serre, cf. [41].

Definition 3.7. A subgroup $D \subset G$ is called *G-completely reducible* (*G-cr* for short) if, whenever D is contained in a parabolic subgroup P of G, it is contained in a Levi subgroup of P.

A G-cr subgroup is reductive. For $G = GL(V)$ this notion agrees with the usual notion of complete reducibility. In fact, if G is any of the classical groups then the notions coincide, although for the symplectic and orthogonal groups this requires the assumption that char $\mathbb{K}$ is a good prime for G. The class of G-cr subgroups is wide. Some conditions which guarantee that certain subgroups satisfy the G-cr condition can be found in [41, 46].

The proof of Theorem 3.2 given above implies:

- if H is not contained in a proper parabolic subgroup of G, then G/H is strongly affinely closed;

- if there exists $\nu : \mathbb{K}^* \to C_G(H)$ such that $\nu(\mathbb{K}^*)$ is not contained in H, then G/H is not affinely closed;

- if H is a G-cr subgroup of G, then the following conditions are equivalent:

1) G/H is affinely closed;
2) G/H is strongly affinely closed;
3) for any one-parameter subgroup $\nu : \mathbb{K}^* \to C_G(H)$ one has $\nu(\mathbb{K}^*) \subseteq H$.

Example 3.8. The following example produced by George J. McNinch shows that the group $W(H)$ may be unipotent even for reductive H. Let L be the space of $(n \times n)$-matrices and H the image of $SL(n)$ in $G = SL(L)$ acting on L by conjugation.

If $p = \text{char } \mathbb{K} \mid n$, then L is an indecomposable $SL(n)$-module with three composition factors, cf. [46, Prop. 4.6.10, a)]. It turns out that $C_G(H)^0$ is a one-dimensional unipotent group consisting of operators of the form $\text{Id}+aT$, where $a \in \mathbb{K}$, and T is a nilpotent operator on L defined by $T(X) = \text{tr}(X)E$. The subgroup H is contained in a quasi-parabolic subgroup of G, hence G/H is not strongly affinely closed.

In the simplest case $n = p = 2$, we have $H \cong PSL(2) \subset SL(4)$, $N_G(H) = HC_G(H)$ (because H does not have outer automorphisms), $C_G(H)$ is connected, and $W(H) \cong (\mathbb{K}, +)$.

It would be very interesting to obtain a complete description of affinely closed spaces in arbitrary characteristic and to answer the following question: is it true that any affinely closed space is strongly affinely closed?

3.3 Affinely closed spaces of non-reductive groups

For non-reductive G the class of affinely closed homogeneous spaces is much wider. For example, it is well-known that an orbit of a unipotent group acting on an affine variety is closed, hence any homogeneous space of a unipotent group is affinely closed. Conversely, if any (quasi-affine) homogeneous space of an affine group G is affinely closed, then the connected component of the identity in G is unipotent [15, 10.1], [20, Th.4.2]. In this subsection we give a complete characterization of affinely closed homogeneous spaces of non-reductive groups.

Let us fix the Levi decomposition $G = LG^u$ of the group G in the semidirect product of a reductive subgroup L and the unipotent radical G^u. By ϕ denote the homomorphism $G \to G/G^u$. We shall identify the image of ϕ with L. Put $K = \phi(H)$.

Theorem 3.9 ([10, Th.2]). *G/H is affinely closed if and only if L/K is.*

Proof. The subgroup H is observable in G if and only if the subgroup K is observable in L [63], [29, Th.7.3].

Suppose that L/K admits a non-trivial affine embedding. Then there are an L-module V and a vector $v \in V$ such that the stabilizer L_v equals K and the orbit boundary $Y = Z \setminus Lv$, where $Z = \overline{Lv}$, is nonempty. Let

$I(Y)$ be the ideal in $\mathbb{K}[Z]$ defining the subvariety Y. There exists a finite-dimensional L-submodule $V_1 \subset I(Y)$ that generates $I(Y)$ as an ideal. The inclusion $V_1 \subset \mathbb{K}[Z]$ defines L-equivariant morphism $\psi : Z \to V_1^*$ and $\psi^{-1}(0) = Y$. Then L-equivariant morphism $\xi : Z \to V_2 = V_1^* \oplus (V \otimes V_1^*)$, $z \to (\psi(z), z \otimes \psi(z))$ maps Y to the origin and is injective on the open orbit in Z. Hence we obtain an embedding of L/K in an L-module such that the closure of the image of this embedding contains the origin. Put $v_2 = \xi(v)$. By the Hilbert-Mumford Criterion, there is a one-parameter subgroup $\lambda : \mathbb{K}^* \to L$ such that $\lim_{t \to 0} \lambda(t)v_2 = 0$. Consider the weight decomposition $v_2 = v_2^{(i_1)} + \cdots + v_2^{(i_s)}$ of the vector v_2, where $\lambda(t)v_2^{(i_k)} = t^{i_k} v_2^{(i_k)}$. Here all i_k are positive.

By the identification $G/G^u = L$, one may consider V_2 as a G-module. Let W be a finite-dimensional G-module with a vector w whose stabilizer equals H. Replacing the pair (W, w) by the pair $(W \oplus (W \otimes W), w + w \otimes w)$, one may suppose that the orbit Gw intersects the line $\mathbb{K}w$ only at the point w. The weight decomposition shows that, for a sufficiently large N, in the G-module $W \otimes V_2^{\otimes N}$ one has $\lim_{t \to 0} \lambda(t)(w \otimes v_2^{\otimes N}) = 0$ ($\lambda(\mathbb{K}^*)$ may be considered as a subgroup of G). On the other hand, the stabilizer of $w \otimes v_2^{\otimes N}$ coincides with H. This implies that the space G/H is not affinely closed.

Conversely, suppose that G/H admits a non-trivial affine embedding. This embedding corresponds to a G-invariant subalgebra $A \subset \mathbb{K}[G/H]$ containing a non-trivial G-invariant ideal I. Note that the algebra $\mathbb{K}[L]$ may be identified with the subalgebra in $\mathbb{K}[G]$ of (left- or right-) G^u-invariant functions, $\mathbb{K}[G/H]$ is realized in $\mathbb{K}[G]$ as the subalgebra of right H-invariants, and $\mathbb{K}[L/K]$ is the subalgebra of left G^u-invariants in $\mathbb{K}[G/H]$. Consider the action of G^u on the ideal I. By the Lie-Kolchin Theorem, there is a non-zero G^u-invariant element in I. Thus the subalgebra $A \cap \mathbb{K}[L/K]$ contains the non-trivial L-invariant ideal $I \cap \mathbb{K}[L/K]$. If the space L/K is affinely closed then we get a contradiction with the following lemma.

Lemma 3.10. *Let L/K be an affinely closed space of a reductive group L. Then any L-invariant subalgebra in $\mathbb{K}[L/K]$ is finitely generated and does not contain non-trivial L-invariant ideals.*

Proof. Let $B \subset \mathbb{K}[L/K]$ be a non-finitely generated invariant subalgebra. For any chain $W_1 \subset W_2 \subset W_3 \subset \dots$ of finite-dimensional L-invariant submodules in $\mathbb{K}[L/K]$ with $\cup_{i=1}^{\infty} W_i = \mathbb{K}[L/K]$, the chain of subalgebras $B_1 \subset B_2 \subset B_3 \subset \dots$ generated by W_i does not stabilize. Hence one may suppose that all inclusions here are strict. Let Z_i be the affine L-variety corresponding the algebra B_i. The inclusion $B_i \subset \mathbb{K}[L/K]$ induces the dominant morphism $L/K \to Z_i$ and Theorem 3.2 implies that $Z_i = L/K_i$, $K \subset K_i$. But $B_1 \subset B_2 \subset B_3 \subset \cdots$, and any K_i is strictly contained in K_{i-1}, a contradiction. This shows that B is finitely generated and, as proved above, L acts transitively on the affine variety Z corresponding to B. But any non-trivial L-invariant ideal in B corresponds to a proper L-invariant subvariety in Z.   □

Theorem 3.9 is proved.   □

Corollary . *Let G/H be an affinely closed homogeneous space. Then for any affine G-variety X and a point $x \in X$ such that $Hx = x$, the orbit Gx is closed.*

Proof. The stabilizer G_x is observable in G, hence $\phi(G_x)$ is observable in L. The subgroup $\phi(G_x)$ contains $K = \phi(H)$, and Theorems 3.2, 3.9 imply that the space $L/\phi(G_x)$ is affinely closed. By Theorem 3.9, the space G/G_x is affinely closed.   □

Corollary . *If X is an affine G-variety and a point $x \in X$ is T-fixed, where T is a maximal torus of G, then the orbit Gx is closed.*

A characteristic-free description of affinely closed homogeneous spaces for solvable groups is given in [67].

3.4 The Slice Theorem

The Slice Theorem due to D. Luna [42] is one of the most important technical tools in modern invariant theory. In this text we need only some corollaries of the theorem which are related to affine embeddings [42, 56].

- Let $G/H \hookrightarrow X$ be an affine embedding with a closed G-orbit isomorphic to G/F, where F is reductive. By the Slice Theorem, we may assume that $H \subseteq F$. Then there exists an affine embedding $F/H \hookrightarrow Y$ with an F-fixed point such that X is G-isomorphic to the homogeneous fibre space $G *_F Y$. This allows one to reduce many problems to affine embeddings with a fixed point. On the other hand, this gives us a G-equivariant projection of X onto G/F.
- Let $G/H \hookrightarrow X$ be a smooth affine embedding with closed G-orbit isomorphic to G/F. Then X is a homogeneous vector bundle over G/F. In particular, if X contains a G-fixed point, then X is a vector space with a linear G-action.

3.5 Fixed-point properties

Here we list some results concerning G-fixed points in affine embeddings.

- If G/H is a quasi-affine non affinely closed homogeneous space, then G/H admits an affine embedding with a G-fixed point [9, Prop. 3].

- A homogeneous space G/H admits an affine embedding $G/H \hookrightarrow X$ such that $X = G/H \cup \{o\}$, where o is a G-fixed point, if and only if H is a quasi-parabolic subgroup of G [53, Th. 4, Cor.5]. In this case the normalization of X is an HV-variety and the normalization morphism is bijective.

- Consider the canonical decomposition $\mathbb{K}[G/H] = \mathbb{K} \oplus \mathbb{K}[G/H]_G$, where the first term corresponds to the constant functions and $\mathbb{K}[G/H]_G$ is the sum of all nontrivial simple G-submodules in $\mathbb{K}[G/H]$. Suppose that H is an observable subgroup of G. The following conditions are equivalent [9, Prop. 6]:

(1) any affine embedding of G/H contains a G-fixed point;
(2) H is not contained in a proper reductive subgroup of G;
(3) $\mathbb{K}[G/H]_G$ is an ideal in $\mathbb{K}[G/H]$.

If H is a Grosshans subgroup, then conditions (1)-(3) are equivalent to

(4) $CE(G/H)$ contains a G-fixed point.

Example 3.11. Let G be a connected semisimple group and P a parabolic subgroup containing no simple components of G. For $H = P^u$ the properties (1)-(4) hold. In fact, (3) follows from the observation that $\mathbb{K}[G/P^u]_G$ is the positive part of a G-invariant grading on $\mathbb{K}[G/P^u]$ defined by the G-equivariant action of a suitable one-parameter subgroup in the centre of the Levi subgroup of P on G/P^u [9].

Proposition 3.12. *Let H be an observable subgroup of G.*

- *If either G/H is affinely closed or H is a quasi-parabolic subgroup of G, then G/H admits only one normal affine embedding (up to G-isomorphism);*
- *if $G = \mathbb{K}^*$ and H is finite, then there exist only two normal affine embeddings, namely $\mathbb{K}^*/H$ and $\mathbb{K}/H$;*
- *in all other cases there exists an infinite sequence*

$$X_1 \xleftarrow{\phi_1} X_2 \xleftarrow{\phi_2} X_3 \xleftarrow{\phi_3} \dots$$

of pairwise nonisomorphic normal affine embeddings $G/H \hookrightarrow X_i$ and equivariant dominant morphisms ϕ_i.

Proof. The statements are obvious for affinely closed G/H and for $G = \mathbb{K}^*$. If H is a quasi-parabolic subgroup, then $\mathbb{K}[G/H]^U = \mathbb{K}[t]$. Suppose that $G/H \hookrightarrow X$ is a normal affine embedding. Then $\mathbb{K}[X]^U \subseteq \mathbb{K}[t]$ is a graded integrally closed subalgebra with $Q(\mathbb{K}[X]^U) = \mathbb{K}(t)$. This implies $\mathbb{K}[X]^U = \mathbb{K}[t]$ and $\mathbb{K}[X] = \mathbb{K}[G/H]$, hence X is G-isomorphic to the canonical embedding of G/H.

In all other cases there exists an integrally closed non-finitely generated invariant subalgebra $\mathfrak{B}$ in $\mathbb{K}[G/H]$ with $Q\mathfrak{B} = \mathbb{K}(G/H)$; see Proposition 6.4. Let $f_1, f_2, \ldots, f_n, f_{n+1}, \ldots$ be a set of generators of $\mathfrak{B}$ such that $\mathbb{K}(f_1, \ldots, f_n) = \mathbb{K}(G/H)$. Define $\mathfrak{B}_k$ as the integral closure of $\mathbb{K}[\langle Gf_1, \ldots, Gf_{n+k}\rangle]$ in $\mathfrak{B}$. The varieties $X_k = \mathrm{Spec}\,\mathfrak{B}_k$ are birationally isomorphic to G/H and hence $G/H \hookrightarrow X_k$. Infinitely many of the X_k are pairwise nonisomorphic. Renumbering, one may suppose that all X_k are nonisomorphic. The chain

$$\mathfrak{B}_1 \subset \mathfrak{B}_2 \subset \mathfrak{B}_3 \ldots$$

corresponds to the desired chain

$$X_1 \leftarrow X_2 \leftarrow X_3 \leftarrow \ldots$$

$\square$

4 Embeddings with a finite number of orbits

4.1 The characterization theorem

Spherical homogeneous spaces admit the following nice characterization in terms of equivariant embeddings.

Theorem 4.1 ([61, 45, 2]). *A homogeneous space G/H is spherical if and only if any embedding of G/H has finitely many G-orbits.*

To be more precise, F. J. Servedio proved that any affine spherical variety contains finitely many G-orbits, D. Luna, Th. Vust and D. N. Akhiezer extended this result to an arbitrary spherical variety and D. N. Akhiezer constructed a projective embedding with infinitely many G-orbits for any homogeneous space of positive complexity.

Now we are concerned with the following problem: characterize all quasi-affine homogeneous spaces G/H of a reductive group G with the property

(AF)   *For any affine embedding $G/H \hookrightarrow X$, the number of G-orbits*
in X is finite.

It follows from the results considered above that

- spherical homogeneous spaces
- affinely closed homogeneous spaces
- homogeneous spaces of the group $SL(2)$

have property (AF). Our main result in some sense gives a unification
of these three classes.

Theorem 4.2 ([11]). *For a reductive subgroup $H \subseteq G$, (AF) holds if
and only if either $W(H) = N_G(H)/H$ is finite or any extension of H by
a one-dimensional torus in $N_G(H)$ is spherical in G.*

Corollary . *For an affine homogeneous space G/H of complexity > 1,
(AF) holds if and only if G/H is affinely closed.*

Corollary . *An affine homogeneous space G/H of complexity 1 satisfies
(AF) if and only if either $W(H)$ is finite, or $\operatorname{rk} W(H) = 1$ and $N_G(H)$
is spherical.*

Corollary . *Let G be a reductive group with infinite center $Z(G)$ and H
a reductive subgroup in G that does not contain $Z(G)^0$. Then property
(AF) holds for G/H if and only if H is a spherical subgroup of G.*

    The proof of Theorem 4.2 is based on the analysis of Akhiezer's con-
struction [2] of projective embeddings and on some results of F. Knop.
We give this proof in Section 4.2, obtaining a more general result, Theorem 4.7.

    Our method applied to an arbitrary quasi-affine space G/H gives a
necessary condition for property (AF) (see the remark on page 31 below), but a characterization of quasi-affine spaces with property (AF) is
not obtained yet. Another open problem is to characterize Grosshans
subgroups H of a reductive group G such that $CE(G/H)$ contains only
a finite number of G-orbits [9].

4.2 Modality

The aim of this subsection is to generalize Theorem 4.2 following the
ideas of [3], and to find the maximal number of parameters in a continuous family of G-orbits over all affine embeddings of a given affine space
G/H.

Definition 4.3. Let $F : X$ be an algebraic group action. The integer

$$d_F(X) = \min_{x \in X} \text{codim}_X \, Fx = \text{tr.deg } \mathbb{K}(X)^F$$

is called the *generic modality* of the action. This is the number of parameters in the family of generic orbits. The *modality* of $F : X$ is the integer $\text{mod}_F \, X = \max_{Y \subseteq X} d_F(Y)$, where Y runs through F-stable irreducible subvarieties of X.

An action of modality zero is just an action with a finite number of orbits. Note that $c(X) = d_B(X)$. E. B. Vinberg [70] proved that $\text{mod}_B(X) = c(X)$ for any G-variety X. This means that if we pass from X to a B-stable irreducible subvariety $Y \subset X$, then the number of parameters for generic B-orbits does not increase. Simple examples show that the inequality $d_G(X) \leq \text{mod}_G(X)$ can be strict. This motivates the following

Definition 4.4. With any G-variety X we associate the integer

$$m_G(X) = \max_{X'} \text{mod}_G(X'),$$

where X' runs through all G-varieties birationally G-isomorphic to X.

For a homogeneous space G/H we have $m_G(G/H) = \max_X \text{mod}_G(X)$, where X runs through all embeddings of G/H.

It is clear that for any subgroup $F \subset G$ the inequality $m_G(X) \leq m_F(X)$ holds. In particular, $m_G(X) \leq c(X)$. The next theorem shows that $m_G(X) = c(X)$.

Theorem 4.5 ([3]). *There exists a projective G-variety X' birationally G-isomorphic to X such that $\text{mod}_G(X') = c(X)$.*

Now we introduce an affine counterpart of $m_G(X)$.

Definition 4.6. With any quasi-affine homogeneous space G/H we associate the integer

$$a_G(G/H) = \max_X \text{mod}_G(X),$$

where X runs through all affine embeddings $G/H \hookrightarrow X$.

Theorem 4.7 ([7]). *Let H be a reductive subgroup of G.*
  (1) *If the group $W(H)$ is finite, then $a_G(G/H) = 0$;*
  (2) *If $W(H)$ is infinite, then*

$$a_G(G/H) = \max_{H_1} \, c(G/H_1),$$

where H_1 runs through all non-trivial extensions of H by a one-dimensional subtorus of $C_G(H)$. In particular,

$$a_G(G/H) = c(G/H) \text{ or } c(G/H) - 1.$$

Proof. **Step 1** – Affine cones. Consider the natural surjection $\kappa : N_G(H) \to W(H)$.

Proposition 4.8. *Let H be an observable subgroup of G. Suppose that there is a non-trivial one-parameter subgroup $\lambda : \mathbb{K}^* \to W(H)$ and put $H_1 = \kappa^{-1}(\lambda(\mathbb{K}^*))$. Then there exists an affine embedding $G/H \hookrightarrow X$ with $\mathrm{mod}_G(X) \geq c(G/H_1)$.*

The idea of the proof is to apply Akhiezer's construction [3] to the homogeneous space G/H_1 and to consider the affine cone over a projective embedding $G/H_1 \hookrightarrow X'$ with $\mathrm{mod}_G(X') = c(G/H_1)$

Lemma 4.9. *In the notation of Proposition 4.8, there exists a finite-dimensional G-module V and an H_1-eigenvector $v \in V$ such that*

1) *the orbit $G\langle v \rangle$ of the line $\langle v \rangle$ in $\mathbb{P}(V)$ is isomorphic to G/H_1;*
2) *H fixes v;*
3) *H_1 acts transitively on $\mathbb{K}^* v$;*
4) *$\mathrm{mod}_G(\overline{G\langle v \rangle}) = c(G/H_1)$.*

Proof (of Lemma 4.9). By Chevalley's theorem, there exist a G-module V' and a vector $v' \in V'$ having property 1). Let χ be the eigenweight of H at v'. Since H is observable in G, each finite-dimensional H-module can be embedded into a finite-dimensional G-module [48]. In particular, there exists a G-module V'' containing H-eigenvectors of the weight $-\chi$. Among them we can choose an H_1-eigenvector v'' and set $V = V' \otimes V''$, $v = v' \otimes v''$. This pair has properties 1)–2).

If H_1 does not act transitively on $\mathbb{K}^* v$, then take an arbitrary G-module W containing a vector with stabilizer H. Take an H_1-eigenvector in W^H with nonzero weight and replace V by $V \otimes W$ and v by $v \otimes w$. Conditions 1)–3) are now satisfied.

By a result of Akhiezer [3], we can find a pair (V', v') with properties 1) and 4). Then we proceed as above obtaining a pair (V, v). The closure $\overline{G\langle v \rangle} \subseteq \mathbb{P}(V)$ lies in the image of the Segre embedding

$$\mathbb{P}(V') \times \mathbb{P}(V'') \times \mathbb{P}(W) \hookrightarrow \mathbb{P}(V),$$

and it projects G-equivariantly onto $\overline{G\langle v' \rangle} \subseteq \mathbb{P}(V')$. Now properties 1)–4) are satisfied for the pair (V, v). $\qquad\square$

Remark . If H is reductive, then one can find v in Lemma 4.9 such that $G_v = H$. This is not possible for an arbitrary observable subgroup, see [11, Remark 2].

Proof (of Proposition 4.8). Let (V, v) be the pair from Lemma 4.9. Put $H' = G_v$ and $\tilde{X} = \overline{Gv}$. By properties 1)-3) and since H_1/H is isomorphic to $\mathbb{K}^*$, H' is a finite extension of H. By 3), the closure of the orbit Gv in V is a cone, therefore 4) implies the inequality $\mathrm{mod}_G(\tilde{X}) \geq c(G/H_1)$.

Consider now the morphism $G/H \to G/H'$. It determines an embedding $\mathbb{K}[G/H'] \subseteq \mathbb{K}[G/H]$. Let A be the integral closure of the subalgebra $\mathbb{K}[\tilde{X}] \subseteq \mathbb{K}[G/H']$ in the field $\mathbb{K}(G/H)$. We have the following commutative diagrams:

$$
\begin{array}{ccccc}
A & \hookrightarrow & \mathbb{K}[G/H] & \hookrightarrow & \mathbb{K}(G/H) \\
\uparrow & & \uparrow & & \uparrow \\
\mathbb{K}[\tilde{X}] & \hookrightarrow & \mathbb{K}[G/H'] & \hookrightarrow & \mathbb{K}(G/H')
\end{array}
\qquad
\begin{array}{ccc}
\mathrm{Spec}\,A & \hookleftarrow & G/H \\
\downarrow & & \downarrow \\
\tilde{X} & \hookleftarrow & G/H'
\end{array}
$$

The affine variety $X = \mathrm{Spec}\,A$ with the natural G-action can be regarded as an affine embedding of G/H. The embedding $\mathbb{K}[\tilde{X}] \subseteq A$ defines a finite (surjective) morphism $X \to \tilde{X}$, therefore

$$\mathrm{mod}_G(X) = \mathrm{mod}_G(\tilde{X}) \geq c(G/H_1).$$

$\square$

Step 2. Here we formulate several results due to F. Knop.

Lemma 4.10 ([34, 7.3.1], see also [11, Lemma 3]). *Let X be an irreducible G-variety, and v a G-invariant valuation of $\mathbb{K}(X)$ over $\mathbb{K}$ with residue field $\mathbb{K}(v)$. Then $\mathbb{K}(v)^B$ is the residue field of the restriction of v to $\mathbb{K}(X)^B$.*

Definition 4.11 ([36, §7]). Let X be a normal G-variety. A discrete $\mathbb{Q}$-valued G-invariant valuation of $\mathbb{K}(X)$ is said to be *central* if it vanishes on $\mathbb{K}(X)^B \setminus \{0\}$. A *source* of X is a non-empty G-stable subvariety $Y \subseteq X$ that is the center of a central valuation of $\mathbb{K}(X)$.

The following lemma is an easy consequence of [36]; for more details see [11, Lemma 4].

Lemma 4.12. *If X is a normal affine G-variety containing a proper source, then there exists a one-dimensional torus $S \subseteq \mathrm{Aut}_G(X)$ such that $\mathbb{K}(X)^B \subseteq \mathbb{K}(X)^S$. (Here $\mathrm{Aut}_G(X)$ is the group of G-equivariant automorphisms of X.)*

Step 3. Assertion (1) of Theorem 4.7 follows from Theorem 3.2. To prove (2) we use Proposition 4.8. Since H is reductive, the group $W(H)$ is reductive and contains a one-dimensional subtorus $\lambda(\mathbb{K}^*)$. Hence $a_G(G/H) \geq c(G/H_1) \geq c(G/H) - 1$. If there exists a one-dimensional torus in $W(H)$ such that $c(G/H) = c(G/H_1)$, we obtain an affine embedding of G/H of modality $c(G/H)$.

Conversely, let $G/H \hookrightarrow X$ be an affine embedding of modality $c(G/H)$. We have to find a one-dimensional subtorus $\lambda(\mathbb{K}^*) \subseteq W(H)$ such that $c(G/H_1) = c(G/H)$. By the definition of modality, there exists a proper G-invariant subvariety $Y \subset X$ such that the codimension of a generic G-orbit in Y is $c(G/H)$, hence $c(Y) = c(G/H)$. Consider a G-invariant valuation v of $\mathbb{K}(X)$ with centre Y. For the residue field $\mathbb{K}(v)$ we have tr.deg $\mathbb{K}(v)^B \geq$ tr.deg $\mathbb{K}(Y)^B$, therefore tr.deg $\mathbb{K}(v)^B =$ tr.deg $\mathbb{K}(X)^B$. If the restriction of v to $\mathbb{K}(X)^B$ is non-trivial, then, by Lemma 4.10, tr.deg $\mathbb{K}(v)^B <$ tr.deg $\mathbb{K}(X)^B$, a contradiction. Thus, v is central and Y is a source of X. Lemma 4.12 provides a one-dimensional subtorus $S \subseteq \mathrm{Aut}_G(X) \subseteq \mathrm{Aut}_G(G/H) = W(H)$ that yields an extension of H of the same complexity. □

Note that Theorem 4.2 is a particular case of Theorem 4.7 with $a_G(G/H) = 0$.

Remark . If H is an observable subgroup and $W(H)$ contains a non-trivial subtorus, then the formula $a_G(G/H) = \max_{H_1} c(G/H_1)$ can be obtained by the same arguments. In particular, Corollary 4.1 holds for observable H. But for non-reductive H the group $W(H)$ can be unipotent [11]: this is the case when $G = SL(3) \times SL(3)$ and

$$H = \left\{ \begin{pmatrix} 1 & a & b + \frac{a^2}{2} \\ 0 & 1 & a \\ 0 & 0 & 1 \end{pmatrix}, \begin{pmatrix} 1 & b & a + \frac{b^2}{2} \\ 0 & 1 & b \\ 0 & 0 & 1 \end{pmatrix} \mid a, b \in \mathbb{K} \right\}.$$

For such subgroups our proof yields only the inequality $a_G(G/H) \leq c(G/H) - 1$.

Let us mention an application of Theorem 4.7 which may be regarded as its algebraic reformulation. Let G be a connected semisimple group. Note that, for the action by left multiplication, one has $c(G) = \frac{1}{2}(\dim G - \mathrm{rk}\, G)$ and $c(G/S) = \frac{1}{2}(\dim G - \mathrm{rk}\, G) - 1$, where S is a one-dimensional subtorus in G. Applying Theorem 4.7 to the case $H = \{e\}$, we obtain

Theorem 4.13 ([8]). *Let $A \subset \mathbb{K}[G]$ be a left G-invariant finitely generated subalgebra and $I \subset A$ a G-invariant prime ideal. Then*

$$\mathrm{tr.deg}\,(Q(A/I))^G \;\leq\; \frac{1}{2}\,(\dim G - \mathrm{rk}\,G) - 1. \qquad (*)$$

Moreover, there exist a subalgebra A and an ideal I such that $(*)$ is an equality.

Example 4.14. The closure of an $SL(3)$-orbit in an algebraic $SL(3)$-variety X may contain at most a 3-parameter family of $SL(3)$-orbits. If X is affine then the maximal number of parameters equals 2.

4.3 Equivariant automorphisms and symmetric embeddings

The group $\mathrm{Aut}_G(G/H)$ of G-equivariant automorphisms of G/H is isomorphic to $W(H)$. The action $W(H) : G/H$ is induced by the action $N_G(H) : G/H$ by right multiplication, i.e. $n * gH = gn^{-1}H$. Let $G/H \hookrightarrow X$ be an embedding. The group $\mathrm{Aut}_G X$ preserves the open orbit, and may be considered as a subgroup of $W(H)$.

Definition 4.15. An embedding $G/H \hookrightarrow X$ is said to be *symmetric* if $W(H)^0 \subseteq \mathrm{Aut}_G(X)$. If $\mathrm{Aut}_G(X) = W(H)$, we say that X is *very symmetric*.

Lemma 4.16. *The following affine embeddings are very symmetric:*

1) *an affine embedding of a spherical homogeneous space;*
2) *the canonical embedding $CE(G/H)$;*
3) *an affine monoid M considered as the embedding $G(M)/\{e\} \hookrightarrow M$.*

Proof. 1) Let G/H be a quasi-affine spherical homogeneous space. By the Schur Lemma, the group $W(H)$ acts on any isotypic component of $\mathbb{K}[G/H]$ by dilation. Hence any G-invariant subspace of $\mathbb{K}[G/H]$ is also $W(H)$-invariant.
2) The group $W(H)$ acts on G/H and on $\mathbb{K}[G/H]$, thus on $\mathrm{Spec}\,\mathbb{K}[G/H]$.
3) The group $W(H) \cong G(M)$ acts on M by right multiplication.   □

Proposition 4.17. *Let H be a reductive subgroup of G. The following conditions are equivalent:*

(1) *there exists a unique symmetric embedding of G/H, namely $X = G/H$;*
(2) *$W(H)^0$ is a semisimple group.*

Proof. The existence of a non-trivial affine embedding $G/H \hookrightarrow X$ with $\dim \operatorname{Aut}_G(X) = \dim W(H)$ means that G/H is not affinely closed as a $(G \times W(H)^0)$-homogeneous space. Denote by L the $(G \times W(H)^0)$-stabilizer of the point eH. Then $L = \{(n, nH) \mid n \in \kappa^{-1}(W(H)^0)\}$ and $N_{G \times W(H)^0}(L)/L$ is finite if and only if $W(H)$ is semisimple. $\quad\square$

Proposition 4.17 implies that in the case of affine $SL(2)$-embeddings only the trivial embedding $X = SL(2)$ is symmetric. In fact, in all other cases with normal X the group $\operatorname{Aut}_{SL(2)} X$ is a Borel subgroup of $SL(2)$ [37, III.4.8, Satz 1]. The theorem below is a partial generalization of this result.

Theorem 4.18 ([12]). *Let $G/H \hookrightarrow X$ be an affine embedding with a finite number of G-orbits and with a G-fixed point. Then the group $\operatorname{Aut}_G(X)^0$ is solvable.*

We begin the proof with the following

Lemma 4.19. *Let X be an affine variety with an action of a connected semisimple group S. Suppose that there is a point $x \in X$ and a one-parameter subgroup $\gamma : \mathbb{K}^* \to S$ such that $\lim_{t \to 0} \delta(t)x$ exists in X for any subgroup δ conjugate to γ. Then x is a $\gamma(\mathbb{K}^*)$-fixed point.*

Proof. Let T be a maximal torus in S containing $\gamma(\mathbb{K}^*)$. One can realize X as a closed S-stable subvariety in V for a suitable S-module V. Let $x = x_{\lambda_1} + \cdots + x_{\lambda_n}$ be the weight decomposition (with respect to T) of x with weights $\lambda_1, \ldots, \lambda_n$. One-parameter subgroups of T form the lattice $\Xi_*(T)$ dual to the character lattice $\Xi(T)$. The existence of $\lim_{t \to 0} \gamma(t)x$ in X means that all pairings $\langle \gamma, \lambda_i \rangle$ are non-negative. Let $\gamma_1, \ldots, \gamma_m$ be all the translates of γ under the action of the Weyl group $W = N_S(T)/T$. By assumption, $\langle \gamma_j, \lambda_i \rangle \geq 0$ for any $i = 1, \ldots, n$, $j = 1, \ldots, m$, hence $\langle \gamma_1 + \cdots + \gamma_m, \lambda_i \rangle \geq 0$. Since $\gamma_1 + \cdots + \gamma_m = 0$, one has $\langle \gamma_j, \lambda_i \rangle = 0$ for all i, j. This shows that the points x_{λ_i} (and x) are $\gamma(\mathbb{K}^*)$-fixed. $\quad\square$

The next proposition is a generalization of [28, Th. 4.3].

Proposition 4.20. *Suppose that $G/H \hookrightarrow X$ is an affine embedding with a non-trivial G-equivariant action of a connected semisimple group S. Then the orbit $S * x$ is closed in X, for every $x \in G/H$.*

Proof. We may assume $x = eH$. If $S * x$ is not closed, then by [31, Th. 1.4] there is a one-parameter subgroup $\gamma : \mathbb{K}^* \to S$ such that the limit

$$\lim_{t \to 0} \gamma(t) * x$$

exists in X and does not belong to $S * x$. Replacing S by a finite cover, we may assume that S embeds in $N_G(H)$ (and thus in G) with a finite intersection with H. By the definition of $*$-action, one has $\gamma(t) * x = \gamma(t^{-1})x$. For any $s \in S$ the limit

$$\lim_{t \to 0}(s\gamma(t)) * x = \lim_{t \to 0} \gamma(t^{-1})s^{-1}x$$

exists. Hence $\lim_{t \to 0} s\gamma(t^{-1})s^{-1}x$ exists too. This shows that for any one-parameter subgroup δ of S, conjugate to $-\gamma$, $\lim_{t \to 0} \delta(t)x$ exists in X. Lemma 4.19 implies that $x = \lim_{t \to 0} \gamma(t)*x$, and this contradiction proves Proposition 4.20. □

Proof (of Theorem 4.18). Suppose that $\mathrm{Aut}_G(X)^0$ is not solvable. Then there is a connected semisimple group S acting on X G-equivariantly. By Proposition 4.20, any $(S, *)$-orbit in the open G-orbit of X is closed in X.

Let X_1 be the closure of a G-orbit in X. Since G has a finite number of orbits in X, the variety X_1 is $(S, *)$-stable. Applying the above arguments to X_1, we show that any $(S, *)$-orbit in X is closed. But in this case all $(S, *)$-orbits have the same dimension $\dim S$. On the other hand, a G-fixed point is an $(S, *)$-orbit, a contradiction. □

Corollary (of the proof). *Let X be an affine G-variety with an open G-orbit. Suppose that*

- *a semisimple group S acts on X effectively and G-equivariantly;*
- *the dimension of a closed G-orbit in X is less than $\dim S$.*

Then the number of G-orbits in X is infinite.

Corollary . *Let M be a reductive algebraic monoid with zero. Then the number of left (right) $G(M)$-cosets in M is finite if and only if M is commutative.*

The following corollary gives a partial answer to a question posed in Subsection 4.1.

Corollary . *The number of G-orbits in $CE(G/P^u)$ is finite if and only if either $P \cap G_i = G_i$ or $P \cap G_i = B \cap G_i$ for each simple factor $G_i \subseteq G$.*

In many cases, Theorem 4.18 may be used to show that the group $\mathrm{Aut}_G(X)$ cannot be very big. On the other hand, the group $\mathrm{Aut}_G(X)$ may be finite (trivial), in particular, for $X = G/H$ with affinely closed G/H. Answering a question from [11], I. V. Losev proposed an example of an observable non-reductive subgroup H in $SL(n)$, where $W(H)$ is finite. (This example is included in the electronic version of [12].) Note that any affine embedding of $SL(n)/H$ gives an example of a locally transitive non-transitive reductive group action on an affine variety with a finite group of equivariant automorphisms.

Finally, we give a variant of Theorem 4.2 for symmetric embeddings.

Theorem 4.21 ([11, Prop. 2]). *Let H be a reductive subgroup of G. Every symmetric affine embedding of G/H has finitely many G-orbits if and only if either (AF) holds or $W(H)^0$ is semisimple.*

5 Application One: Invariant algebras on homogeneous spaces of compact Lie groups

5.1 *Invariant algebras and self-conjugate algebras*

For any compact topological space M the set $C(M)$ of all continuous $\mathbb{C}$-valued functions on M is a commutative Banach algebra with respect to pointwise addition, multplication, and the uniform norm. We shall consider the case where $M = K/L$ is a homogeneous space of a compact connected Lie group K. Let us recall that A is an *invariant algebra* on M if A is a K-invariant uniformly closed subalgebra with unit in $C(M)$. In this section G, H denote the complexifications of K, L respectively. The group G is a complex reductive algebraic group with a reductive subgroup H.

The main problem is to describe all invariant algebras on a given space M and to study their properties. Let us start with a particular class of invariant algebras.

Definition 5.1. An invariant algebra A is *self-conjugate* if $f \in A$ implies $\overline{f} \in A$, where the bar denotes the complex conjugation.

The classification of self-conjugate invariant algebras is based on the Stone-Weierstrass Theorem. Here we follow [39].

The Stone-Weierstrass Theorem. *Let R be a compact topological space and A a subalgebra with unit in $C(R)$ such that*

- *A separates points on R, i.e. for any $x_1 \neq x_2 \in R$ there exists $f \in A$ such that $f(x_1) \neq f(x_2)$;*
- *A is invariant under complex conjugation.*

Then A is dense in $C(R)$.

Given a self-conjugate invariant algebra A, define an equivalence relation on M: $x \sim y$ if and only if $f(x) = f(y)$ for any $f \in A$. The space M' of equivalence classes is a homogeneous K-space, hence $M' = K/L'$, where L' is a closed subgroup containing L. By construction, the self-conjugate algebra A separates points on M' and thus $A = C(M')$. Conversely, for any $L \subseteq L' \subseteq K$ the inverse image of $C(K/L')$ under the projection $K/L \to K/L'$ determines a self-conjugate invariant algebra on M. This shows that self-conjugate invariant algebras on M are in one-to-one correspondence with closed subgroups L', $L \subseteq L' \subseteq K$.

5.2 Spherical functions

The space $M = K/L$ may be considered as a compact subset of the affine homogeneous space $X_0 = G/H$. Moreover, M is a real form of X_0 in the natural sense. In particular, the restriction of polynomial functions to M determines an embedding $\mathbb{C}[X_0] \hookrightarrow C(M)$. Denote the image of this embedding by $\mathbb{C}[M]$.

Definition 5.2. A function $f \in C(M)$ is called *spherical* if the linear span $\langle Kf \rangle$ is finite-dimensional. More generally, for a linear action of a Lie group K on vector space V, a vector $v \in V$ is *spherical* if $\langle Kv \rangle$ is finite-dimensional.

Denote by V_{sph} the subspace of all spherical vectors in V.

Proposition 5.3. *The algebra $\mathbb{C}[M]$ coincides with $C(M)_{sph}$.*

Proof. Any regular function is contained in a finite-dimensional invariant subspace. Conversely, any complex finite-dimensional representation of K is completely reducible and any irreducible component may be considered as a simple G-module. Hence the matrix entries of such a module are in $\mathbb{C}[M]$. If $f \in C(M)$ is spherical and $V = \langle Kf \rangle$, then f is a linear combination of the matrix entries of the dual representation $K : V^*$. Indeed, let $f_1, \ldots, f_k$ be a basis in V. For any $f \in V$, $g \in K$ one has $f_i(g^{-1}eL) = \sum a_{ij}(g)f_j(eL)$ and $f_i(gL) = \sum c_j a_{ij}(g^{-1})$, where $c_j = f_j(eL)$ are constants. $\qquad \square$

By the Peter-Weyl Theorem, the matrix entries (with respect to some orthonormal basis) over all irreducible finite-dimensional representations of K form an orthonormal basis in space $L^2(K)$. Spherical functions are finite linear combinations of the basic elements. They form a uniformly dense subspace in $C(K)$. The following generalization of this result plays a key role in this section.

Proposition 5.4 ([51, Th. 5.1], [47, 2.16]). *Given a continuous linear representation of a compact Lie group K in a Fréchet space E, the subspace E_{sph} is dense in E.*

In particular, in any invariant algebra, spherical functions form a dense subalgebra. Moreover, if S is K-invariant subspace in $C(M)_{sph}$ and $\overline{S}$ is its uniform closure in $C(M)$, then $\overline{S} \cap C_{sph}(M) = S$. (For the proof see [26, Lemma 14].) Finally, we get

Theorem 5.5. *There is a natural bijection ψ between invariant algebras on the space M and invariant subalgebras in $\mathbb{C}[M]$. More precisely, $\psi(A) = \mathfrak{A} = A_{sph} = A \cap \mathbb{C}[M]$ and $\psi^{-1}(\mathfrak{A}) = \overline{\mathfrak{A}}$.*

This result provides nice connections between functional and algebraic problems. To make this link really useful we need to reformulate functional properties in algebraic terms and conversely. For this purpose we are going to use the geometric language of affine embeddings.

5.3 Finitely generated invariant algebras and affine embeddings

Definition 5.6. An invariant algebra A is *finitely generated* if it is generated (as a Banach algebra) by a K-invariant finite-dimensional subspace.

An invariant algebra A is finitely generated if and only if A_{sph} is a finitely generated algebra. It is clear that $C(M)$ is finitely generated. As follows from the discussion above, any self-conjugate invariant algebra is finitely generated. The question as to when any invariant subalgebra in $\mathbb{C}[M]$ is finitely generated will be considered in the last section.

Any finitely generated subalgebra $\mathfrak{A} \subset \mathbb{C}[G/H]$ defines an affine G-variety $X = \operatorname{Spec}\mathfrak{A}$ with an open orbit isomorphic to G/F, where F is an observable subgroup containing H. The inclusion $\mathfrak{A} \subset \mathbb{C}[G/H]$ defines the morphism $\phi : G/H \to X$ and the base point $x_0 = \phi(eH)$. If

we look at $\mathfrak{A}$ as at an abstract G-algebra, then there may exist different equivariant inclusion homomorphisms $\mathfrak{A} \to \mathbb{C}[G/H]$ with the same
image. Two different base points $x_0 \in X$ and $x_0' \in X$ determine the
same subalgebra $\mathfrak{A} \subset \mathbb{C}[G/H]$ if and only if there exists $n \in \mathrm{Aut}_G(X)$
such that $x_0 = nx_0'$. (Corresponding inclusions $\mathfrak{A} \subset \mathbb{C}[G/H]$ differ by a
G-equivariant automorphism of $\mathfrak{A}$.) Let us denote the subalgebra $\mathfrak{A}$ by
$\mathfrak{A}(X, x_0)$ and the corresponding invariant algebra $\overline{\mathfrak{A}(X, x_0)}$ by $A(X, x_0)$.
We have proved:

Theorem 5.7. *Invariant finitely generated algebras on the space* $M = K/L$ *are in one-to-one correspondence with the following data:*

- *an affine embedding* $G/F \hookrightarrow X$, *where* $F \subseteq G$ *is an observable sub-group containing* H;
- *an H-fixed point* x_0 *in the open G-orbit on X, which is defined up to the action of* $\mathrm{Aut}_G(X)$.

It is natural to classify invariant algebras up to some equivalence.
The group of K-equivariant automorphisms of M is the group $N = N_K(L)/L$, acting as $n*kL = kn^{-1}L$. This action defines a K-equivariant
action $N : C(M)$. The group N acts transitively on the set M^L.

Definition 5.8. Two invariant algebras A_1 and A_2 on M are *equivalent*
if there exists $n \in N$ such that $n * A_1 = A_2$.

Clearly, this equivalence preserves all reasonable properties of invariant algebras. In terms of Theorem 5.7, it is reasonable to expect that
base points from the same K-orbit in X determine equivalent invariant
algebras.

Definition 5.9. Two invariant algebras $A(X, x_0)$ and $A(X', x_0')$ on M
are *weakly equivalent* if $X \cong_G X'$ and there exist $n \in \mathrm{Aut}_G(X)$ and
$k \in K$ such that $x_0 = n * kx_0'$.

An invariant algebra A on M may be regarded as an invariant algebra
$\tilde{A}$ on K such that every element $f \in \tilde{A}$ is fixed by right L-multiplication.
Two such subalgebras A_1 and A_2 are weakly equivalent if A_1 may be
shifted to A_2 by the map $R(k) : f(x) \to f(xk)$ for some $k \in K$.

Clearly, equivalent invariant algebras are weakly equivalent, but the
converse is not always true. One may suppose that $x_0 = kx_0'$ ($\mathrm{Aut}_G(X)$-
action does not change the subalgebra). Consider the subgroups $L_1 = K_{x_0}$, $L_2 = K_{x_0'}$, and the map $\phi : K/L \to X$, $\phi(eL) = x_0$. Denote
by $\mathrm{Aut}(X, x_0)$ the subgroup of $\mathrm{Aut}_G(X)$ that preserves Kx_0. (In fact,
$\mathrm{Aut}(X, x_0) \subset N_K(L_1)/L_1$.)

Definition 5.10. A closed subgroup $L \subset K$ is an *A-subgroup* if any two weakly equivalent finitely generated invariant algebras on $M = K/L$ are equivalent.

Proposition 5.11. *A subgroup $L \subset K$ is an A-subgroup if and only if for any affine embedding $G/F \hookrightarrow X$, $H \subset F$, and any base point $x_0 \in (G/F)^H$ one has* $\mathrm{Aut}(X, x_0)\phi((K/L)^L) = (Kx_0)^L$.

Proof. Let $x_0' = kx_0$ be an L-fixed point. The equivalence of invariant algebras $\mathfrak{A}(X, x_0)$ and $\mathfrak{A}(X, x_0')$ means that there is an element $n \in N_K(L)$ such that $\mathfrak{A}(X, nx_0) = \mathfrak{A}(X, x_0')$, i.e. nx_0 and x_0' are in the same $\mathrm{Aut}_G(X)$-orbit. If $m \in \mathrm{Aut}_G(X)$ and $m * nx_0 = x_0'$, then $m \in \mathrm{Aut}(X, x_0)$. But the set of points nx_0, $n \in N_K(L)$, coincides with $\phi((K/L)^L)$. $\qquad\square$

If for any $L \subseteq L_1$ the natural map $(K/L)^L \to (K/L_1)^L$ is surjective, then L is an A-subgroup. In particular, the unit subgroup and any maximal subgroup in K are A-subgroups.

Corollary . *If L is an A-subgroup, two subgroups L_1 and L_2 contain L and are K-conjugate, then they are $N_K(L)$-conjugate.*

Proof. On K/L_1 any point fixed by L has the form $m * nL_1$, where $m \in N_K(L_1)$ and $n \in N_K(L)$. In particular, for $L_2 = kL_1k^{-1}$, $k \in K$, one has $kL_1 = m * nL_1$ and $L_2 = nm^{-1}L_1mn^{-1} = nL_1n^{-1}$. $\qquad\square$

Example 5.12. Put $K = SU(5)$, $L = \{e\} \times \{e\} \times \{e\} \times SU(2)$, $L_1 = SU(2) \times SU(3)$, $L_2 = SU(3) \times SU(2)$ as shown on the picture. Here L_1 and L_2 are K-conjugate, contain L, but are not $N_K(L)$-conjugate. This proves that L is not an A-subgroup.

5.4 Some classes of invariant algebras

The results of Subsection 5.1 and Theorem 5.5 imply:

Proposition 5.13 ([39]). *An invariant algebra $A = A(X, x_0)$ is self-conjugate if and only if $X = Gx_0$ and G_{x_0} is the complexification of K_{x_0}.*

Remark. There is one more characterization of this class of K-orbits obtained by V. M. Gichev and I. A. Latypov. Consider any G-equivariant embedding of X into a G-module V. Then the conditions of Proposition 5.13 are equivalent to the polynomial convexity of the orbit Kx_0 in V; see [26] for details.

The following theorem due to I. A. Latypov may be regarded as a variant of Luna's theorem (see 3.1) for compact groups.

Theorem 5.14 ([38]). *Any invariant algebra on M is self-conjugate if and only if the group $N = N_K(L)/L$ is finite.*

In this case any invariant algebra on M is finitely generated. It follows from the results of Section 6 that any invariant algebra on M is finitely generated if and only if either N is finite or $K = U(1)$. (Here we assume that the action $K : M$ is effective.)

Now we introduce a class of invariant algebras, which are in some sense opposite to self-conjugate algebras.

Definition 5.15. An invariant algebra A is said to be *antisymmetric* if the set $\{f \in A \mid \overline{f} \in A\}$ coincides with the set of constant functions.

It is easy to see that antisymmetry is equivalent to either of the following conditions:

- any real-valued function in A is a constant;
- A contains no non-trivial self-conjugate invariant subalgebra.

Hence an invariant algebra $A = A(X, x_0)$ is antisymmetric if and only if there exists no G-equivariant map $\phi : X \to G/H'$, where G/H' is an affine homogeneous space of positive dimension and $G_{\phi(x_0)}$ is the complexification of $K_{\phi(x_0)}$. In particular, if X contains a G-fixed point, then $A(X, x_0)$ is antisymmetric.

Example 5.16. Let $K = SU(2)$, $G = SL(2)$, and $L = H = \{e\}$. Consider $X = SL(2)/T$. Any point $x_0 \in X$ may be regarded as a base point for some invariant algebra $A(X, x_0)$ on $M = K$. If the stabilizer of x_0 contains a torus from K, then $A(X, x_0)$ is self-conjugate, and any two such invariant algebras are equivalent. Other base points determine antisymmetric algebras: we obtain a 1-parameter family of mutually non-equivalent antisymmetric invariant algebras on $SU(2)$. In particular, this example shows that the property '$A(X, x_0)$ separates points on M' depends on the choice of the base point x_0 on X. For more information on invariant algebras on $SU(2)$, see [40].

Finally we consider one more natural class of invariant algebras.

Definition 5.17. An invariant algebra A on M is called a *Dirichlet algebra* if the real parts of functions from A are uniformly dense in the algebra of real-valued continuous functions on M.

Any Dirichlet algebra separates the points of M, but the converse is not true. Some results on Dirichlet invariant algebras on compact groups can be found in [58]. In particular, it is proved there that there exists a biinvariant antisymmetric Dirichlet algebra on K if and only if K is connected and commutative. It would be interesting to characterize Dirichlet algebras $A(X, x_0)$ in terms of affine embeddings.

5.5 Biinvariant algebras and invariant algebras on spheres.

A *biinvariant* algebra on K is a uniformly closed subalgebra with unit in $C(K)$ invariant with respect to both left and right translations (here $M = (K \times K)/\Delta(K)$).

Suppose that F is a subgroup in $G \times G$ containing $\Delta(G)$. Then the subgroup $F_0 = \{g \in G \mid (g, e) \in F\}$ is normal in G. This shows that F is the preimage of $\Delta(\tilde{G})$ for the homomorphism $G \times G \to \tilde{G} \times \tilde{G}$, where $\tilde{G} = G/F_0$. Moreover, $\Delta(G)$-fixed points in $(\tilde{G} \times \tilde{G})/\Delta(\tilde{G})$ correspond to central elements of $\tilde{G}$. These elements form an orbit of the center $Z(\tilde{G})$, and $Z(\tilde{G})$ acts $(\tilde{G} \times \tilde{G})$-equivariantly on any affine embedding of $(\tilde{G} \times \tilde{G})/\Delta(\tilde{G})$. Hence different base points on such embeddings define the same invariant algebras. An affine embedding of the space $(\tilde{G} \times \tilde{G})/\Delta(\tilde{G})$ is nothing else but an algebraic monoid $\tilde{S}$ with $G(\tilde{S}) = \tilde{G}$ (Proposition 1.12).

Let us summarize all these observations in the following one-to-one correspondences (all biinvariant algebras are supposed to be finitely generated):

- $\left\{ \begin{array}{c} \text{self-conjugate biinvariant} \\ \text{algebras on } K \end{array} \right\} \Longleftrightarrow \left\{ \begin{array}{c} \text{quotient groups } \tilde{G} \\ \text{of the group } G \end{array} \right\};$

- $\{\text{biinvariant algebras on } K\} \Longleftrightarrow \left\{ \begin{array}{c} \text{algebraic monoids } \tilde{S} \\ \text{with } G(\tilde{S}) = \tilde{G} \end{array} \right\};$

- $\left\{ \begin{array}{c} \text{biinvariant algebras separating} \\ \text{points on } K \end{array} \right\} \Longleftrightarrow \left\{ \begin{array}{c} \text{algebraic monoids } S \\ \text{with } G(S) = G \end{array} \right\};$

$$\bullet \left\{ \begin{array}{c} \text{antisymmetric biinvariant} \\ \text{algebras on } K \end{array} \right\} \Longleftrightarrow \left\{ \begin{array}{c} \text{algebraic monoids } \tilde{S} \text{ with zero} \\ \text{and } G(\tilde{S}) = \tilde{G} \end{array} \right\}.$$

To explain the last equivalence, we note that $\tilde{S}$ has a zero if and only if the closed $(\tilde{G} \times \tilde{G})$-orbit in $\tilde{S}$ is a point. Embeddings with a G-fixed point correspond to antisymmetric invariant algebras (see 5.4). If the closed orbit has positive dimension, it is isomorphic to $(\tilde{G}_1 \times \tilde{G}_1)/\Delta(\tilde{G}_1)$ for a non-trivial quotient $\tilde{G}_1$ of the group $\tilde{G}$, and the corresponding projection (see Section 3.4) determines a non-trivial self-conjugate subalgebra in our invariant algebra.

Theorem 5.14 (or Proposition 3.4) shows that any biinvariant algebra on K is self-conjugate if and only if K is semisimple. This result was proved by R. Gangolli [24] and J. Wolf [73].

Our final remark concerns invariant algebras on spheres S^n. The classification of transitive actions of compact Lie groups on spheres was obtained by A. Borel, D. Montgomery and H. Samuelson (see [50]). All corresponding homogeneous spaces are spherical with a unique exception: there is a transitive action of $Sp(n) = GL(n, \mathbb{H}) \cap U(2n)$ on S^{4n-1} with stabilizer $Sp(n-1)$, and the complexification of $Sp(n)/Sp(n-1)$ is a homogeneous space of complexity one. (This is the reason why the clasification of invariant algebras on spheres was not completed in this case only, see [39].)

The complexification of $Sp(n)/Sp(n-1)$ satisfies the conditions of Theorem 4.2. This implies the following general result: the number of radical invariant ideals in any invariant algebra on a sphere (with respect to any transitive action) is finite.

6 Application Two: G-algebras with finitely generated invariant subalgebras

6.1 The reductive case

In this section by $\mathfrak{A}$ we denote a finitely generated G-algebra without zero divisors. Let us introduce three special types of G-algebras.

Type C.   Here $\mathfrak{A}$ is a finitely generated domain of Krull dimension $\mathrm{Kdim}\,\mathfrak{A} = 1$ (i.e. the transcendence degree of the quotient field $Q\mathfrak{A}$ equals one) with any (for example, trivial) G-action. Such algebras may be considered as the algebras of regular functions on irreducible affine curves.

Type HV. Let λ be a dominant weight of the group G (with respect to some fixed Borel subgroup) and $V(\lambda)$ be a simple finite-dimensional G-module with highest weight λ. Let λ^* be the highest weight of the dual module $V(\lambda)^*$. Consider a subsemigroup P in the additive semigroup of non-negative integers (it is automatically finitely generated), and put

$$\mathfrak{A}(P, \lambda) = \oplus_{p \in P} V(p\lambda).$$

There exists a unique structure (up to G-isomorphism) of a G-algebra on $\mathfrak{A}(P, \lambda)$ such that $V(p\lambda)V(m\lambda) = V((p+m)\lambda)$. In fact, consider the closure $X(\lambda) = \overline{Gv}$ of the orbit of a highest weight vector v in $V(\lambda^*)$. The algebra $\mathbb{K}[X(\lambda)]$ of regular functions on $X(\lambda)$ as a G-module has the isotypic decomposition $\mathbb{K}[X(\lambda)] = \oplus_{k \geq 0} \mathbb{K}[X(\lambda)]_{k\lambda}$, any $\mathbb{K}[X(\lambda)]_{k\lambda}$ is a simple G-module, and (see Section 1.3)

$$\mathbb{K}[X(\lambda)]_{k\lambda}\mathbb{K}[X(\lambda)]_{m\lambda} = \mathbb{K}[X(\lambda)]_{(k+m)\lambda}.$$

This allows us to realize $\mathfrak{A}(P, \lambda)$ as a subalgebra in $\mathbb{K}[X(\lambda)]$. The proof of uniqueness of such multiplication is left to the reader. Further we shall say that the algebra $\mathfrak{A}(P, \lambda)$ is an algebra of type HV.

Example 6.1. Let $G = SL(n)$ and $\omega_1, \ldots, \omega_{n-1}$ be its fundamental weights. The natural linear action $G : \mathbb{K}^n$ induces an action on regular functions

$$G : \mathfrak{A} = \mathbb{K}[x_1, \ldots, x_n], \quad (gf)(v) := f(g^{-1}v).$$

The homogeneous polynomials of degree m form an (irreducible) isotypic component corresponding to the weight $m\omega_{n-1}$. The algebra $\mathfrak{A}$ is of type HV with $\lambda = \omega_{n-1}$ and $P = \mathbb{Z}_+$. The variety $X(\omega_{n-1})$ is the original space $\mathbb{K}^n$.

Type N. Let H be a closed subgroup of G and

$$\mathfrak{A}(H) = \mathbb{K}[G]^H = \mathbb{K}[G/H]$$
$$= \{f \in \mathbb{K}[G] \mid f(gh) = f(g) \text{ for any } g \in G, \ h \in H\}.$$

If H is reductive, then $\mathfrak{A}(H)$ is finitely generated. We say that a G-algebra $\mathfrak{A}$ is of type N if there exists a reductive subgroup $H \subset G$ with $|N_G(H)/H| < \infty$ and $\mathfrak{A}$ is G-isomorphic to $\mathfrak{A}(H)$.

Example 6.2. The algebra

$$\mathfrak{A}(T) = \{f \in \mathbb{K}[G] \mid f(gt) = f(g) \text{ for any } t \in T\}$$

is a G-algebra of type N with respect to the left G-action.

Now we are ready to formulate the main result.

Theorem 6.3 ([9]). *Let $\mathfrak{A}$ be a finitely generated G-algebra without zero divisors. Then any G-invariant subalgebra of $\mathfrak{A}$ is finitely generated if and only if $\mathfrak{A}$ is an algebra of one of the types* C, HV *or* N.

We start the proof of Theorem 6.3 with a method of constructing a non-finitely generated subalgebra. Let X be an irreducible affine algebraic variety and Y a proper closed irreducible subvariety. Consider the subalgebra

$$\mathfrak{A}(X,Y) = \{f \in \mathbb{K}[X] \mid f(y_1) = f(y_2) \text{ for any } y_1, y_2 \in Y\} \subset \mathfrak{A} = \mathbb{K}[X].$$

Proposition 6.4. *The algebra $\mathfrak{A}(X,Y)$ is finitely generated if and only if Y is a point.*

Proof. If Y is a point, then $\mathfrak{A}(X,Y) = \mathbb{K}[X]$. Suppose that Y has positive dimension and $\mathfrak{I} = \mathfrak{I}(Y) = \{f \in \mathbb{K}[X] \mid f(y) = 0 \text{ for any } y \in Y\}$. Then $\mathfrak{A}/\mathfrak{I}$ is infinite-dimensional vector space. By the Nakayama Lemma, we can find $i \in \mathfrak{I}$ such that in the local ring of Y the element i is not in $\mathfrak{I}^2$. Then for any $a \in k[X] \setminus \mathfrak{I}$ the element ia is in $\mathfrak{I} \setminus \mathfrak{I}^2$. Hence the space $\mathfrak{I}/\mathfrak{I}^2$ has infinite dimension.

On the other hand, suppose that $f_1, \ldots, f_n$ are generators of $\mathfrak{A}(X,Y)$. Subtracting constants, one may assume that all f_i are in $\mathfrak{I}$. Then $\dim \mathfrak{A}(X,Y)/\mathfrak{I}^2 \leq n + 1$, a contradiction. $\qquad\qquad\square$

Proposition 6.5. *Let $\mathfrak{A}$ be a finitely generated domain. Then any subalgebra in $\mathfrak{A}$ is finitely generated if and only if* $\operatorname{Kdim}\mathfrak{A} \leq 1$.

Proof. If $\operatorname{Kdim}\mathfrak{A} \geq 2$, then the statement follows from the previous proposition. The case $\operatorname{Kdim}\mathfrak{A} = 0$ is obvious. It remains to prove that if $\operatorname{Kdim}\mathfrak{A} = 1$, then any subalgebra is finitely generated. By taking the integral closure, one may suppose that $\mathfrak{A}$ is the algebra of regular functions on a smooth affine curve C_1. Let C be the smooth projective curve such that $C_1 \cong C \setminus \{P_1, \ldots, P_k\}$. The elements of $\mathfrak{A}$ are rational functions on C that may have poles only at points P_i.

Let $\mathfrak{B}$ be a subalgebra in $\mathfrak{A}$. By induction on k, we may suppose that the subalgebra $\mathfrak{B}' \subset \mathfrak{B}$ consisting of functions regular at P_1 is finitely generated, say $\mathfrak{B}' = \mathbb{K}[s_1, \ldots, s_m]$. (Functions that are regular at any point P_i are constants.) Let $v(f)$ be the order of the zero/pole of $f \in \mathfrak{B}$ at P_1. The set $V = \{v(f), \ f \in \mathfrak{B}\}$ is an additive subsemigroup of integers. Such a subsemigroup is finitely generated. Let $f_1, \ldots, f_n$ be elements of $\mathfrak{B}$ such that the $v(f_i)$ generate V. Then for any $f \in \mathfrak{B}$

there exists a polynomial $P(y_1, \ldots, y_n)$ with $v(f - P(f_1, \ldots, f_n)) \geq 0$, and thus $f - P(f_1, \ldots, f_n) \in \mathfrak{B}'$. This shows that $\mathfrak{B}$ is generated by $f_1, \ldots, f_n, s_1, \ldots, s_m$. $\qquad\square$

Let $\mathfrak{A}$ be a finitely generated G-algebra with $\operatorname{Kdim}\mathfrak{A} \geq 2$. Consider the affine variety $X = \operatorname{Spec}\mathfrak{A}$. The action $G : \mathfrak{A}$ induces a regular action $G : X$.

Suppose that there exists a proper irreducible closed invariant subvariety $Y \subset X$ of positive dimension. Then $\mathfrak{A}(X, Y)$ is an invariant subalgebra, which is not finitely generated. In particular, this is the case if G acts on X without a dense orbit. Hence we may assume that either

(i) the action $G : X$ is transitive, or
(ii) X consists of an open orbit and a G-fixed point p.

In case (i), $X = G/H$ and H is reductive. If G/H is not affinely closed then there exists a non-trivial affine embedding $G/H \hookrightarrow X'$, and the complement in X to the open affine subset G/H is a union of irreducible divisors. Let Y be one of these divisors. The algebra $\mathfrak{A}(X', Y)$ is a non-finitely generated invariant subalgebra in $\mathbb{K}[X']$ and the inclusion $G/H \hookrightarrow X'$ defines an embedding $\mathbb{K}[X'] \subset \mathbb{K}[X] = \mathfrak{A}$. On the other hand,

Lemma 6.6. *If $X = G/H$ is affinely closed, i.e. $\mathfrak{A}$ is of type N, then any invariant subalgebra in $\mathfrak{A}$ is finitely generated.*

Proof. Suppose that there exists an invariant subalgebra $\mathfrak{B} \subsetneq \mathfrak{A}$ that is not finitely generated. Let $f_1, f_2, \ldots$ be a system of generators of $\mathfrak{B}$. Consider the finitely generated subalgebras $\mathfrak{B}_i = \mathbb{K}[\langle Gf_1, \ldots, Gf_i \rangle]$. Infinitely many of them are pairwise different. For the corresponding varieties $X_i := \operatorname{Spec}\mathfrak{B}_i$ one has natural dominant G-morphisms

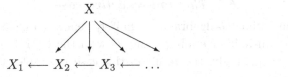

By Theorem 3.2, any X_i is an affine homogeneous space G/H_i, $H \subseteq H_i$. The infinite sequence of algebraic subgroups

$$H_1 \supset H_2 \supset H_3 \supset \ldots$$

leads to a contradiction. $\qquad\square$

Remark . As is obvious from what has been said, any invariant subalgebra in the algebra $\mathfrak{A}(H)$ of type N has the form $\mathfrak{A}(H')$, where $H \subseteq H' \subseteq G$ and also has type N. Algebras of type N can be characterized by the following equivalent properties:

- any invariant subalgebra contains no proper invariant ideals;
- the algebra contains no proper invariant ideals and the group of equivariant automorphisms is finite.

Now consider case (ii). Let us recall the following theorem due to F. Bogomolov.

Theorem 6.7 ([16], see also [29, Th.7.6]). *Let X be an irreducible affine variety with a non-trivial G-action and with a unique closed orbit, which is a G-fixed point. Then there exists a G-equivariant surjective morphism $\phi : X \to X(\mu)$ for some dominant weight $\mu \neq 0$.*

In our case the preimage $\phi^{-1}(0)$ is the point p, and thus all fibres of ϕ are finite. This shows that X is a spherical variety of rank one (see [17] for definitions), i.e.

$$\mathbb{K}[X] = \oplus_{m \geq 0} \mathbb{K}[X]_{m\lambda},$$

where $\mathbb{K}[X]_{m\lambda}$ is either zero or irreducible, and $\mu = k\lambda$ for some $k > 0$. On the other hand, the stabilizer of any point on $X(\mu)$ contains a maximal unipotent subgroup of G, and the same is true for X. By Theorem 1.9, this implies $\mathbb{K}[X]_{m_1\lambda}\mathbb{K}[X]_{m_2\lambda} = \mathbb{K}[X]_{(m_1+m_2)\lambda}$. Hence $\mathfrak{A} = \mathbb{K}[X]$ is an algebra of type HV.

Conversely, any subalgebra of the $\mathfrak{A}(P, \lambda)$ is finitely generated because it corresponds to some subsemigroup $P' \subset P$ and P' is finitely generated. This completes the proof of Theorem 6.3. $\qquad\square$

6.2 The non-reductive case

Let us classify affine G-algebras with finitely generated invariant subalgebras for a non-reductive affine group G with the Levi decomposition $G = LG^u$. Surprisingly, the result in this case is simpler than in the reductive case.

In the previous subsection we assumed that a G-algebra $\mathfrak{A}$ has no zero divisors. In fact, this restriction is inessential.

Lemma 6.8 ([10]). *Let $\mathrm{rad}(\mathfrak{A})$ be the ideal of all nilpotents in $\mathfrak{A}$. The following conditions are equivalent:*

- *any G-invariant subalgebra in $\mathfrak{A}$ is finitely generated;*
- *any G-invariant subalgebra in $\mathfrak{A}/\mathrm{rad}(\mathfrak{A})$ is finitely generated and $\dim \mathrm{rad}(\mathfrak{A}) < \infty$.*

Proof. Any finite-dimensional subspace in $\mathrm{rad}(\mathfrak{A})$ generates a finite-dimensional subalgebra in $\mathfrak{A}$. Hence if $\dim \mathrm{rad}(\mathfrak{A}) = \infty$, then the subalgebra generated by $\mathrm{rad}(\mathfrak{A})$ is not finitely generated. On the other hand, the preimage in $\mathfrak{A}$ of any non-finitely generated subalgebra in $\mathfrak{A}/\mathrm{rad}(\mathfrak{A})$ is not finitely generated.

Conversely, assume that the second condition holds. Then any subalgebra in $\mathfrak{A}$ is generated by elements whose images generate the image of this subalgebra in $\mathfrak{A}/\mathrm{rad}(\mathfrak{A})$, and by a basis of the radical of the subalgebra. □

If $\mathfrak{A}$ contains non-nilpotent zerodivisors, then the proof of Theorem 6.3 goes through with small technical modifications, see [10]. The same proof also goes well for a non-reductive G. The only difference is that case HV is excluded by the result of V. L. Popov.

Proposition 6.9 ([53, Th. 3]). *If G acts on an affine variety X with an open orbit, and*

- *the induced action $G^u : X$ is non-trivial and*
- *the complement to the open G-orbit in X does not contain a component of positive dimension,*

then the action $G : X$ is transitive.

These arguments prove

Theorem 6.10 ([10, Th. 3]). *Let $\mathfrak{A}$ be a G-algebra without nilpotents with the non-trivial induced G^u-action. The following conditions are equivalent:*

- *any G-invariant subalgebra in $\mathfrak{A}$ is finitely generated;*
- *there is no G-invariant subalgebra in $\mathfrak{A}$ with non-trivial G-invariant ideals;*
- *there is no L-invariant subalgebra in $\mathfrak{A}^{G^u}$ with non-trivial L-invariant ideals;*
- *$\mathfrak{A} = \mathbb{K}[G/H]$, where G/H is an affinely closed homogeneous space.*

Bibliography

[1] A. A'Campo-Neuen. Note on a counterexample to Hilbert's fourteenth problem given by P. Roberts. *Indag. Math.*, 5(3):253–257, 1994.

[2] D. N. Akhiezer. Actions with a finite number of orbits. *Func. Anal. Appl.*, 19(1):1–4, 1985.

[3] D. N. Akhiezer. On modality and complexity of reductive group actions. *Russ. Math. Surveys*, 43(2):157–158, 1988.

[4] I. V. Arzhantsev. Invariant ideals and Matsushima's criterion. arXiv: math.AG/0506430.

[5] I. V. Arzhantsev. On actions of reductive groups with one-parameter family of spherical orbits. *Math. Sbornik*, 188(5):639–655, 1997.

[6] I. V. Arzhantsev. On $SL(2)$-actions of complexity one. *Izvestiya RAN Ser. Math.*, 61(4):685–698, 1997.

[7] I. V. Arzhantsev. On modality and complexity of affine embeddings. *Math. Sbornik*, 192(8):1133–1138, 2001.

[8] I. V. Arzhantsev. Invariant subalgebras and affine embeddings of homogeneous spaces. In *Recent advances in Lie theory (Vigo, 2000)*, volume 25 of *Res. Exp. Math.*, pages 121–126. Heldermann, Lemgo, 2002.

[9] I. V. Arzhantsev. Algebras with finitely generated invariant subalgebras. *Ann. Inst. Fourier*, 53(2):379–398, 2003.

[10] I. V. Arzhantsev and N. A. Tennova. On affinely closed homogeneous spaces. *J. Math. Sciences (Springer)*, 131(6):6133–6139, 2005. arXiv:math.AG/0406493.

[11] I. V. Arzhantsev and D. A. Timashev. Affine embeddings with a finite number of orbits. *Transformation Groups*, 6(2):101–110, 2001.

[12] I. V. Arzhantsev and D. A. Timashev. On the canonical embedding of certain homogeneous spaces. In E. B. Vinberg, editor, *Lie groups and invariant theory: A. L. Onishchik's jubilee volume*, volume 213 of *Amer. Math. Soc. Transl. Ser. 2*, pages 63–83. Amer. Math. Soc., Providence, RI, 2005. arXiv:math.AG/0308201.

[13] D. Bartels. Quasihomogene affine Varietäten für $SL(2, \mathbb{C})$. In *Séminaire d'algèbre Paul Dubreil et Marie-Paule Malliavin, 36ème année (Paris, 1983–1984)*, volume 1146 of *Lecture Notes in Math.*, pages 1–105. Springer-Verlag, 1985.

[14] F. Bien, A. Borel, and J. Kollár. Rationally connected homogeneous spaces. *Invent. Math.*, 124:103–127, 1996.

[15] D. Birkes. Orbits of linear algebraic groups. *Ann. of Math. (2)*, 93:459–475, 1971.

[16] F. A. Bogomolov. Holomorphic tensors and vector bundles on projective manifolds. *Math. USSR-Izv*, 13(3):499–555, 1979.

[17] M. Brion. On spherical varieties of rank one. In *Group actions and invariant theory (Montreal, PQ, 1988)*, volume 10 of *CMS Conf. Proc.*, pages 31–41. Amer. Math. Soc., Providence, RI, 1989.

[18] M. Brion. Variétés sphériques. Notes de la session de la S.M.F. 'Opérations hamiltoniennes et opérations de groupes algébriques', Grenoble, 1997. Available at http://www-fourier.ujf-grenoble.fr/~mbrion/spheriques.ps.

[19] S. Donkin. Invariants of unipotent radicals. *Math. Z.*, 198:117–125, 1988.

[20] W. Ferrer Santos. Closed conjugacy classes, closed orbits and structure of algebraic groups. *Comm. Algebra*, 19(12):3241–3248, 1991.

[21] G. Freudenburg. A survey of counterexamples to Hilbert's fourteenth problem. *Serdica Math. J.*, 27:171–192, 2001.

[22] G. Freudenburg and D. Daigle. A counterexample to Hilbert's fourteenth problem in dimension 5. *J. Algebra*, 221:528–535, 1999.

[23] W. Fulton. *Introduction to toric varieties*, volume 131 of *Ann. Math. Stud.* Princeton Univ. Press, Princeton, 1993.

[24] R. Gangolli. Invariant function algebras on compact semisimple Lie groups. *Bull. Amer. Math. Soc.*, 71:634–637, 1965.

[25] V. M. Gichev. Domains with homogeneous skeletons and invariant algebras. In *'Harmonic analysis on homogeneous spaces'*, volume 3 of *Vestnik Tambovskogo Univ.*, pages 38–44. 1998.

[26] V. M. Gichev and I. A. Latypov. Polynomially convex orbits of compact Lie groups. *Transformation Groups*, 6(4):321–331, 2001.

[27] F. D. Grosshans. Observable groups and Hilbert's fourteenth problem. *Amer. J. Math.*, 95(1):229–253, 1973.

[28] F. D. Grosshans. The invariants of unipotent radicals of parabolic subgroups. *Invent. Math.*, 73:1–9, 1983.

[29] F. D. Grosshans. *Algebraic Homogeneous Spaces and Invariant Theory*, volume 1673 of *Lecture Notes in Math.* Springer-Verlag, Berlin, 1997.

[30] J. E. Humphreys. *Linear algebraic groups*, volume 21 of *Graduate Texts in Mathematics*. Springer-Verlag, New York, 1975.

[31] G. Kempf. Instability in invariant theory. *Ann. of Math. (2)*, 108(2):299–316, 1978.

[32] B. N. Kimel'feld and E. B. Vinberg. Homogeneous domains on flag manifolds and spherical subgroups of semisimple Lie groups. *Func. Anal. Appl.*, 12(3):168–174, 1978.

[33] F. Knop. The Luna-Vust theory of spherical embeddings. In S. Ramanan, editor, *Proc. Hyderabad Conf. on Algebraic Groups*, pages 225–249. Manoj Prakashan, Madras, 1991.

[34] F. Knop. Über Bewertungen, welche unter einer reductiven Gruppe invariant sind. *Math. Ann.*, 295:333–363, 1993.

[35] F. Knop. Über Hilberts vierzehntes Problem für Varietäten mit Kompliziertheit eins. *Math. Z.*, 213:33–35, 1993.

[36] F. Knop. The asymptotic behavior of invariant collective motion. *Invent. Math.*, 116:309–328, 1994.

[37] H. Kraft. *Geometrische Methoden in der Invariantentheorie*. Vieweg Verlag, Braunschweig-Wiesbaden, 1985.

[38] I. A. Latypov. Homogeneous spaces of compact connected Lie groups which admit nontrivial invariant algebras. *J. Lie Theory*, 9:355–360, 1999.

[39] I. A. Latypov. *Invariant algebras of continuous functions on homogeneous spaces of compact Lie groups*. PhD thesis, Omsk State University, 1999.

[40] I. A. Latypov. Invariant function algebras on $SU(2)$. *Siberian Adv. Math.*, 10(4):122–133, 2000.

[41] M. W. Liebeck and G. M. Seitz. Variations on a theme of Steinberg. *J. Algebra*, 260:261–297, 2003.

[42] D. Luna. Slices étales. In *Sur les groupes algébriques*, pages 81–105. Bull. Soc. Math. France, Paris, Mémoire 33. Soc. Math. France, 1973.

[43] D. Luna. Adhérences d'orbite et invariants. *Invent. Math.*, 29:231–238, 1975.

[44] D. Luna and R. W. Richardson. A generalization of the Chevalley restriction theorem. *Duke Math. J.*, 46(3):487–496, 1979.

[45] D. Luna and Th. Vust. Plongements d'espaces homogènes. *Comment. Math. Helv.*, 58:186–245, 1983.

[46] G. J. McNinch. Dimensional criteria for semisimplicity of representations. *Proc. London Math. Soc. (3)*, 76:95–149, 1998.

[47] G. D. Mostow. Cohomology of topological groups and solvmanifolds. *Ann. of Math. (2)*, 73(1):20–48, 1961.

[48] G. D. Mostow, A. Bialynicki-Birula, and G. Hochschild. Extensions of representations of algebraic linear groups. *Amer. J. Math.*, 85:131–144, 1963.

[49] M. Nagata. On the fourteenth problem of Hilbert. In *Proc. Internat. Congress Math. 1958*, pages 459–462. Cambridge Univ. Press, 1960.

[50] A. L. Onishchik. Transitive compact transformation groups. *Amer. Math. Soc. Transl.*, 55:153–194, 1966.

[51] R. S. Palais and T. E. Stewart. The cohomology of differentiable transformation groups. *Amer. J. Math.*, 83(4):623–644, 1961.

[52] V. L. Popov. Quasihomogeneous affine algebraic varieties of the group $SL(2)$. *Math. USSR-Izv.*, 7:793–831, 1973.

[53] V. L. Popov. Classification of three-dimensional affine algebraic varieties that are quasihomogeneous with respect to an algebraic group. *Math. USSR-Izv.*, 9:535–576, 1975.

[54] V. L. Popov. Contractions of actions of reductive algebraic groups. *Math. USSR-Sb.*, 58(2):311–335, 1987.

[55] V. L. Popov and E. B. Vinberg. A certain class of quasihomogeneous affine algebraic varieties. *Math. USSR-Izv.*, 6:743–758, 1972.

[56] V. L. Popov and E. B. Vinberg. *Invariant theory*, volume 55 of *Encyclopædia of Mathematical Sciences*, pages 123–278. Springer-Verlag, Berlin-Heidelberg, 1994.

[57] R. W. Richardson. Affine coset spaces of reductive algebraic groups. *Bull. London Math. Soc.*, 9:38–41, 1977.

[58] D. Rider. Translation-invariant Dirichlet algebras on compact groups. *Proc. Amer. Math. Soc.*, 17(5):977–985, 1966.

[59] A. Rittatore. Algebraic monoids and group embeddings. *Transformation Groups*, 3(4):375–396, 1998.

[60] P. Roberts. An infinitely generated symbolic blow-up in a power series ring and a new counterexample to Hilbert's fourteenth problem. *J. Algebra*, 132(2):461–473, 1990.

[61] F. J. Servedio. Prehomogeneous vector spaces and varieties. *Trans. Amer. Math. Soc.*, 176:421–444, 1973.

[62] R. Steinberg. Nagata's example. In *Algebraic groups and Lie groups*, volume 9 of *Austral. Math. Soc. Lect. Ser.*, pages 375–384. Cambridge Univ. Press, Cambridge, 1997.

[63] A. A. Sukhanov. Description of the observable subgroups of linear algebraic groups. *Math. USSR-Sb.*, 65(1):97–108, 1990.

[64] L. Tan. On the Popov-Pommerening conjecture for the groups of type G_2. *Algebras Groups Geom.*, 5(4):421–432, 1988.

[65] L. Tan. On the Popov-Pommerening conjecture for groups of type A_n. *Proc. Amer. Math. Soc.*, 106(3):611–616, 1989.

[66] L. Tan. Some recent developments in the Popov-Pommerening conjecture. In *Group actions and invariant theory (Montreal, PQ, 1988)*, volume 10 of *CMS Conf. Proc.*, pages 207–220. Amer. Math. Soc., Providence, RI, 1989.

[67] N. A. Tennova. A criterion for affinely closed homogeneous spaces of a solvable algebraic group. *Vestnik Moskov Univ., Ser. 1*, 5:67–70, 2005. (Russian). English transl. *Moscow University Math. Bulletin*, 2005, *in print*.

[68] D. A. Timashev. Classification of G-varieties of complexity 1. *Math. USSR-Izv.*, 61(2):363–397, 1997.

[69] D. A. Timashev. Equivariant compactifications of reductive groups. *Math. Sbornik*, 194(4):589–616, 2003.

[70] E. B. Vinberg. Complexity of actions of reductive groups. *Func. Anal. Appl.*, 20(1):1–11, 1986.

[71] E. B. Vinberg. On reductive algebraic semigroups. In S. Gindikin and E. Vinberg, editors, *Lie Groups and Lie Algebras: E. B. Dynkin Seminar*, volume 169 of *AMS Transl.*, pages 145–182. Amer. Math. Soc., 1995.

[72] Th. Vust. Sur la théorie des invariants des groupes classiques. *Ann. Inst. Fourier*, 26:1–31, 1976.

[73] J. A. Wolf. Translation-invariant function algebras on compact groups. *Pacif. J. Math.*, 15:1093–1099, 1965.

Department of Higher Algebra
Faculty of Mechanics and Mathematics
Moscow State University
Leninskie Gory, 119992 Moscow
Russia
arjantse@mccme.ru

Formal groups over local fields: a constructive approach

Mikhail V. Bondarko

Introduction

This paper is dedicated to the work of St. Petersburg mathematicians on formal groups over local fields. This work can be divided into three main topics.

The first consists of explicit reciprocity formulas on formal groups over local fields. This activity in St. Petersburg was started by S. Vostokov in the 1970s and was carried on by his students.

The second topic is the explicit classification of formal groups over local fields. It was started recently by M. Bondarko and S. Vostokov. One should also mention an earlier work of O. Demchenko. We also describe briefly some recent results on the classification of finite local commutative group schemes over complete discrete valuations fields (obtained by Bondarko).

We note that recently both groups of results were carried over to complete discrete valuation fields with imperfect residue field; see [9] and [11].

The third topic is the connection between formal groups and Galois module structure. This subject was studied by M. Bondarko.

One of the main features of this work is that the results do not require any sophisticated knowledge. An interested reader may find the proofs in the literature which we cite and see that the proofs are relatively simple (though a little technical sometimes).

In Section 1 we give the most basic definitions: a formal group, its logarithm, a homomorphism between two formal groups and the notion of formal module. We also give the definition of Lubin-Tate formal groups, which are very important in local algebraic number theory.

We begin Section 2 by recalling the notion of a p-typical formal group.

For simplicity, we shall mostly deal only with formal groups of this type, although our methods work for general formal groups. Next we recall the results of Honda on the classification of formal groups over rings of integers of unramified local fields. All these results can be found in the book [23]. We also describe a canonical representative in each strict isomorphism class of formal group laws. This result appears to be new. At the end of the section we define Honda formal group modules.

Section 3 is dedicated to explicit formulas for the (generalized) Hilbert symbol on formal groups. First we describe the formula for Lubin-Tate formal groups. Next we formulate the more general result for arbitrary Honda formal group laws. The proofs can be found in the book [21] and in the papers [16, 17, 18, 19].

Sections 4–8 are dedicated to the classification of formal groups. More can be found in the papers [12] and [9].

In Section 4 we describe the main classification method. By restriction of scalars we apply Honda's classification of formal groups over unramified fields to the classification of formal groups over general local fields. The main idea here is to replace an m-dimensional group over $\mathfrak{O}_K$ by an me-dimensional one over an unramified ring $\mathfrak{o}$ by means of restriction of scalars. Hence the problem is reduced to studying certain matrices. Using the matrix method S. V. Vostokov was able to state and verify the classification result for one-dimensional groups up to $e = 2p - 2$.

In Section 5 we prove that the operator corresponding to a p-typical logarithm can be presented as a fraction. Next we define the fractional part invariant. We state and prove the main classification result (Theorem 5.5).

In Section 6 we define the invariant Cartier-Dieudonné modules for formal groups. The difference from Cartier's definition is that we take logarithms of p-typical curves. That gives us a canonical imbedding of our module in $K[[\Delta]]^m$. In order to illustrate the usefulness of such a modification we describe which formal groups are isogenous to a group defined over a subfield of the base field. Next we define the module invariant M_F. We prove that, together with the fractional part invariant it classifies formal groups up to a strict isomorphism.

In Section 7 we apply our methods to the classification of formal groups for $e \leq p^2/2$. We distinguish the cases $e < p$ and $e \geq p$. In the latter case we describe the classification in the more simple case of dimension 1 and height > 1.

In Section 8 we describe canonical representatives of isogeny classes for one-dimensional formal groups. We also discuss the relation between formal groups, finite group schemes and reduction of Abelian varieties. An interested reader should consult [4] for a survey of the author's results on this subject.

Note that the classification results described can easily be generalized to formal groups over multidimensional fields.

In Section 9 we state the main result on the links between formal groups and associated Galois modules. Several more statements on this topic can be found in the papers [5, 6, 8].

The author is deeply grateful to Professor M. J. Taylor for the invitation to Manchester and the possibility to publish this work; also to Professor S. V. Vostokov for his useful advice on this manuscript.

1 Basic definitions

Notation . Throughout the paper $M_m(\mathfrak{A})$ will denote the ring of $m \times m$ matrices over a (possibly non-commutative) ring $\mathfrak{A}$, I_m will denote the unit matrix of size m; e_i will denote the m-vector $(0, 0, \ldots, 1, 0, \ldots, 0)$ (1 is in the ith place).

K is a finite extension of $\mathbb{Q}_p$, v_K is the normalized valuation on K, $e = v_K(p)$ is the absolute ramification index of K, $\mathfrak{M}$ is the maximal ideal of K.

m will usually denote the dimension of a formal group F, $X = (X_i) = X_1, \ldots X_m$ will denote formal variables, x is one variable.

For local number fields N and K their rings of integers are denoted by $\mathfrak{o}$ and $\mathfrak{O}_K$, respectively.

The definitions and the results of the first two sections, except Subsection 2.3, can be found in the book [23]. Another source is [21].

1.1 Formal groups and their logarithms

Let A be a commutative ring with unity. An $m$tuple of formal power series $F_i(X_1, \ldots, X_m, Y_1, \ldots, Y_m)$ over A is said to determine a commutative *formal group* (or a formal group law) F of *dimension* m over A if for $F = (F_i)$, $X = (X_i)$, etc. for $1 \leq i \leq m$ one has

$$F(X, 0) = F(0, X) = X,$$
$$F(F(X, Y), Z) = F(X, F(Y, Z)) \qquad \text{(associativity)},$$
$$F(X, Y) = F(Y, X). \qquad \text{(commutativity)}.$$

Natural examples of formal groups are the additive formal group

$$F_+(X,Y) = X + Y$$

and the multiplicative formal group

$$F_\times(X,Y) = X + Y + XY = (1+X)(1+Y) - 1.$$

The definition implies that $F_i(X,Y) = X_i + Y_i + \sum_{\#I, \#J \geq 1} a_{iIJ} X^I Y^J$, $a_{iIJ} \in A$; here I, J are multi-indices.

The most simple example of a multidimensional formal group is a sum of one-dimensional groups, i.e. for one-dimensional formal groups F^i we put $F_i(X,Y) = F^i(X_i, Y_i)$.

An $s$tuple of formal power series $f = (f_i(X))$, where $f_i \in A[[X]]_0$ (the constant terms are zero), $1 \leq i \leq s$, is called a *homomorphism* from an m-dimensional formal group F to an s-dimensional formal group G if

$$f(F(X,Y)) = G(f(X), f(Y)).$$

f is called an isomorphism if there exists a series $g = f^{-1}$ inverse to it with respect to composition, i.e. such that $(f \circ g)(X) = (g \circ f)(X) = X$. The set of homomorphisms from F to G will be denoted by $\mathrm{Hom}(F, G)$; it has the structure of an abelian group defined by

$$f(X) + g(X) = G(f(X), g(X)).$$

Moreover, the set $\mathrm{End}_A(F)$ of all homomorphisms of F to F has the structure of a ring:

$$f(X) \cdot g(X) - f(g(X)).$$

Lemma 1.1. *There exists a uniquely determined homomorphism*

$$\mathbb{Z} \to \mathrm{End}_A(F): \quad n \to [n]_F.$$

satisfying $[1]_i = X_i$.

It is easily seen that a homomorphism f of F is an isomorphism if and only if $f(X) \equiv BX \mod \deg 2$, where B is an invertible matrix over A. If $B = I_m$, then the isomorphism f is called *strict*.

Now let $A = K$ be a field of characteristic 0.

Proposition 1.2. *For any two formal groups of the same dimensions there exists a unique strict isomorphism between them. More generally, let M be an $s \times m$ matrix over A. Then for an m-dimensional formal group F and an s-dimensional formal group G there exists a unique $f \in \mathrm{Hom}(F, G)$ such that $f(X) \equiv MX \mod \deg 2$.*

In particular, any m-dimensional formal group F over K is strictly isomorphic to F_+^m. Hence there exists a $\lambda(X) = (\lambda_i)$ for $\lambda_i(X) \in (X_1, \ldots, X_m) A[[X_j]]$, $1 \le i, j \le m$, $\lambda_i(X) \equiv X_i \mod \deg 2$ such that

$$F(X, Y) = \lambda^{-1}(\lambda(X) + \lambda(Y)).$$

The $m$tuple $\lambda(X)$ is called the *logarithm* of the formal group F. We will denote it by $\log_F(X)$. The series inverse to it with respect to composition is denoted by $\exp_F(X)$. Then $F(X, Y) = \exp_F(\log_F(X) + \log_F(Y))$.

The theory of formal groups is presented in [23].

1.2 Formal modules

Obviously, for $a \in \mathbb{Z}$ we have $[a]_i \equiv a X_i \mod \deg 2$.

One says that F is a *B-formal module* for some ring $B \subset A$ if for any $b \in B$ there exists a $[b]_F \in \operatorname{End}_A(F)$, $[b]_F \equiv bX \mod \deg 2$.

Remark 1.3. Sometimes one demands that $[b]_F \equiv \phi(b)X$ where ϕ is some ring homomorphism from B into A.

If A is not a characteristic zero ring, one also has to demand that the map $b \to [b]_F$ be a ring homomorphism. For char $A = 0$ this condition follows immediately from the second part of Proposition 1.2. Here the addition for $\operatorname{End}_A(F)$ is F and multiplication is given by composition.

If A is a p-complete ring then any F is canonically a $\mathbb{Z}_p$-module. Usually one says that F is a B-formal module if B is an integral extension of $\mathbb{Z}_p$.

1.3 Lubin-Tate formal group laws

From now on we assume that K is a local number field. For such a field the Lubin-Tate formal groups play an important role. Let $\mathfrak{F}_\pi$ denote the set of formal power series $f(X) \in \mathfrak{O}_K[[X]]$ such that $f(X) \equiv \pi X \mod \deg 2$, $f(X) \equiv X^q \mod \pi$, where π is a prime element in K and q is the cardinality of the residue field $\overline{K}$. The following assertion holds.

Proposition 1.4. *Let $f(X) \in \mathfrak{F}_\pi$. Then there exists a unique formal group $F = F_f$ over $\mathfrak{O}_K$ such that*

$$F_f(f(X), f(Y)) = f(F_f(X, Y)).$$

For each $\alpha \in \mathfrak{O}_K$ there exists a unique $[\alpha]_F \in \operatorname{End}_{\mathfrak{O}_K}(F)$ such that

$$[\alpha]_F(X) \equiv \alpha X \mod \deg 2.$$

The map $\mathfrak{O}_K \to \operatorname{End}_{\mathfrak{O}_K}(F): \quad \alpha \to [\alpha]_F$ is a ring homomorphism, and $f = [\pi]_F$. If $g(X) \in \mathfrak{F}_\pi$ and F_g is the corresponding formal group, then F_f and F_g are isomorphic over $\mathfrak{O}_K$, i.e. there is a series $\rho(X) \in K[[X]]$, $\rho(X) \equiv X \mod \deg 2$, such that

$$\rho(F_f(X, Y)) = F_g(\rho(X), \rho(Y)).$$

The formal group F_f is called a *Lubin-Tate formal group*. It is an example of an $\mathfrak{O}_K$-formal module.

Note that the multiplicative formal group $F_\times$ is a Lubin-Tate group for $\pi = q = p$.

2 Classification over unramified fields and related topics

2.1 *p-typical groups*

Since throughout this paper we will work with commutative formal group laws over characteristic zero rings, our formal group laws will have logarithms. The following statement is well known (see [23]).

Proposition 2.1. *Let $\mathfrak{A}$ be a commutative $\mathbb{Z}_p$-algebra. Then a formal group law with logarithm $\lambda = (\lambda_i)$, $1 \le i \le m$, where $\lambda_i = \sum a^i_{i_1 \dots i_m} X_1^{i_1} \dots X_m^{i_m}$, is strictly isomorphic to a formal group law whose logarithm is equal to (λ'_i). Here $\lambda'_i = \sum a'^i_{i_1 \dots i_m} X_1^{i_1} \dots X_m^{i_m}$ for $a'^i_{i_1 \dots i_m} = a^i_{i_1 \dots i_m} X_1^{i_1}$ if and only if $i_s = 0$ for all s except one, the remaining i_s is a power of p, all other $a'^i_{i_1 \dots i_m}$ are 0.*

Thus we can assume that the logarithm λ of the formal group F has the form $\Lambda(\Delta)(X)$, where $\Lambda(\Delta) = (\Lambda_i)$ is in the matrix ring $M_m(K[[\Delta]])$ and $\Delta(aX_i^b) = aX_i^{pb}$ for any i, $1 \le i \le m$.

2.2 *The results of Honda*

We consider an unramified extension $N/\mathbb{Q}_p$. Let $\sigma \in \operatorname{Gal}(N/\mathbb{Q}_p)$ be the Frobenius, i.e. let it satisfy $\sigma(x) - x^p \in p\mathfrak{o}$ for any $x \in \mathfrak{o}$.

We introduce a non-commutative ring W, which is equal to $\mathfrak{o}[[\Delta]]$ as a left $\mathfrak{o}$-module and satisfies the the relation $\Delta a = \sigma(a)\Delta$ for any $a \in \mathfrak{o}$. We will often identify W and $\mathfrak{o}[[\Delta]]$ as sets.

The following statement is the main tool for proving classification results.

Theorem 2.2.(i) $\Lambda(\Delta)(X)$, $\Lambda(\Delta) \in \mathbb{Q}_p W$, *is the logarithm of a formal group over* $\mathfrak{o}$ *if and only if* $\Lambda = pU^{-1}$ *for some* $U \in M_m(\mathfrak{A}[[\Delta]]')$, $U \equiv pI_m \mod (\Delta)$.

(ii) Λ *and* Λ' *give strictly isomorphic formal group laws if and only if* $U' = \mathfrak{E}U$ *for some* $\mathfrak{E} \in M_m(\mathfrak{A}[[\Delta]]')$, $\mathfrak{E} \equiv I_m \mod (\Delta)$.

(iii) *More generally, there exists a homomorphism* f *from* F *to* F' *of dimensions* m *and* m', $f \equiv AX \mod \deg 2$, A *is some* $m' \times m$ *matrix over* $\mathfrak{A}$, *if and only if there exist* $C \in M_{m' \times m}(\mathfrak{A}[[\Delta]]')$ *such that* $CU = U'A$.

(iv) *If* $\Lambda \equiv \Lambda' \mod p$, *then* $\Lambda'(X)$ *is the logarithm of a formal group* F' *over* $\mathfrak{A}$ *if and only if* $\Lambda(X)$ *satisfies the same property. In this case* F' *is strictly isomorphic to* F.

Following Honda, we will call matrices that satisfy the conditions for U in the theorem *special* elements of $M_m(W)$.

Example: For the multiplicative formal group law we have $U = p - \Delta$.

2.3 Canonical representatives

The previous theorem gives a classification of formal groups over $\mathfrak{A}$ in terms of equivalence classes of certain matrices U. In the one-dimensional case one can choose in any class a polynomial of the minimal possible degree, thus obtaining a canonical representative. This method cannot be extended to multi-dimensional groups. However, canonical representatives can be chosen in a completely different way.

We fix a system of representatives $\theta : \mathfrak{o}/p\mathfrak{o} \to \mathfrak{o}$.

Proposition 2.3. *Each equivalence class of special matrices contains a unique representative* U *such that* $U = pI_m + r\Delta$ *where the coefficients of* r *belong to* $\theta(\mathfrak{o}/p\mathfrak{o})$.

The proof is very easy. It can be found in [12].

A natural choice of θ is the Teichmüller representative map. This choice is very convenient when one studies the reduction of formal group laws.

2.4 Honda formal modules

Let L be a local field, Fr be an automorphism of L satisfying $\mathrm{Fr}(x) \equiv x^q \mod \pi = \pi_L$.

The set of operators of the form $\sum_{i \geq 0} a_i \Delta^i$, where $a_i \in \mathfrak{O}_L$, forms a noncommutative ring $\mathfrak{O}'_L[[\Delta]]$ of series in Δ in which $\Delta\, a = a^{\mathrm{Fr}}\, \Delta$ for $a \in \mathfrak{O}_L$.

In order to extend the action of $\mathfrak{O}'_L[[\Delta]]$ onto $L[[X]]$ we define $\Delta \circ X_i = X_i^q$.

A (not necessarily p-typical) formal group $F \in \mathfrak{O}_L[[X, Y]]$ with logarithm $\log_F(X) \in L[[X]]^m$ is called a *Honda formal group* if

$$u \circ \log_F \equiv 0 \quad \mathrm{mod}\ \pi$$

for some operator $u = \pi I_m + A_1 \Delta + \cdots \in M_m \mathfrak{O}'_L[[\Delta]]$. The operator u is called the *type* of the formal group F.

Every Honda formal group is a formal module over $\mathfrak{O}_L$.

The main statements on Honda formal modules are very similar to those on Honda formal groups for $L = \mathbb{Q}_p$, $\mathrm{Fr} = \mathrm{id}$.

Every formal $\mathfrak{O}_L$-module over an unramified extension of L is a Honda formal group (cf. [24] or [23]). In particular, a Lubin-Tate formal group is a Honda group with $u = \pi - \Delta$.

Types u and v of a formal group F are called *equivalent* if $u = \varepsilon \circ v$ for some $\varepsilon \in \mathfrak{O}'_L[[\Delta]]$, $\varepsilon(0) = I_m$.

Let F be of type u. Then $v = \pi I_m + B_1 \Delta + \cdots \in M_m(\mathfrak{O}_L[[X]])$ is a type of F if and only if v is left equivalent to u.

Using the Weierstrass preparation theorem for the ring $\mathfrak{O}_L[[\Delta]]$, one can prove [24] that for every one-dimensional formal Honda group F there is a unique *canonical type*

$$u - \pi - a_1 \wedge - \cdots - a_h \Delta^h, \quad a_1, \ldots, a_{h-1} \in \mathfrak{M}_L, a_h \in \mathfrak{O}_L^*.$$

This type determines the group F uniquely up to isomorphism. Here h is called the *L-height* of F; the product $h[L : \mathbb{Q}_p]$ is equal to the height of F.

3 Explicit reciprocity formulas

3.1 The setting

We assume in this section that $p > 2$. All formal groups are one-dimensional. The results of this section can be found in [21].

The problem of obtaining explicit formulas for the Hilbert pairing originates from the ninth Hilbert problem. Symbols on formal groups over local fields were defined by Fröhlich. Let F be a formal group defined over the ring of integers of a local field N. Suppose that K/N is a finite extension and K contains the kernel of $[p^n]_F$. We can define the pairing $(\; , \;)_{F,n} : K^* \times F(\mathfrak{M}) \to \ker[p^n]_F$ by

$$(\alpha, \beta)_{F,n} = \Psi(\alpha)(\tilde{\beta}) \underset{F}{-} \tilde{\beta}, \qquad (3.1)$$

where $\Psi : K^* \to \mathrm{Gal}(K^{\mathrm{ab}}/K)$ is the (local) reciprocity map and

$$[p^n]_F(\tilde{\beta}) = \beta \; .$$

The usual Hilbert pairing is obtained if, in the above formula, one takes F to be the mutliplicative formal group law. Thus in the setting of formal groups one should take $F(X, Y) = X + Y + XY$. The formula for this case was obtained in the paper [28].

3.2 The formula for Lubin-Tate groups

A method similar to that in the multiplicative case allowed Vostokov to prove the formula for the case of Lubin-Tate formal groups in the papers [29] and [30]. In this case the Hilbert pairing takes on values in the group $\ker[\pi^n]_F$, which is naturally isomorphic to $\mathfrak{o}/\pi^n\mathfrak{o}$. Hence the answers may be indexed by elements of $\mathfrak{o}/\pi^n\mathfrak{o}$.

Let K be a local number field with residue field F_q, π a prime element in K and $F = F_f$ a Lubin-Tate formal group over $\mathfrak{O}_K$ for $f \in \mathfrak{F}_\pi$. Let L/K be a finite extension such that the $\mathfrak{O}_K$-module κ_n of π^n-division points is contained in L. We have $\kappa_n = \langle \zeta_n \rangle$, where ζ_n satisfies $[\pi^n]_F(\zeta_n) = 0$ and $[\pi^{n-1}]_F(\zeta_n) \neq 0$.

For α_i in the completion of the maximal unramified extension K^{ur} of K put

$$\Delta \left(\sum \alpha_i X^i \right) = \sum \mathrm{Fr}_K(\alpha_i) X^{qi},$$

where Fr_K is the continuous extension of the Frobenius automorphism over K and q is the cardinality of $\overline{K}$. Let $\mathfrak{O}$ be the ring of integers in the completion of K^{ur}. Let $F(X\mathfrak{O}[[X]])$ denote the $\mathfrak{O}_K$-module of formal power series $X\mathfrak{O}[[X]]$ with respect to operations

$$f +_F g = F(f, g), \quad a \cdot f = [a]_F(f), \quad a \in \mathfrak{O}_K.$$

We introduce the map $l_F = l_{F,X}$:

$$l_F(g(X)) = \left(1 - \frac{\Delta}{\pi} \right) (\log_F(g(X))), \qquad g(X) \in X\mathfrak{O}[[X]].$$

For $a_i \in \mathfrak{o}_0$ put

$$\delta \left(\sum a_i X^i \right) = \sum \mathrm{Fr}(a_i) X^{pi},$$

where $\mathrm{Fr} = \mathrm{Fr}_{\mathbb{Q}_p}$ is the Frobenius automorphism over $\mathbb{Q}_p$.

For the series $\alpha(X) = \theta X^m \varepsilon(X)$, where θ is a lth root of unity, l is relatively prime to p and $\varepsilon(X) \in 1 + X\mathfrak{o}_0[[X]]$, put

$$l(\alpha(X)) = l(\varepsilon(X)) = \left(1 - \frac{\Delta}{q}\right)(\log \varepsilon(X)) = \frac{1}{q}\log\left(\frac{\alpha(X)^q}{\alpha(X)^\Delta}\right)$$

and

$$L(\alpha(X)) = (1 + \delta + \delta^2 + \dots)l(\alpha(X)).$$

Let $\alpha \in L^*$, $\beta \in F(\mathfrak{M}_L)$. Let $\alpha = \alpha(X)|_{X=\Pi}$, $\beta = \beta(X)|_{X=\Pi}$, where $\alpha(X)$ is as above, $\beta(X) \in X\mathfrak{o}_T[[X]]$. Put

$$\Phi_{\alpha(X),\beta(X)} = \frac{\alpha(X)'}{\alpha(X)}l_F(\beta(X)) - l(\alpha(X))\left(\frac{\Delta}{q}\log_F(\beta(X))\right)'.$$

If $p > 2$ then

$$(\alpha,\beta)_F = [\mathrm{Tr}\,\mathrm{Res}_X\,\Phi_{\alpha(X),\beta(X)}/s(X)](\xi_n).$$

In the case $p = 2$ the formula was found by the participants of the St. Petersburg number theory seminar organized by I. B. Fesenko (see [20, 31]).

O. Demchenko, one of the participants of this seminar, obtained an explicit formula for the relative Lubin-Tate formal groups (cf. [18]).

3.3 Formulas for Honda groups

Let K be a local field with residue field of cardinality $q = p^f$ and L be a finite unramified extension of K. Let π be a prime element of K and let $\mathfrak{o}_L$ be the ring of integers of L. We put $\mathrm{Fr} = \mathrm{Fr}_K$ (the Frobenius fixing K).

D. Benois, also a participant of the St. Petersburg number theory seminar, found the formulas for the Hilbert pairing for $n - 1$ in his joint work with S. V. Vostokov. These results were applied to Galois representations corresponding to the Tate module (cf. [2] and [3]).

There were some difficulties in the case of arbitrary n. They were overcome by O. Demchenko. He proved that Honda formal groups satisfy a property that generalizes the definition of Lubin-Tate groups. See the papers [16] and [17] for the proofs.

Theorem 3.1. *Let F be a Honda formal group of type*

$$\tilde{u} = \pi - a_h\,\Delta^h - a_{h+1}\,\Delta^{h+1} - \dots, \quad a_i \in \mathfrak{o}_L,$$

where a_h is invertible in $\mathfrak{O}_L$. Let $u = \pi - a_1 \triangle - \cdots - a_{h-1} \triangle^{h-1} - a_h \triangle^h$ be the canonical type of F, $a_1, \ldots, a_{h-1} \in \mathfrak{M}_L$. Let $\lambda = \log_F$ be the logarithm of F. Put $\lambda_1 = B_1 \lambda^{\mathrm{Fr}^h}$, where

$$B_1 = 1 + \frac{a_{h+1}}{a_h} \triangle + \frac{a_{h+2}}{a_h} \triangle^2 + \ldots$$

(i.e. $\tilde{u} = \pi - a_h B_1 \triangle^h$). Then

(i) λ_1 is the logarithm of the Honda formal group F_1 of type $\tilde{u}_1 = a_h^{-1} \tilde{u} a_h$ and of canonical type $u_1 = a_h^{-1} u a_h$;

(ii) $f = \left[\dfrac{\pi}{a_h} \right]_{F,F_1} \in \mathrm{Hom}_{\mathfrak{O}_L}(F, F_1)$ and $f(X) \equiv X^{q^h} \mod \pi$.

Remark 3.2. Some examples:

(i) for a formal Lubin-Tate group $F_1 = F$;
(ii) for a relative Lubin-Tate group $F_1 = F^{\mathrm{Fr}}$.

A certain converse result is also valid.

Theorem 3.3 (the converse to Theorem 3.1). *Let $f \in \mathfrak{O}_L[[X]]$ be a series satisfying relations*

$$f(X) \equiv X^{q^h} \mod \pi, \quad f(X) \equiv \frac{\pi}{a_h} X \mod \deg 2,$$

where a_h is an invertible element of $\mathfrak{O}_L$. Let $u = \pi - a_1 \triangle - \cdots - a_h \triangle^h$, where $a_1, \ldots, a_{h-1} \in \mathfrak{M}_L$. Let

$$C = 1 - \frac{a_1}{\pi} \triangle - \cdots - \frac{a_{h-1}}{\pi} \triangle^{h-1}$$

and $\tilde{u} = C^{-1} u = \pi - a_h \triangle^h - a_{h+1} \triangle^{h+1} - \ldots$. Then there exists a unique Honda formal group F of type $\tilde{u}$ and of canonical type u such that $f = \left[\dfrac{\pi}{a_h} \right]_{F,F_1}$ is a homomorphism from F to the formal group F_1. Here F_1 is the group defined in Theorem 3.1.

These results allowed Demchenko and Vostokov to study the arithmetic of a Honda formal module.

We define on the set of Honda formal groups over the ring $\mathfrak{O}_L$ the invertible operator $\mathfrak{A}: F \to F_1$. Define the sequence of Honda formal groups by iterating $\mathfrak{A}$, i.e. $F_m = \mathfrak{A}(F_{m-1})$. We denote by f_{m-1} the homomorphism from F_{m-1} into F_m that is defined in Theorem 3.1. We have $f_0 = f$.

Let $\lambda_m = \log_{F_m}$ be the logarithm of F_m and let u_m be the canonical type of F_m. Put

$$\pi_1 = \pi/a_h, \pi_m = \pi/a_h^{\mathrm{Fr}^{h(m-1)}},$$

$$\pi_1^{(m)} = \prod_{i=1}^{m} \pi_i = \pi^m/a_h^{1+\varphi^h+\cdots+\varphi^{h(m-1)}}.$$

Then $u_m \circ \pi_1^{(m)} = \pi_1^{(m)} u$.

Denote $f^{(m)} = f_{m-1} \circ f_{m-2} \circ \cdots \circ f_1 \circ f$. From Theorem 3.1 one can deduce that

$$f_{m-1}(X) \equiv \pi_m X \quad \mathrm{mod \ deg \ 2}, \qquad f^{(m)}(X) \equiv \pi_1^{(m)} X \quad \mathrm{mod \ deg \ 2}.$$

Let E be a finite extension of L which contains all elements of π^n-division points $\kappa_n = \ker [\pi^n]_F$.

Along with the *generalized Hilbert pairing*

$$(\cdot, \cdot)_F = (\cdot, \cdot)_{F,n} \colon E^* \times F(\mathfrak{M}_E) \to \kappa_n, \quad (\alpha, \beta)_F = \Psi_E(\alpha)(\gamma) -_F \gamma,$$

where Ψ_E is the reciprocity map and γ is such that $[\pi^n]_F(\gamma) = \beta$, we also need another generalization that uses the homomorphism $f^{(n)}$:

$$\{\cdot, \cdot\}_F = \{\cdot, \cdot\}_{F,n} \colon E^* \times F(\mathfrak{M}_E) \to \kappa_n, \quad \{\alpha, \beta\}_F = \Psi_E(\alpha)(\delta) -_F \delta,$$

where δ is such that $f^{(n)}(\delta) = \beta$. Then

$$(\alpha, \beta)_F = \{\alpha, [\pi_1^{(n)}/\pi^n]_{F,F_n}(\beta)\}_F.$$

We introduce a generalization for Honda formal modules of the map l_F defined above. Let T be the maximal unramified extension of K in E. Denote by $F(X\mathfrak{O}_T[[X]])$ the $\mathfrak{O}_K$-module with underlying set $X\mathfrak{O}_T[[X]]$ and operations given by

$$f +_F g = F(f, g); \quad a \cdot f = [a]_F(f), \quad a \in \mathfrak{O}_K.$$

The class of isomorphic Honda formal groups F contains the canonical group F_{ah} of type

$$u = \pi - a_1 \Delta - \cdots - a_h \Delta^h, \quad a_i, \ldots, a_{h-1} \in \mathfrak{M}_L, \quad a_h \in \mathfrak{O}_L^*$$

with Artin-Hasse type logarithm

$$\log_{F_{ah}} = (u^{-1}\pi)(X) = X + \alpha_1 X^q + \alpha_2 X^{q^2} + \ldots, \quad \alpha_i \in L.$$

Define the map l_F as follows: for $g \in X\mathfrak{O}_T[[X]]$, set

$$l_F(g) = \left(1 - \frac{a_1}{\pi} \Delta - \cdots - \frac{a_h}{\pi} \Delta^h\right)(\log_F \circ g).$$

We also need similar maps for the formal group F_n. Let

$$u_n = \pi - b_1 \Delta - \cdots - b_h \Delta^h$$

be the canonical type of F_n. Consider the canonical formal group F_b of type u_n whose logarithm is

$$\lambda_b = (u_n^{-1}\pi)(X) = X + \beta_1 X^q + \beta_2 X^{q^2} + \ldots, \quad \beta_i \in L.$$

The groups F_n and F_b are isomorphic because they have the same type u_n. Now we define the function

$$l_{F_n}(g) = (u_n\pi^{-1})(\lambda_n \circ \psi) = \left(1 - \frac{b_1}{\pi} \Delta - \cdots - \frac{b_h}{\pi} \Delta^h\right)(\lambda_n \circ g).$$

For a monomial $d_i X^i \in T((X))$ put $\nu(d_i X^i) = v_T(d_i) + i/q^h$ where v_T is the discrete valuation of T. Denote by $\mathfrak{L}$ the T-algebra of series

$$\mathfrak{L} = \left\{\sum_{i \in \mathbb{Z}} d_i X^i : d_i \in T, \quad \inf_i \nu(d_i X^i) > -\infty, \quad \lim_{i \to +\infty} \nu(d_i X^i) = +\infty\right\}.$$

Since the $\mathfrak{O}_K$-module κ_n has h generators, we are naturally led to work with $h \times h$ matrices. Denote the ring of integers of the maximal unramified extension of $\mathbb{Q}_p$ in E by $\mathfrak{O}_0$.

Fix a set of generators $\xi_1, \ldots, \xi_h$ of κ_n. Let $z_i(X) \in \mathfrak{O}_T[[X]]$ be the series corresponding to an expansion of ξ_i into a power series in Π, i.e. $z_i(\Pi) = \xi_i$.

Theorem 3.4. *For $\alpha \in E^*$ let $\alpha(X)$ be a series in*

$$\{X^i\theta\varepsilon(X) : \theta \text{ is a Teichmüller representative}, \varepsilon \in 1 + X\mathfrak{O}_0[[X]]\}.$$

For $\beta \in F(\mathfrak{M}_E)$ let $\beta(X)$ be a series in $X\mathfrak{O}_T[[X]]$ such that $\beta(\Pi) = \beta$.

The generalized Hilbert symbol $(\cdot, \cdot)_F$ is given by the following explicit formula:

$$(\alpha, \beta)_F = \sum_{j=1}^{h} {}_{(F)}[\operatorname{Tr} \operatorname{Res} \Phi V_j]_F(\xi_j),$$

where $\Phi(X)V_j(X)$ belongs to $\mathfrak{L}$, $V_j = A_j/\det A$, $1 \leq j \leq h$,

$$A = \begin{pmatrix} \pi^n \lambda \circ z_1(X) & \cdots & \pi^n \lambda \circ z_h(X) \\ \pi^n \Delta (\lambda \circ z_1(X)) & \cdots & \pi^n \Delta (\lambda \circ z_h(X)) \\ \cdots & \cdots & \cdots \\ \pi^n \Delta^{h-1} (\lambda \circ z_1(X)) & \cdots & \pi^n \Delta^{h-1} (\lambda \circ z_h(X)) \end{pmatrix},$$

A_j is the cofactor of the $(j,1)$-element of A,

$$\Phi = \frac{\alpha(X)'}{\alpha(X)} l_F(\beta(X)) - \frac{1}{\pi} \sum_{i=1}^{h} a_i \left(1 - \frac{\Delta^i}{q^i}\right) (\log \varepsilon(X)) \; \Delta^i \, (\lambda \circ \beta(X)),$$

and the formula means that we add (with respect to F) ξ_j, multiplied in the formal sense by the numbers $[\mathrm{Tr}\,\mathrm{Res}\,\Phi V_j]$.

See [19] for the proof.

Remark 3.5.(i) One can easily deduce the formula for the Lubin-Tate groups from the formula above.

(ii) The formula above can be simplified in the case of $n = 1$, see [2].

(iii) The first explicit formula for the generalized Hilbert pairing for formal Honda groups and arbitrary n in the case of odd p under some additional assumptions on the field E was obtained by V A. Abrashkin [1] using the link between the Hilbert pairing and the Witt pairing via an auxiliary construction of a crystalline symbol as a generalization of the method developed in his previous papers.

(iv) A similar result for the case of an arbitrary residue field was proved in [11].

4 The matrix method

All results of Sections 4–8 can be found in [12]. In [9] they were extended to the case of imperfect residue fields.

4.1 Restriction of scalars

Let N be the inertia subfield of a complete discrete valuation field K, and let $\mathfrak{o} = \mathfrak{o}_N$ be its ring of integers. We have $e = [K : N]$. As in Subsection 2.2 we consider the ring W. We put W' equal to $N[\Delta]$ as a set with multiplication defined as in W.

The ring $K[[\Delta]]$ has a natural structure of a right W-module. Note that in order to define it one does not need to extend σ to K (since we do not consider the products of the type Δx for $x \in K \setminus N$).

In order to apply the results of Honda to ramified local fields we replace an m-dimensional group over $\mathfrak{O}_K$ by an me-dimensional one over $\mathfrak{o}$. To this end one needs the Weil restriction of scalars.

We fix a basis $w = (w_1, \ldots, w_s)$ of $\mathfrak{O}_K$ over $\mathfrak{o}$.

Consider an operator $S = S_w : N^s \to K$ that maps (n_i) into $\sum n_i w_i$ and extend it to various formal power series rings over N.

Suppose that we have an m-dimensional formal group law $F = (F_l)$ over $\mathfrak{O}_K$, i.e. an $m$tuple of series in variables $X_1, \ldots, X_m, Y_1, \ldots, Y_m$. We introduce the variables X_l^r, Y_l^r for $1 \leq l \leq m$ and $1 \leq r \leq s$. Assume that the new variables 'take their values in N' and satisfy the relations $X_l = S((X_l^r))$, $Y_l = S((Y_l^r))$. Then the standard properties of the Weil restriction of scalars imply that there exists a unique $m$tuple (F_l^r) of formal power series in $X_1, \ldots, X_m, Y_1, \ldots, Y_m$ over $\mathfrak{o} = \mathfrak{o}_N$ satisfying

$$F_i(S((X_l^r)), S((X_l^r))) = S((w_r F_i^r(X_1, \ldots, X_m, Y_1, \ldots, Y_m)) \qquad (4.1)$$

for all i, $1 \leq i \leq m$. Moreover, $F_w = (F_l^r)$ is a formal group law. We can apply a similar process to homomorphisms of formal groups. We denote the functor obtained in this way by ϕ. In particular, $\phi(\lambda)$ is the logarithm of $\phi(F)$.

An $n \times m$ matrix A can be regarded as a homomorphism from G_a^m into G_a^n. Thus it corresponds to a unique $ns \times ms$ matrix A' over N. We will denote A' by $\phi(A)$.

We formulate the main properties of ϕ.

Proposition 4.1.(i) *Let A be an $m_2 \times m_1$ matrix over $\mathfrak{O}_K$. There exists a homomorphism f from F_1 of dimension m_1 into F_2 of dimension m_2, $f \equiv AX \mod \deg 2$, if and only if there exists a homomorphism f' from $\phi(F_1)$ into $\phi(F_2)$), $f' \equiv A'X \mod \deg 2$.*

(ii) *F_1 and F_2 are strictly isomorphic if and only if so are $\phi(F_1)$ and $\phi(F_2)$.*

4.2 The logarithmic matrix

The goal of this section is to apply Honda's classification of formal groups over unramified fields to the classification of formal groups over general local fields. We apply the functor ϕ and consider a matrix that corresponds to the p-typical part of the logarithm of $\phi(F)$.

We define an operator $\langle \alpha \rangle$ on $K[[\Delta]]$: $\langle \alpha \rangle(\sum c_i \Delta^i) = \sum c_i \alpha^{p^i} \Delta^i$. Thus for any $h \in K[[\Delta]]$ we have

$$\langle \alpha \rangle(h)(x) = h(y),$$

where $y = ax$.

In what follows we always assume that N is the inertia subfield of K and w is a $\mathfrak{o}$-base of $\mathfrak{O}_K$. Sometimes we will impose some extra restrictions on w.

Definition 4.2. For an $m$tuple $f = (f_i)$ of p-typical series in X_i over K we define the matrix $T_w(f)$ consisting of the columns (t_j^i), where the coefficients of t_j^i are $t_{jk}^{il} \in W'$, by the equality

$$f(w_i x e_j)_k = \left(\sum_l w_l t_{jk}^{il} \right) (x e_j).$$

Thus $S_l(t_{jk}^{il}) = \langle w_i \rangle \Lambda_j$.

If $\Lambda(X)$ is the logarithm of a formal group F we call $T_w(\Lambda)$ the *logarithmic matrix* for F.

4.3 Main properties of the logarithmic matrix

We study the connection of $\phi(F)$ with the formal group obtained from $T(\Lambda)$.

Proposition 4.3.(i) *Let $\Lambda \in K[[\Delta]]^m$. Then $\lambda = \Lambda(X)$ is the logarithm of a formal group F over $\mathfrak{O}_K$ if and only if $\lambda_e = T(\Lambda)((X_j^i))$ is the logarithm of a formal group over $\mathfrak{o}$.*

ii) *If the conditions of the first part are fulfilled, then $T(\Lambda)((X_j^i))$ is the logarithm of a formal group law that is strictly isomorphic to $\phi(F)$.*

iii) *There exists a homomorphism f from F into F' of dimensions m and m' respectively, of the form $f \equiv AX$ mod deg 2 for some $m' \times m$ matrix A over $\mathfrak{O}_K$, if and only if there exists a $C \in M_{m' e \times m e}(W)$ such that $\phi(A)T(F) = T(F')C$.*

Proof (Sketch). Part (i) is verified by relatively simple calculations. Part (ii) follows immediately from Proposition 2.1. Part (iii) is obtained by application of Theorem 2.2 to the groups $\phi(F_1)$ and $\phi(F_2)$. □

Using very simple linear algebra, one can state the previous result in a more convenient form.

Proposition 4.4.(i) *A p-typical $m$tuple $\lambda = \Lambda(X)$, $\Lambda \equiv I_m$ mod Δ, is the logarithm of a formal group law over $\mathfrak{O}_K$ if and only if $pW^{me} \subset T(\Lambda)W^{me}$.*

ii) *There exists a homomorphism f from F to F' of dimensions m and m' respectively, $f \equiv AX$ mod deg 2, A is some $m' \times m$-matrix over $\mathfrak{O}_K$, if and only if $\phi(A)T(F)(W^{me}) \subset T(F')(W^{me})$.*

iii) *There exists a homomorphism f from F to F' of dimensions m and m' respectively, $f \equiv Ap^s X$ mod deg 2, A is some $m' \times m$ matrix over $\mathfrak{O}_K$, for some $s \in \mathbb{Z}$ if and only if $\phi(A)T(F)(W^{me}) \subset T(F')(W^{me})N$.*

5 Fractional part invariant: classification up to a strict isogeny

In this section we define the fractional part of elements of $K[[X]]$. We prove that the fractional part of the logarithm of a formal group classifies formal groups up to an isogeny of an explicitly described sort.

We will consider the algebra NW, which is equal to $\cup_s p^{-s}W$.

5.1 Fractional parts

We consider the ring $R_X = \mathfrak{O}_K[[X_i]]\mathbb{Q}_p$, i.e. the ring of series with bounded denominators. We denote by DR_X the K-module of series whose partial derivatives belong to R_X.

Definition 5.1. We denote the residue of an element $f \in K[[X_i]]$ modulo R_X by $\{f\}$ and call it the _fractional part_ of f.

Thus $\{f\} = \{g\}$ if and only if $f - g \in R_X$.

We introduce the valuation v_X on R_X as the minimum of the valuations of coefficients.

We list the basic properties of the fractional part. The proof is quite easy.

Lemma 5.2. _Let_ $f, g \in DR_X$ _and_ $h_1, h_2 \in \mathfrak{O}_K[[X_i]]_0^m$ _(i.e. the constant terms are zero)._

(i) _If_ $\{f\} = \{g\}$ _and_ $h_1 \equiv h_2 \mod (\pi)$, _then_ $\{f(h_1)\} = \{g(h_2)\}$.

(ii) _Let_ $\lambda \in DR_X^m$ _be the logarithm of a finite height formal group. If_ $\{\lambda(h_1)\} = \{\lambda(h_2)\}$ _then_ $h_1 \equiv h_2 \mod (\pi)$.

(iii) _There exists a constant_ c _(c depends on e) such that_

$$v_X(g(h_1) - g(h_2)) \geq \min v_X(\frac{dg}{dX_i}) - c.$$

(iv) _For the obvious right action of NW on $K[[X]]$, $\{fr\} = \{f\}r$ for any_ $r \in NW$, $f \in K[[X]]$.

5.2 Presentation of a logarithm as a 'fraction'

Theorem 5.3.(i) _Suppose that_ $\Lambda \in M_m(K[[\Delta]])$ _comes from a formal group F (i.e. $\Lambda(\Delta)(X)$ is the logarithm of a p-typical group F). Then Λ can be presented in the form v/u, where $vp^l\pi^{-p^l} \in M_m(\mathfrak{O}_K[[\Delta]])$, $l = [\log_p(e/(p-1))]$ and u is a special element of $M_m(W)$._

(ii) *Let F, F' be finite height groups. Suppose that $\Lambda = v/u$ and $\Lambda' = v'/u'$, where $v, v' \in M_m(R)$, u, u' are special elements. Then the reductions of F and F' modulo π are equal if and only if $u' = u\varepsilon$ for $\varepsilon \in GL_m(W)$.*

Proof (Sketch). The idea is to present $\mathfrak{D}_K$ as a factor of a ring $\mathfrak{A} = \mathfrak{o}[[t]]$ (the homomorphism maps t to π). We can raise F to a p-typical formal group law $F_{\mathfrak{A}}$ over $\mathfrak{A}$. We have $\Lambda_{\mathfrak{A}} = pU^{-1}$. Here $U = u + tu'$ in a certain non-commutative ring that is defined similarly to W (with $\mathfrak{A}$ instead of $\mathfrak{o}$, $\sigma(t) = t^p$, see subsection 2.2) and $u \in W$. Then it can be easily checked that Λu has bounded denominators.

If the reductions of F and F' modulo π are equal then one can choose U' that is congruent to U modulo t. In order to check that for finite height formal groups F, F' we always have $u' = u\varepsilon$ one should use Theorem 5.5 below. $\square$

Remark 5.4.(i) Thus we can take the same denominator in the presentation of Λ and Λ' if and only if the reductions are equal.

(ii) F is a formal group of finite height if and only if the minimal possible rank of the reduction of u over W mod p is equal to m.

(iii) If F is a formal module over some $B \supset \mathbb{Z}_p$ then one can take a smaller l in part (i) of the theorem.

5.3 Main theorem about 'fractional parts'

For a formal group law Γ, define $r(F)$ to be the residue of Λ mod $M_m(R)$. We have $r(F) \in M_m(R)u^{-1}/M_m(R)$ where $\Lambda = vu^{-1}$.

Theorem 5.5.(i) *Let A belong to $M_{m' \times m}(K)$ and let the dimensions of F and F' be m and m' respectively. Then there exists a homomorphism f from F into F' such that $f(X) \equiv Ap^s X$ mod deg 2 for some $s \in \mathbb{Z}$, if and only if $Ar = r'\varepsilon$ for some $\varepsilon \subset M_{m'}(NW)$. Here $r = r(F)$ and $r' = r(F')$.*

(ii) *Suppose that there exists a homomorphism f from F into F' such that $f \equiv AX$ mod deg 2 for some $A \in M_{m' \times m}(\mathfrak{D}_K)$. Then we can present f as*

$$f(X) = \sum_{(F'), i, j, l} (a_{ijl} X_i^{p^j} e_l) \tag{5.1}$$

for some $a_{ijl} \in \mathfrak{D}_K$. For this a_{ijl} we have $Ar = r'B$, where $B_{il} = \sum \theta(\overline{a_{ijl}})\Delta^j$ and θ denotes the Teichmüller representative.

Proof.(i) We take $w_1 = 1$, $\pi \mid w_j$ for $j > 1$. Then we obtain

$$S(T(F)(NW^{me})) = vu^{-1}(NW^{me}) + R^m.$$

Thus the condition on the fractional parts is equivalent to

$$AT(F)(NW^{me}) \subset T(F')(NW^{m'e}).$$

Now Proposition 4.4 gives us the statement of the first part.
(ii) According to the definition of a homomorphism, we have

$$\lambda'(f(X)) = A\lambda(X). \tag{5.2}$$

Hence we have to prove $\{\lambda'(f(X))\} = r'B$.

If θ is a Teichmüller representative, for any $s > 0$, $1 \le i \le m$ we have $\lambda'(\theta x_i^{p^s}) = \Lambda\theta\Delta^s e_i(X)$.

Suppose that $f(X)$ cannot be presented in the form (5.1). Obviously, $f(X)$ can be presented as

$$\sum_{(F'),i,J,l} (a_{iJl}X^J e_l),$$

where J runs through all multi-indices. If J_0 is the least non-p-typical multi-index (i.e. $J \ne e_i p^s$) such that $a_{iJl} \ne 0$, then the coefficient of $\lambda'(f(X)) = \sum_{i,J,l} f(a_{iJl}X^J e_l)$ at $X^J e_l$ is non-zero. We deduce that $A\lambda$ is not p-typical, therefore λ is not p-typical.

Hence $f(X)$ can be presented in the form (5.1). We have

$$\{\lambda'(f(X))\} = \sum_{i,l}\{\lambda'(a_{ijl}x_i^{p^j} e_l)\} = r'B$$

according to part (ii) of Lemma 5.2.

$\square$

We call two formal groups *rationally isogenous* if they satisfy the conditions of the first part of the theorem for $A = I_m$.

5.4 *Implications of the theorem*

Theorem 5.5 can be used for explicit calculations of the fractional part invariant. We describe the most natural implications.

Proposition 5.6. *Let* F, F' *be finite height groups.*

(i) *For fixed r-invariants of F and F', the knowledge of f mod deg 2 fixes f modulo π.*

ii) *In particular, $r = r'$ if and only if there is a homomorphism $f \equiv p^s I_m X$ mod deg 2 from F into F' for some $s \in \mathbb{Z}$ such that $f(x) \equiv [p^s]_F(X)$ mod $\pi \mathfrak{O}_K[[X]]^m$.*

iii) *If F is strictly isomorphic to F' then $r = r'$ if and only if the strict isomorphism between this groups is congruent X modulo π.*

iv) *If F is strictly isomorphic to F', then*

    a.) $u' = u\eta$ for $\eta \in I_m + M_m(W)\Delta$
    b.) $r' = r\varepsilon$ for $\varepsilon \in I_m + M_m(W)\Delta$.

v) *Let F and F' have dimension m and be of finite height. Then F and F' are isogenous if and only if there exists an $A \in GL_m(K)$ and $\varepsilon \in M_m(NW)$ such that $Ar = r'\varepsilon$.*

According to Theorem 5.5, strict isomorphisms multiply r by $\varepsilon \in I_m + M_m(W)\Delta$. Moreover, any such ε is possible (for any r). Hence it seems natural to describe homomorphisms between F and F' fixing r and r' only modulo $I_m + M_m(W)\Delta$.

We state the corresponding result for one-dimensional groups. We expand u as $p - \sum u_i \Delta^i$, $u_i \in \mathfrak{o}$. We have $u_h \in \mathfrak{o}^*$, $u_i \in p\mathfrak{o}$ for $i < h$, where h is the height of F (see the remark below).

The following result follows immediately from Theorem 5.5.

Proposition 5.7. *Suppose that F and F' are finite height formal groups. Let $a \in K$, $b \in \mathfrak{o}$ and $m \in \mathbb{Z}$. Then we have $ar(F) = r(F')b\varepsilon\Delta^m$ for some $\varepsilon \in 1 + W\Delta$ if and only if there is an isogeny $f = \sum a_i x^i \equiv ap^s x$ mod deg 2 from F into F' for some $s \in \mathbb{Z}$ such that the height of f equals $sh + m$ and*

$$a_{p^{sh+m}} \equiv b u_h^{p^m + p^{h+m} + \ldots p^{(s-1)h+m}} \qquad \text{mod } \pi.$$

Remark 5.8.(i) Unfortunately, in order to describe the corresponding multi-dimensional result one has to introduce many technical definitions and results.

ii) A certain invariant that is similar to $r(F)$ was defined by Fontaine (see [22]). Our definition is much more explicit. The main disadvantage of Fontaine's functor is that it is formulated in terms of modules, so it gives $r(F)$ only modulo a $GL_m(W)$ multiplier instead of $I_m + \Delta M_m(W)$. Thus one cannot recover $a_{p^{sh+m}}$ mod π from Fontaine's invariant. One can construct non-isomorphic (here we also consider non-strict isomorphisms) formal groups that are glued together by Fontaine's functor but have distinct r-invariants.

Parts (i)–(iii) of Proposition 5.6 are also completely new.

We note that by fixing a base for the formal group law one can formulate substantially more precise results than in the base-independent approach.

6 Invariant Cartier-Dieudonné modules; the module invariant of a formal group

6.1   D-modules; two definitions of D_F

Definition 6.1. We denote by $\mathfrak{D}$ the category of (left) W-submodules D of $K[[\Delta]]^m$ satisfying the following conditions.

(i) D is a free W-module of rank em;
(ii) $D \mod \Delta = \mathfrak{O}_K^m$;
(iii) $p\mathfrak{O}_K^m \subset D$.

One can easily verify the following statements.

Lemma 6.2.(i)   *The elements $v_1, \dots, v_{me} \in D$ form a W-base of $D \in \mathfrak{D}$ if and only if $v_i \mod \Delta$ form an $\mathfrak{o}$-base of $\mathfrak{O}_K^m$.*
(ii)   *If $D \subset D'$ for $D, D' \in \mathfrak{D}$, then $D = D'$.*

Sometimes it can be useful to replace the axioms of $\mathfrak{D}$ (especially the first one) by slightly different ones.

Proposition 6.3. *If D, satisfying conditions (ii) and (iii) of Definition 6.1, satisfies also one of the following conditions:*

1) *D is generated by at most me elements;*
2) *any $v \in W^m \Delta \cap D$ also satisfies $v \in D\Delta$,*

then it belongs to $\mathfrak{D}$. Conversely, any $D \in \mathfrak{D}$ satisfies 1) and 2).

Definition 6.4. For a formal group F with logarithm $\lambda = \Lambda(X)$ one defines

$$D_F = S(T_F(W^m)) = \langle\langle w_i \rangle \Lambda_j \rangle.$$

We describe the functorial properties of D_F.

Proposition 6.5. *For formal groups F_1 and F_2 of dimensions m_1 and m_2, whose D-modules are equal to D_1 and D_2 respectively, the following statements are valid.*

(i)   *Let A be an $m_2 \times m_1$ matrix over $\mathfrak{O}_K$. There exists a homomorphism f from F_1 into F_2, $F(X) \equiv AX \mod \deg 2$ if and only if $AD_1 \subset D_2$.*

(ii) *For $m_1 = m_2$ the groups F_1 and F_2 are strictly isomorphic if and only if $D_1 = D_2$.*

iii) *For any F we have $D_F \in \mathfrak{D}$.*

The proof is a direct application of Proposition 4.4.

It turns out that D_F is the logarithm of the module of p-typical curves (i.e. of the classical Cartier-Dieudonné module).

Proposition 6.6.

$$D_F = \{f \in K[[\Delta]]^m : \ \exp_F(f(x)) \in \mathfrak{O}_K[[x]]\}$$

where $\exp_F$ is the composition inverse of λ_F.

Proof. First we verify that $D_F \subset D'_F$.

We have $\exp_F(\langle w_i \rangle \Lambda(xe_i)) = w_i x e_i$, hence D'_F contains a W-base of D_F.

We check that D'_F is an W-module. For $c_i \in \mathbb{Z}_p$ and $h \in K[[\Delta]]^m$ one has

$$\exp_F(h \sum c_i \Delta^i)(x) = \sum_{(F)} [c_i]_F \exp_F(h(X^{p^i})),$$

therefore D'_F is an $\mathbb{Z}_p[[\Delta]]$-module. It remains to check that for any Teichmüller representative θ we have $D'F\theta \subset D'F$. This assertion follows immediately from the relation $\exp_F(h\theta(x)) = \exp_F(h(y))$, where $y = \theta x$, for any $h \in K[[\Delta]]^m$.

Now, according to Lemma 6.2, it is sufficient to prove $D'_F \in \mathfrak{D}$. We have $\exp_F(h(x)) \equiv h(x) \mod \deg 2$, hence $D'_F \mod \Delta \subset \mathfrak{O}_K^m$. Thus properties (ii) and (iii) of the definition of $\mathfrak{O}$ follow from the fact $D_F \subset D'_F$. Obviously, D'_F also satisfies condition 2) of Proposition 6.3. Therefore $D'_F \in \mathfrak{D}$ and the proposition is proved. $\square$

6.2 Basic properties of D_F

A module $D \in \mathfrak{D}$ corresponds to a formal group over $\mathfrak{O}_K$ if and only if it is $\langle \pi \rangle$-stable. It is an analogue of the classification result of Cartier.

Proposition 6.7. *For $D \in \mathfrak{D}$ the following conditions are equivalent:*

(i) *$D = D_F$ for some F;*

ii) *for any $a \in \mathfrak{O}_K$ we have $\langle a \rangle D \subset D$;*

iii) *$\langle \pi \rangle D \subset D$;*

iv) *there exist $\Lambda_i \in D$, $\Lambda_i \equiv e_i \mod \Delta$ such that $\langle w_j \rangle \Lambda_i \in D$, $1 \leq i \leq m$, $1 \leq j \leq e$, where (w_i) is some $\mathfrak{o}$-base of $\mathfrak{O}_K$.*

Instead of π we could take any other polynomial generator of $\mathfrak{O}_K$ over $\mathfrak{o}$, for example, $1 + \pi$.

Next we describe the logarithms of formal groups isomorphic to F in terms of D_F.

Proposition 6.8. *For $f \in M_m K[[\Delta]]$ the series $f(\Delta)(X)$ is the logarithm of a formal group strictly isomorphic to F if and only if $f \equiv I_m$ mod Δ and $f(\Delta)(xe_i) \in D_F$.*

Thus we can choose any set of $\Lambda_i \equiv e_i$ mod Δ in D.

We also note that the intersection of D_F with K^m always contains a module that is slightly larger than $p\mathfrak{O}_K^m$.

Remark 6.9.(i) For $s = -[e/(1 - p)]$ we have $\pi^s \mathfrak{O}_K W^m \subset D_F$.

(ii) Let $a \in \mathfrak{O}_K$, $v_K(a) = l$, $l \leq s$. Then for the formal group law $F_a = a^{-1} F(aX, aY)$ one has $\pi^{s-l} \mathfrak{O}_K W^m \subset D_{F_a}$.

6.3 The change of the base field

We can easily describe how the module D_F behaves if we replace K by $L \supset K$. The proof is very easy.

We denote the ring of integers of L by $\mathfrak{o}_L$.

Proposition 6.10. *Let F be a formal group over $\mathfrak{O}_K$, $\{s_i\}$ be an $\mathfrak{O}_K$-base of $\mathfrak{o}_L$. Then $D_L(F) = \langle\langle s_i \rangle D_K(F)\rangle$ (the envelope can be understood in the Abelian group sense).*

Now suppose that K' is a subfield of K and π' is a uniformizing element of K'. We prove the following result in order to demonstrate the advantages of our classification methods when compared with the methods of Breuil (cf. [13]).

Proposition 6.11. *A finite height formal group F is isogenous to a formal group defined over $\mathfrak{O}'_K$ if and only if $r = Ar'\varepsilon$ for some r' defined over K', $A \in GL_m(K)$ and $\varepsilon \in GL_m(W)$.*

Proof. The only if part follows immediately from Theorem 5.5.

Now suppose that $r = Ar'\varepsilon$. We consider the module $D_1 = A^{-1}D \cap K'[[\Delta]]^m$. We have $D_1 \mod \Delta \subset A^{-1}\mathfrak{O}_K^m \cap K'^m$ and $pA^{-1}\mathfrak{O}_K^m \cap K'^m \subset D'$. Hence if we choose $A' \in GL_m(K')$ so that $A'D_1 \mod \Delta = \mathfrak{O}'_K{}^m$, then the module $A'D_1$ satisfies properties (ii) and (iii) of the definition of $\mathfrak{D}$ (over K'). Obviously, D' satisfies property 2) of Proposition 6.3.

Thus $D' \in \mathfrak{D}'$. Since for any $a \in \mathfrak{O}'_K$ we have $\langle a \rangle D' \subset D'$, it corresponds to some formal group F' over $\mathfrak{O}'_K$.

It remains to prove that F' is isogenous to F. Using Proposition 6.10, we obtain $AA'^{-1}D_{F'} \subset D_F$. Hence there exists a homomorphism $f \equiv AA'^{-1}X \mod \deg 2$ from F' into F.

On the other hand, the condition on the fractional parts implies that there exists an $s \in \mathbb{Z}$ such that for all i, $1 \leq i \leq m$, $p^s \Lambda_i \in AA'^{-1}D_{F'}$. Hence a homomorphism with nonzero Jacobian in the inverse direction also exists. $\qquad\qquad\qquad\qquad\qquad\qquad\qquad\qquad\qquad\qquad\qquad\qquad\square$

Note that when the conditions of the proposition are not fulfilled one still obtains a formal group F' over K' and a canonical (up to an isomorphism) homomorphism from F into F'.

6.4 The module invariant

The goal of this section is an explicit description of a module invariant M_F. It completes the V-invariant to a classification of formal groups up to a strict isomorphism; furthermore, the set of possible M_F for finite height formal groups of fixed dimension over a fixed field K is finite.

For any local field L we denote by $L\{\{\Delta\}\}$ the two dimensional field of series

$$L\{\{\Delta\}\} = \left\{ \sum_{i \in \mathbb{Z}} a_i \Delta^i \ : \ \begin{array}{l} a_i \in L, \ a_i \to 0 \text{ for } i \to -\infty, \\ v(a_i) > s \text{ for some } s \in \mathbb{Z} \end{array} \right\}.$$

As proved above, the module D_F lies in $(R^m)u^{-1}$. Hence it lies in $w^{-1}R^m$, where $w \in \mathbb{Z}_p[[\Delta]]$ is the determinant of u as an $\mathbb{Q}_p[[\Delta]]$-operator. We can consider w^{-1} as a matrix over $\mathbb{Q}_p\{\{\Delta\}\}$. Therefore we can canonically embed D_F in $K\{\{\Delta\}\}$.

Note that we can define on $K\{\{\Delta\}\}$ the natural right action of

$$\Omega = \left\{ \sum_{i \in \mathbb{Z}} a_i \Delta^i \ : \ a_i \in \mathfrak{o}, \ a_i \to 0 \text{ as } i \to -\infty \right\},$$

where the multiplication, as in W, is defined by the relation $\Delta a = \sigma(a)\Delta$ for any $a \in \mathfrak{o}$. We have $w \subset \Omega$.

Definition 6.12. For a formal group F we define M_F as $D_F\Omega \subset K\{\{\Delta\}\}$.

Note that M_F is a Ω-module.

We describe the main properties of M_F that allow its use in classification questions as a complement of the invariant $r(F)$.

Theorem 6.13.(i) *Suppose that F and F' satisfy the conditions of part (i) of Theorem 5.5, i.e. that for some $A \in M_{m' \times m}(K)$ and $\varepsilon \in M_{m'}(NW)$ we have $Ar = r'\varepsilon$. Then there exists a homomorphism f from F to F' given by $f(X) \equiv Ap^s x \mod \deg 2$ for a fixed $s \in \mathbb{Z}$, if and only if $Ap^s M_F \subset M_{F'}$.*

(ii) *Suppose that F is a finite height group and for some $A \in M_{m' \times m}(K)$ and $\varepsilon \in M_{m'}(NW)$ we have $Ar = r'\varepsilon$. If $p^s A \in M_{m' \times m}(\mathfrak{M}^l)$, where*

$$s = p^{[\log_p(e/(p-1))]} - e[\log_p(e/(p-1))] - [e/(1-p)],$$

$\varepsilon \in M_{m'}(NW)$, then there exists a homomorphism f from F to F', with $f(X) \equiv Ap^s x \mod \deg 2$.

(iii) *F and F' are strictly isomorphic if and only if they are rationally isogenous and $M_F = M_{F'}$.*

The proof is easy and uses the fact that $NW \cap \Omega = W$.

7 Classification for small ramification index

7.1 Formal groups for $e < p$

Suppose now that the ramification index of K is less than p.

Theorem 7.1.(i) *$\lambda = \Lambda(\Delta)(X)$ is the logarithm of a p-typical formal group if and only if $\Lambda = vu^{-1}$ for some $u \in M_m(W)$, $u \equiv pI_m \mod \Delta$ and $v \in M_m(\mathfrak{O}_K[\Delta])$, $v \equiv pI_m \mod \pi\Delta$.*

(ii) *The formal group F corresponding to vu^{-1} is strictly isomorphic to F_1 corresponding to $v_1 u_1^{-1}$ if and only if $u_1 = \varepsilon u$, $v_1 = v + gu$, where $\varepsilon \in I_m + \Delta M_m(W)$, $g \in \pi M_m(\mathfrak{O}_K[\Delta])\Delta$.*

Remark 7.2.(i) One can also prove a result about homomorphisms between formal groups similar to part (iii) of 2.2. One should take v into account in the same way as in Theorem 5.5.

(ii) One can choose a canonical representative in each strict isomorphism class by demanding the coefficients of u at positive degrees of Δ to be Teichmüller representatives (see Proposition 2.3), the coefficients of v at positive degrees of Δ to have zero trace into $\mathfrak{o}$.

(iii) One can easily prove that in the one-dimensional case the height of F is equal to the index of the first invertible coefficient of u.

iv) In a similar way one can describe formal groups of height > 1 for $e = p$. For a multidimensional group the height should be understood in a multidimensional sense. It is possible to classify one-dimensional formal groups for $e \leq 2p-2$. To this end one should take into account the first two columns of T. The method described below seems to be more useful, as it easily gives the classification of formal groups of height > 1 for $e \leq p^2/2$.

7.2 M_F for finite height groups. Algorithm for classification of formal groups

Proposition 7.3.(i) *If the logarithm of a formal group law F belongs to R_X^m then $D_F = M_F \cap R^m$.*

(ii) *If F is a finite height formal group then $M_F = \pi M_{F_\pi}$, where F_π is the formal group law $\pi^{-1}F(\pi X, \pi Y)$.*

(iii) *If F is a finite height formal group then $D_F \cap R^m = \pi D_{F_\pi}$.*

One can easily check that $\lambda \in R_X^m$ if the reduction of the formal group law F modulo π is equal to G_a^m.

Now we give a result that is crucial for the construction and the classification of formal group laws.

Theorem 7.4.(i) *$\lambda = \Lambda(X)$ is the logarithm of a formal group law F over $\mathfrak{D}_K$ if and only if $\langle \pi \rangle \Lambda_i \in \pi D_{F'}$ for $1 \leq i \leq m$ and $(\Lambda u)_i \in \pi D_{F'}$ for some special u and a formal group F' over $\mathfrak{D}_K$ satisfying $\pi^{s-1}\mathfrak{D}_K W^m \subset D_{F'}$.*

If this happens then F' is strictly isomorphic to F_π.

(ii) *Let $s = -[e/(1-p)]$, $a \in \mathfrak{D}_K$, and $v_K(a) = l \leq s$. Suppose that for some Λ, a special u, and a formal group law F' over $\mathfrak{D}_K$ satisfying $a^{-1}\pi^s \mathfrak{D}_K W^m \subset D_{F'}$ we have $\langle a \rangle \Lambda_i \in a D_{F'}$ and $(\Lambda u)_i \in \pi^s D_{F'}$. Then $\Lambda(X)$ is the logarithm of a formal group law F over $\mathfrak{D}_K$ and F' is strictly isomorphic to F_a.*

Remark 7.5. In particular, the conditions of assertion 2 are fulfilled if $\langle a \rangle \Lambda_i \in a \mathfrak{D}_K^m W$ and $\Lambda u_i \in a \mathfrak{D}_K^m W$, since in this case one can take F' equal to the m-th power of the additive formal group law.

Hence one can obtain all Honda formal group laws (Lubin-Tate ones, for example) by means of this statement. The canonical representatives of the isogeny classes of formal groups given in [25] for the case of an algebraically closed residue field can also be constructed.

Now we describe a general algorithm for classifying formal group laws of fixed dimension m over a fixed field K.

First one describes all possible πD_{F_π}. To this end one can can use the universal p-typical formal group law to construct the logarithms of the form $\pi^{-1}\lambda(\pi X)$. Since D_F depends only on the residues of the coefficients of Λ modulo π^l, where $l = -[e/(1-p)]$ and only the first few coefficients of $\langle \pi \rangle \Lambda$ may not divide π^l, the number of distinct possible πD_{F_π} is finite.

Next one fixes πD_{F_π}, and for each u describes Λ satisfying the conditions of the previous proposition. Thus one obtains the description of all p-typical logarithms of formal group laws.

In order to check which of them give strictly isomorphic formal group laws, one can check which residues modulo $R^m u$ have elements $(v_i) \in \pi M_{F_\pi}^m$, $v \equiv pI_m \mod \Delta$ such that

$$\langle \pi \rangle (vu^{-1}) \in \pi D_{F_\pi}^m. \tag{$*$}$$

After doing this one recovers $r(F)$. Next it is possible to calculate M_F (using Proposition 7.4 for finite height groups). Hence one obtains the pair $r(F), M(F)$ and is able to apply Theorem 6.13.

Note that the condition $(*)$ depends only on the first few coefficients of u and v. Thus if we take u equal to a canonical representative, the calculation will terminate in finitely many steps.

7.3 Classification of formal groups for $e \le p^2/2$

We illustrate the method described in the previous subsection by classifying one-dimensional formal groups of height > 1 for $e \le p^2/2$. This result can easily be extended to multidimensional groups if one defines the height as a certain vector (i.e. in this case we should demand $u \equiv 0 \mod (p, \Delta^2)$).

Calculation of πD_{F_π}

According to Remark 6.9, it is sufficient to calculate $\lambda_\pi = \pi^{-1}\lambda(\pi x) = \Lambda_\pi(\Delta)(x)$ modulo $\pi^{1-[e/(1-p)]}$.

The universal p-typical group law (see [23]) is obtained by inverting the power series $u_c = 1 - \sum c_i \Delta^i p^{-i}$ in the noncommutative ring, that is isomorphic to $\mathfrak{A}[[\Delta]]$ as a left $\mathfrak{A}$-module, and applying it to x. Here $\mathfrak{A} = \mathbb{Z}_p[c_i]$, $i > 0$, $\sigma(c_i) = c_i^p$, $\Delta c_i = c_i^p \Delta$. We have

$$u_c^{-1} = 1 + \frac{c_1}{p}\Delta + \left(\frac{c_2}{p} + \frac{c_1^{p+1}}{p^2}\right)\Delta^2 + \dots. \tag{7.1}$$

For the formal group λ_π we have $\pi^{p^i-1} \mid c_i$. Since the height of F is > 1 we also have $\pi^p \mid c_1$. Hence for $e \leq \frac{p^2}{2}$ we obtain $\Lambda_\pi \equiv 1 + \frac{c}{p}\Delta$ mod $\pi^{1-[e/(1-p)]}$, $\pi^{p-1} \mid c = c_1$. Therefore,

$$\pi D_{F_\pi} = \langle \pi^i + \frac{\pi^{(i-1)p+1}c}{p}\Delta \rangle, \; i \geq 1.$$

We denote $v_K(c)$ by d.

From this one easily obtains that $a \in \pi D_{F_\pi}$ (which is equivalent to $a\Delta \in \pi D_{F_\pi}$) for $a \in \mathfrak{M}$ if and only if $v_K(a) \geq 1 + \frac{e-d}{p-1}$.

Checking (*)

If $h > 1$ then, according to Proposition 2.3, we can assume $u \equiv p$ mod Δ^2. For such u and $e \leq \frac{p^2}{2}$ one easily sees that

$$\langle \pi \rangle (v/u) \equiv \langle \pi \rangle (v/p) \quad \text{mod } \pi^{-[e/(p-1)]},$$

Next, the module πD_{F_π} was constructed for

$$\Lambda = \sum_{i \geq 0} \frac{c}{\pi^{p-1}}^{1+\cdots+p^{i-1}} \left(\frac{\Delta}{p}\right)^i.$$

Since $h > 1$, we have $\frac{c}{\pi^{p-1}} \in \mathfrak{M}$, hence $v' = p + c\frac{\Delta}{p\pi^{p-1}} \in \pi D_{F_\pi}$ and v' satisfies (*) for $u = p$. We deduce that v satisfies (*) if and only if $\langle \pi \rangle \frac{v-v'}{p} \subset D_F$. On the other hand, we have $\langle \pi \rangle \frac{\Delta^2}{p} D_{F_\pi} \in D_{F_\pi}$. Therefore, v satisfies (*) for $v \in \pi D_{F_\pi}$ if and only if $v \equiv v' + v_1\Delta \mod \deg 2$, where $v_1 \in \mathfrak{M}$, $v_K(v_1) \geq e - p + 1 + \frac{e-d}{p-1}$. Furthermore, we obtain

$$v_1\Delta + \frac{v_1^p c\Delta^2}{p\pi^{p-1}} = \langle v/\pi \rangle v'\Delta \in \pi D_{F_\pi}.$$

Hence the final answer for $e > p$ is $d \geq p$,

$$v = p + c\frac{\Delta}{p\pi} + v_1\Delta + \frac{v_1^p c\Delta^2}{p\pi^{p-1}} + v_2\Delta^2$$

for any $v_2 \in \pi D_{F_\pi}$ and $v_1 \in \mathfrak{M}$ satisfying $v_K(v_1) \geq e - p + 1 + \frac{e-d}{p-1}$. One can easily calculate $r(F)$.

For a finite height group one has $M_F = \pi D_{F_\pi}\Omega$. Since πD_{F_π} is generated over $W[\langle \pi \rangle]$ by $\pi + \frac{\pi c}{p}\Delta$, we obtain that $\pi D_{F_\pi}\Omega$ corresponding to c' coincides with πD_{F_π} for c if and only if $v(c - c') \geq \max(e + 1, e + 1 + \frac{e-d}{p-1})$ (see Subsection 7.3). For an infinite height group law one can calculate M_F directly using $D_F \subset R$.

8 Some other classification results

8.1 Representatives of isogeny classes

Using Theorem 7.4 one can construct a representative in each isogeny class of one-dimensional formal groups. See the paper [10] for a stronger and a more general result.

Theorem 8.1.(i) *Every one-dimensional formal group is isogenous to a group with logarithm* $(v(\Delta)/u(\Delta))(x)$, *where* $v = p + \sum_{i=1}^{h-1} v_i \Delta^i$ *and* $v_K(v_i) \geq e(1 - \frac{i}{h})$.

(ii) *Every* $(v(\Delta)/u(\Delta))(x)$ *for* v *as in part (i), is the logarithm of a formal group over* $\mathfrak{O}_K$ *(we call it a 'nice' group)* .

Remark 8.2.(i) It was also shown that if v' mod u has the form described in the theorem, then the group F' is strictly isomorphic to F. Moreover, all homomorphisms between 'nice' formal groups can be described in terms of $V(F)$. The Newton polygon of a 'nice' group and the 'residues' of the torsion elements of $F(\mathfrak{o}_K^{\mathrm{alg}})$ can be calculated.

Using this, one can calculate the Newton polygons and certain invariants of the Tate module $T(F)$ for all groups isogenous to a given 'nice' group. Unfortunately, it is quite difficult to determine which of them are defined over $\mathfrak{O}_K$.

(ii) The result was extended in [10] to the case of an arbitrary residue field, thus giving a vast generalization of an earlier result of Laffaille for the algebraically closed residue field case (see [25]).

(iii) Probably these results can be extended to multidimensional formal groups. An analogue of the first part is not completely clear yet.

8.2 Cartier modules for finite group schemes and reduction of Abelian varieties

Using the invariant Cartier-Dieudonné modules, one can easily define and calculate the (usual) Cartier-Dieudonné module structure on the kernels of isogenies of formal groups (cf. [26]). In particular, by using the theory for formal groups a complete classification of finite local flat commutative group schemes was obtained.

In [4] a natural definition of the tangent space of a finite flat commutative group scheme was given. It was used to calculate the minimal dimension of a finite height formal group F such that a fixed local commutative group scheme S can be embedded into F. The 'almost fullness' for the generic fibre functor for finite commutative group schemes was proven. This generalizes the corresponding fullness result of Raynaud for the small ramification index case (see [27]).

This generic fibre result allowed an explicit answer to the question of Nicholas Katz: whether one can be sure that an Abelian variety that acquires good reduction over K and is defined over $K' \subset K$ has good reduction over K' knowing that some level of p-torsion for this variety (depending only on e) gives a flat group scheme over $o_{K'}$. A first partial result in this direction was proved in [15, §5]. This result was generalized both to p-divisible groups and to Abelian varieties of potentially semistable reduction. See [4] for a comprehensive account on this subject.

9 Connection of formal groups with associated Galois modules

In this section we try to give an idea of the links of formal group laws with associated Galois modules. Many more details can be found in in [5, 6, 8].

9.1 Triviality of cocycles

Let K/N be a finite Galois extension of complete discrete valuation fields with group G. Let v be the (normalized) valuation on K. As above we assume that o is the ring of integers of N and $\mathfrak{O}_K$ is the ring of integers of K. Let $\mathfrak{M}$ be the maximal ideal of K.

We define

$$\mathfrak{C}_i = \{ f \in K[G] : \min_{x \in K^*} v(f(x)) - v(x) \geq i \}.$$

Let d denote the depth of ramification of K/N with respect to K, i.e. the minimum of

$$v(\mathrm{Tr}_{K/N}(x)) - v(x)$$

for $x \in K^*$. The relation $d = v(\mathfrak{D}_{K/N}) - e(K/N) + 1$ is well known. For $N \subset L \subset K$ we put $d(K/L) = v(\mathfrak{D}_{K/L}) - e(K/L) + 1$. We put $d(f) = s$ for $s \in K[G]$ if $f \in \mathfrak{C}_{d+s} \setminus \mathfrak{C}_{d+s+1}$.

Let F be an m-dimensional not necessarily commutative formal group law over o.

Let $Z^1(F(K))$ and $B^1(F(K))$ be the modules (or sets) of 1-cocycles and 1-coboundaries for $F(K)$ as a G-module corresponding to the standard inhomogeneous resolution. For an Abelian F the factor group $H^1(G, F(K)) = Z^1(F(K))/B^1(F(K))$ is the (local) Weil-Chatelet group.

For a vector x we define $v(x) = \min_i(v(x_i))$.

Theorem 9.1. *Let the map A belong to $Z^1(F(K))$,*

$$A : G \to \mathfrak{M}^m, \quad \sigma \to a_\sigma = (a_{1\sigma}, \ldots a_{m\sigma}).$$

(i) *A splits if and only if all f_i belong to $\mathfrak{C}_{d+1}$.*
(ii) *Suppose that A splits. Then for x that split A, i.e. such that*

$$x \underset{F}{-} \sigma(x) = a_\sigma \quad \text{for all } \sigma \in G,$$

we have

$$\max_x(v(x)) = \max\{s : \ f_i \in \mathfrak{C}_{d+s} \text{ for all } i, 1 \le i \le m\}.$$

Proof (Idea). There are two completely different ways to prove this statement. Both use the identification $\phi : K \otimes_N K \to K[G]$, defined by the formula

$$\phi(x \otimes y) = x \sum_{\sigma \in G} \sigma(y)\sigma. \tag{9.1}$$

With use of ϕ all associated Galois modules can be easily described. Several nice properties of ϕ were proved in [5] and [6].

The first way to prove the theorem was used in [8].

We consider a tower of intermediate normal subextensions in K/N:

$$k = N \subset k_1 \subset k_2 \subset \cdots \subset k_l = K, \ [k_i : k_{i-1}] = p$$

. Let (f_i) equal $\phi(\alpha)$, $\alpha \in K \otimes_N K^m$. One can prove by induction (inverse on i) for each i, $0 \le i \le l$, that there exist x and y satisfying

$$\alpha = x \otimes 1 \underset{F}{+} y, \ x \in F(\mathfrak{M}), \ y \in F(k_i \otimes K),$$

where $F(k_i \otimes K)$ is defined in a natural way.

The second way is to generalize ϕ to higher dimensions. In this way one obtains an identification of tensor powers with so-called higher dimensional associated modules. A very natural argument proves that the sheaf that is defined by F on the tensor powers of K is flabby. $\qquad\square$

Note that in order to determine whether a cocycle splits or not one doesn't have to know F – it is sufficient to know the values of A.

9.2 Deeply ramified extensions

Theorem 9.1 does not usually imply triviality of $H^1(F(\mathfrak{M}))$. In order to determine whether H^1 is trivial, one should compute the associated modules. One of the cases when the structure of associated modules is quite simple is the case of unramified extensions. Another one is the case of deeply ramified extensions defined by Coates and Greenberg in the paper [14].

Suppose that K is an injective limit of complete discrete valuation fields, L is a finite Galois extension of K, $G = \text{Gal}(L/K)$. Let F be a formal group (possibly multidimensional) defined over some complete discrete valuation field $k_0 \subset K$. We can compute $H^1(G, F(\mathfrak{M}_L))$; in the case of non-Abelian F it is only a pointed set.

It is easily seen that the extension L/K is defined over some complete discrete valuation field $k \subset K$. It means that there exists a Galois extension E/k such that $L = KE$ and $\text{Gal}(E/k) = G$.

Theorem 9.2. *Suppose that there exists a family of intermediate subfields $k \subset k_i \subset K$ such that*

$$\lim \frac{d(Ek_i/k_i)}{e(k_i/k)} = 0. \tag{9.2}$$

Then $H^1(G, F(L)) = \{0\}$.

Proof. This is immediate from the fact that $\mathfrak{M}^i[G] \subset \mathfrak{C}_i$ for any finite Galois extension K/N of complete discrete valuation fields. $\qquad\square$

In the paper of Coates and Greenberg K was an algebraic extension of $\mathbb{Q}_p$. They also introduced the notion of a deeply ramified extension. K is called a *deeply ramified extension* of $\mathbb{Q}_p$ if for any L a sequence of k_i fulfilling (9.2) exists.

Remark 9.3. Theorem 3.1 considerably generalizes the results of Coates and Greenberg since F is not required to be Abelian.

A certain converse result was proved in [7].

Proposition 9.4. *Suppose that the height of F is greater than m and $v(K^*)$ is non-discrete. If $H^1(K, F(\mathfrak{M}_{K^{\text{alg}}}^m)) = 0$, then K is deeply ramified.*

9.3 Cocycles whose values belong to N

If $A(G) \subset N^m$, then A is a homomorphism.

We define the notion of a Kummer equation for a formal group.

Definition 9.5. Let T be a finite subgroup in $F(N)$ and A be a homomorphism from G to T. We call $x \in F(\mathfrak{M}^m)$ *a solution of a Kummer equation* F, T, A if for any $\sigma \in G$ the relation

$$x \underset{F}{-} \sigma(x) = A(\sigma)$$

is fulfilled.

Let $\mathfrak{A}_i = \mathfrak{C}_i \cap N[G]$. We obtain an immediate corollary from Theorem 9.1.

Proposition 9.6. *Let L be a subextension in K/N and $H \subset G$ be the Galois group for K/L. We define*

$$f_i = \sum_{\sigma \in H} A(\sigma)\sigma.$$

A Kummer equation $F, T, A : H \to T$ has a root x in $F(K)$, $v(x) \geq s > 0$, if and only if

$$f_i \in \mathfrak{A}_{s+d(K/L)}, \ 1 \leq i \leq m.$$

We illustrate Proposition 9.6 by giving two simple examples: $G = \mathbb{Z}/p\mathbb{Z}$ and F is the multiplicative formal group law.

$$G = \mathbb{Z}/p\mathbb{Z}$$

It is not difficult to verify that for $G = \mathbb{Z}/p\mathbb{Z} = \langle \sigma \rangle$ a cocycle splits if and only if $v(a_{i\sigma}) > h$, where h is the ramification jump for the group G.

This condition is obviously necessary. The proof of sufficiency is also easy, see [7].

Multiplicative F

Suppose that F is the multiplicative formal group law, i.e. $F(X, Y) = X + Y + XY$.

An $x \in \mathfrak{M}$ splits a cocycle A if and only if $(x+1)/\sigma(x+1) = A(\sigma)$ for each $\sigma \in G$. According to Hilbert's Theorem 90, for every multiplicative cocycle there exists a splitting element. Hence A splits for F if and only if $1 + A$ in the multiplicative group is split by a principal unit. A does not split for F if and only if $d(f) = 0$, i.e. $f \in \mathfrak{C}_d \setminus \mathfrak{C}_{d+1}$.

Now suppose that $G = \mathbb{Z}/p^l\mathbb{Z} = \langle \sigma \rangle$ is a cyclic p-group and $\zeta_{p^l} \in k$. Then $A(\sigma^i) = \zeta_{p^l}^i - 1$ is a cocycle. An $x \in K$ splits $[s]_F A$ for some $s \in \mathbb{Z}$

prime to p if and only if $k(x) = K$ and $(x+1)^{p^l} \in k$. The maximal possible valuation of x is equal to $d(f)$. In this case $f = \sum_{i=1}^{n} \zeta_{p^l}^i \sigma^i - \text{Tr}$.

For $l = 1$ the knowledge of the maximal possible valuation of x is equivalent to the knowledge of the ramification jump for G.

Let $l = 2$, let $k \subset K_1 \subset K$, $[K : K_1] = [K_1 : k] = p$. The knowledge of the ramification jumps for G is equivalent to the knowledge of ramification jumps in K_1/k and K/K_1. One can easily show that the maximal possible valuation of x cannot be recovered from the jumps; one cannot even know for sure whether x can be chosen to be a principal unit.

9.4 Galois structure of extensions that are Kummer with respect to formal groups

For a one-dimensional F let $x \in \mathfrak{M}$ be a solution of a Kummer equation F, T, A, where A is an isomorphism. Then $P(x) \in N$, where $P(X) = \prod_{t \in T}^{F} (X - t)$.

Let v_0 be the valuation on N, $n = \#T$.

We construct a class of extensions whose Galois module structures can be completely described.

Theorem 9.7. *Let x be a root of $P(X) = y$, where $y \in \mathfrak{M}_N$, $v(y)$ is prime to p, $v_0(y) < n \min_{t \in T} v_0(t)$. We define $K = N(x)$, $c = v(y)$. Let ξ^i denote $\sum_{\sigma \in G} A(\sigma)^i \sigma$.*

(i) *K/N is a Galois extension with group G isomorphic to $F(T)$.*

ii) *All ramification numbers of K/N are congruent to $-c$ modulo n.*

iii)
$$\mathfrak{A}_i = \langle \pi_0^{c(s,i)} \xi^s \rangle,$$

where $c(s,i) = [\frac{i-d-cs-1}{n}] + 1$.

iv) *For any $z \in K$, $v(z) = -c$, we have $\mathfrak{A}_i(z) = \mathfrak{M}^{i-c}$.*

Proof (Idea). One easily checks that $\xi = \phi(\alpha)$, where $\alpha = x \otimes 1 - 1 \otimes x + \alpha'$ and $\xi(\alpha') \in \mathfrak{C}_{d+c+1}$; see the definition of ϕ in (9.1). Then for $0 \leq i < n$ we have $\xi^i = \phi(\alpha_i)$, where $\alpha_i = (x \otimes 1 - 1 \otimes x)^i + \alpha'_i$ and $\phi(\alpha'_i) \in \mathfrak{C}_{d+ic+1}$. The properties of ϕ easily imply the desired statement. $\square$

Remark 9.8.(i) For any two ideals $I, J \subset K$ a basis of $\text{Hom}_{\mathfrak{o}[G]}(I, J)$, similar to the one in the third part of the theorem can be easily found.

ii) In [6] the area where F is defined was extended (for some F). Using that definition a slightly wider class of extensions was constructed; they were called semistable. In [5] and [6] some sets of conditions were given under which K/N is semistable if it contains an ideal that is free over its associated order (i.e. over $\mathrm{Hom}_{o[G]}(I, I)$).

(iii) Any totally ramified extension of degree p that is not maximally ramified (i.e. the ramification jump is less than $pe/(p-1)$) is semistable.

(iv) Considering multidimensional formal groups one can construct extensions that need be neither Abelian nor totally ramified, yet have similar properties.

Bibliography

[1] V. Abrashkin. Explicit formulas for the Hilbert symbol of a formal group over Witt vectors. *Izv. Ross. Akad. Nauk Ser. Mat.*, 61(3):3–56, 1997. (Russian).

[2] D. G. Benois and S. V. Vostokov. Norm pairing in formal groups and Galois representations. *Leningrad Math. J.*, 2:69–92, 1991.

[3] D. G. Benois and S. V. Vostokov. Galois representaions in Honda's formal groups. Arithmetic of the group of points. In *Proceedings of the St. Petersburg Mathematical Society. Vol. II.*, volume 159 of *Amer. Math. Soc. Transl. Ser. 2*, pages 1–14. American Mathematical Society, 1994.

[4] M. V. Bondarko. Finite flat commutative group schemes over complete discrete valuation fields: classification, structural results; application to reduction of Abelian varieties. Preprint, available on http://arxiv.org/abs/math.NT/0412524.

[5] M. V. Bondarko. Local Leopoldt's problem for rings of integers in Abelian p-extensions of complete discrete valuation fields. *Doc. Math.*, 5:657–693, 2000.

[6] M. V. Bondarko. Local Leopoldt's problem for ideals in p-extensions of complete discrete valuation fields. In *Algebraic Number Theory and Algebraic Geometry: Papers Dedicated to A. N. Parshin on the Occasion of his Sixtieth Birthday*, Contemporary Mathematics, pages 27–57. 2002.

[7] M. V. Bondarko. Cohomology of formal group moduli and deeply ramified extensions. *Math. Proc. Cambridge Phil. Soc.*, 135:19–24, 2003.

[8] M. V. Bondarko. Links between associated additive Galois modules and computation of H^1 for local formal group modules. *J. Number Theory*, 101:74–104, 2003.

[9] M. V. Bondarko. Explicit classification of formal groups over complete discrete valuation fields with imperfect residue field. *Trudy St. Peterburgskogo Matematicheskogo Obsh'estva*, 11:1–36, 2005. (Russian).

[10] M. V. Bondarko. Isogeny classes of formal groups over complete discrete valuation fields with arbitrary residue fields. *Algebra i Analiz*, 17(6):105–124, 2005. (Russian).

[11] M. V. Bondarko, F. Lorenz, and S. V. Vostokov. Hilbert pairing for formal groups over σ-rings. *Zap. Nauchn. Sem. Pomi*, 319:5–58, 2004.

[12] M. V. Bondarko and S. V. Vostokov. Explicit classification of formal groups over local fields. *Trudy MIAN*, 241(2):43–67, 2003.

[13] C. Breuil. Groupes p-divisibles, groupes finis et modules filtrés. *Ann. of Math. (2)*, 152(2):489–549, 2000.

[14] J. Coates and R. Greenberg. Kummer theory for Abelian varieties over local fields. *Invent. Math.*, 124:129–174, 1996.

[15] B. Conrad. Finite group schemes over bases with low ramification. *Compositio Math.*, 119:239–320, 1999.

[16] O. V. Demchenko. New relationships between formal Lubin-Tate groups and formal Honda groups. *St. Petersburg Math. J.*, 10:785–789, 1999.

[17] O. V. Demchenko. Formal Honda groups: the arithmetic of the group of points. *Algebra i Analiz*, 12(1):132–149, 2000.

[18] O. V. Demchenko and S. V. Vostokov. Explicit form of Hilbert pairing for relative Lubin-Tate formal groups. *J. Math. Sci. (2)*, 89:1105–1107, 1996.

[19] O. V. Demchenko and S. V. Vostokov. Explicit formula of the Hilbert symbol for a Honda formal group. *Zap. Nauchn. Sem. POMI*, 272:86–128, 2000.

[20] I. B. Fesenko. The generalized Hilbert symbol in the 2-adic case. *Vestnik Leningrad Univ. Math.*, 18:75–87, 1985.

[21] I. B. Fesenko and S. V. Vostokov. *Local fields and their extensions*, volume 121 of *Translations of Mathematical Monographs*. American Mathematical Society, Providence, RI, 1993. With a foreword by I. R. Shafarevich.

[22] J. M. Fontaine. *Groupes p-divisibles sur les corps locaux*. Soc. Math. France, Paris, 1977. Astérisque, n. 47–48.

[23] M. Hazewinkel. *Formal groups and applications*. Academic Press Inc. [Harcourt Brace Jovanovich Publishers], New York, 1978.

[24] T. Honda. On the theory of commutative formal groups. *J. Math. Soc. Japan, 22*, pages 213–246, 1970.

[25] G. Laffaille. Construction de groupes p-divisibles, le cas de dimension 1. In *Journées de Géométrie Algébrique de Rennes. (Rennes, 1978), Vol. III*, volume 65 of *Astérisque*, pages 103–123. Soc. Math. France, Paris, 1979.

[26] F. Oort. Dieudonné modules of finite local group schemes. *Indag. Math.*, 37:103–123, 1975.

[27] M. Raynaud. Schemas en groupes de type $(p, \ldots, p)$. *Bull. Soc. Math. France*, 102:241–280, 1974.

[28] S. V. Vostokov. Explicit form of the reciprocity law. *Math. USSR-Izv.*, 13(3):557–588, 1979.

[29] S. V. Vostokov. A norm pairing in formal modules. *Math. USSR-Izv.*, 15(1):25–51, 1980.

[30] S. V. Vostokov. Symbols on formal groups. *Math. USSR-Izv.*, 19(2):261–284, 1982.

[31] S. V. Vostokov and I. B. Fesenko. The Hilbert symbol for Lubin-Tate formal groups. *J. Soviet Math.*, 30(1):85–96, 1988.

Department of Mathematics and Mechanics
St. Petersburg State University
198904 Staryi Peterhof
Bibliotechnaya pl. 2
Russia
mbondarko@hotmail.com

Classification problems and mirror duality

Vasily V. Golyshev

Introduction

In this paper we make precise, in the case of rank 1 Fano 3-folds, the following programme:

Given a classification problem in algebraic geometry, use mirror duality to translate it into a problem in differential equations; solve this problem and translate the result back into geometry.

The paper is based on the notes of the lecture series the author gave at the University of Cambridge in 2003. It expands the announcement [13], providing the background for and discussion of the modularity conjecture.

We start with basic material on mirror symmetry for Fano varieties. The quantum D–module and the regularized quantum D–module are introduced in Section 1. We state the mirror symmetry conjecture for Fano varieties. We give more conjectures implying, or implied by, the mirror symmetry conjecture. We review the algebraic Mellin transform of Loeser and Sabbah and define hypergeometric D–modules on tori.

In Section 2 we consider Fano 3-folds of Picard rank 1 and review Iskovskikh's classification into 17 algebraic deformation families. We apply the basic setup to Fano 3-folds to obtain the so called counting differential equations of type $D3$. We introduce DN equations as generalizations of these, discuss their properties and take a brief look at their singularities.

In Section 3, motivated by the Dolgachev-Nikulin-Pinkham picture of mirror symmetry for $K3$ surfaces, we introduce (N, d)-modular families; these are pencils of $K3$ surfaces whose Picard-Fuchs equations are the counting $D3$ equations of rank 1 Fano 3-folds. The (N, d)-modular

88

family is the pullback of the twisted symmetric square of the universal elliptic curve over $X_0(N)^W$ to a cyclic covering of degree d.

Our mirror dual problem is stated in Section 4: for which pairs (N, d) is it possible for the Picard-Fuchs equation of the corresponding (N, d)-modular family to be of type $D3$? Through a detailed analysis of singularities, we get a necessary condition on (N, d), bringing the list down to 17 possibilities.

Identifying certain odd Atkin-Lehner, weight 2, level N Eisenstein series (that appear in Section 5) with the sections of the bundle of relative differential 2-forms in our modular family, we compute the corresponding Picard-Fuchs equations and show them to be of type $D3$, recovering the matrix coefficients.

It turns out that the pairs (N, d) that we get are exactly those for which there exists a rank 1 Fano 3-fold of index d and anticanonical degree $2d^2N$. The Iskovskikh classification is revisited in Section 6. We sketch a proof that the matrices we have recovered in Section 5 via modular computations are, up to a scalar shift, the counting matrices of the corresponding Fanos.

In Section 7 we briefly discuss further classification problems to which our approach may be applied.

We refer the reader to [2] for more on the quantum cohomology of *minimal* Fano 3-folds.

1 Conjectures on mirror symmetry for Fano varieties.

There exist two different approaches to differential systems built from Gromov-Witten invariants of a variety. The full Frobenius manifold underlies vector bundles with connections whose construction requires knowledge of the *big* quantum cohomology and therefore of the whole system of multiple-pointed correlators (see Chapter 2 in [19]). On the other hand, if we are content to restrict our study to the divisorial subdirection of the Frobenius manifold, only the *small* quantum cohomology is needed. It is still a strong invariant of a variety but it only requires knowledge of the three-pointed correlators. For this reason, it is easier to compute. The small quantum differential system has the additional advantage of being representable as an algebraic D–module on the 'Neron-Severi-dual' torus.

Given a smooth scheme $X/\mathbb{C}$ we denote by $D^{b,\mathrm{holo}}$ the full subcategory of cohomologically bounded cohomologically holonomic complexes of sheaves of left $\mathcal{D}_X$-modules, where as usual $\mathcal{D}_X$ denotes the sheaf of rings of differential operators on X. For morphisms $f\colon X \longrightarrow Y$ between smooth varieties, the 'six operations' exist and provide a convenient language for the constructions that we are going to need. If $f\colon X \longrightarrow Y$ is smooth of relative dimension d, then $f_*K = Rf_*(K \otimes_{\mathcal{O}_X} \Omega^{\bullet}_{X/Y})[d]$, so that $\mathcal{H}^{i-d}(f_*K) = H^i_{DR}(X/Y, K)$ with its Gauss-Manin connection.

We will need the following notion of pullback: if M is a flat $\mathcal{D}_Y$-module, then $f^!M = f^+M[\dim X - \dim Y]$, where f^+M is the naive pullback of M as module with integrable connection. If G is a separated smooth group scheme over $\mathbb{C}$ with group law $\mu\colon G \times G \longrightarrow G$, then the convolution of objects of $D^{b,\mathrm{holo}}(G)$ is defined by $(K, L) \to K * L = \mu_*(K \boxtimes L)$ where $K \boxtimes L$ is the external tensor product.

Definition 1.1 (Three-point correlators). Let X be a Fano variety. Let $T_{\mathrm{NS}^\vee}$ be the torus dual to the lattice $\mathrm{NS}^\vee(\mathrm{X})$. Define a trilinear functional $\langle \ , \ , \ \rangle$ on the space $H(X)$ by setting

$$\langle \alpha, \beta, \gamma \rangle = \sum_{\chi \in \mathrm{NS}^\vee(X)} \langle \alpha, \beta, \gamma \rangle_\chi \cdot \chi$$

where $\langle \alpha, \beta, \gamma \rangle_\chi$ is 'the expected number of maps'† from $\mathbb{P}^1$ to X in the cohomology class χ such that 0 maps into a general enough representative of α, 1 maps into a representative of β, ∞ maps into a representative of γ. The functional $\langle \ , \ , \ \rangle$ takes values in $\mathbb{C}[\chi]$.

Consider the trivial vector bundle $\mathcal{H}(X)$ over $T_{\mathrm{NS}^\vee}$ with fibre $H(X)$. Extend the Poincaré pairing $[,]$ to the vector space of its sections $H(X) \otimes \mathbb{C}[\chi]$. Raising an index, we turn the trilinear form into a multiplication law on $H(X) \otimes \mathbb{C}[\chi]$:

$$[\alpha \cdot \beta, \gamma] = \langle \alpha, \beta, \gamma \rangle .$$

Identify elements f in the lattice $\mathrm{NS}(X)$ with invariant derivations ∂_f on $T_{\mathrm{NS}^\vee}$ by the rule

$$\partial_f(\chi) = f(\chi)\chi .$$

---

† Technically, a Gromov-Witten invariant, [19, VI-2.1]. Let $\overline{M}_n(X, \chi)$ denote the compactified moduli space of maps of rational curves of class $\chi \in \mathrm{NS}^\vee$ with n marked points, and let $[\overline{M}_n(X, \chi)]^{\mathrm{virt}}$ be its virtual fundamental class of virtual dimension $\mathrm{vdim}\, \overline{M}_n(X, \chi) = \dim X - \deg_{K_X} \chi + n - 3$. Let $ev_i\colon \overline{M}_n(X, \chi) \longrightarrow X$ denote the evaluation map at the i-th marked point. Then

$$\langle \alpha, \beta, \gamma \rangle_\chi = ev_1^*(\alpha) \cdot ev_2^*(\beta) \cdot ev_3^*(\gamma) \cdot [\overline{M}_n(X, \chi)]^{\mathrm{virt}}$$

if $\mathrm{codim}\,\alpha + \mathrm{codim}\,\beta + \mathrm{codim}\,\gamma = \mathrm{vdim}\, \overline{M}_3(X, \chi)$, and $\langle \alpha, \beta, \gamma \rangle_\chi = 0$ otherwise.

(In the left-hand side of the formula χ is a function on a torus; on the right, it is an element of the lattice $NS(X)$ and as such is paired with f.) Define a connection

$$\nabla_{T_{NS^\vee}} : \Omega^0(\mathcal{H}(X)) \longrightarrow \Omega^1(\mathcal{H}(X))$$

in the vector bundle $\mathcal{H}(X)$ by setting for any constant section $\bar{\alpha} = \alpha \otimes 1 \in H(X) \otimes \mathbb{C}[\chi]$

$$\langle \partial_f, \nabla_{T_{NS^\vee}} \bar{\alpha} \rangle = (f \otimes 1) \cdot \bar{\alpha}$$

(the derivation ∂_f is coupled with the vector-valued 1-form $\nabla_{T_{NS^\vee}} \bar{\alpha}$ on the left hand side).

Theorem 1.2. *The connection ∇ is flat.*

In view of this fact the space $H(X) \otimes \mathbb{C}[\chi]$ has the structure of a $\mathcal{D} = \mathcal{D}_{T_{NS^\vee}}$-module. We will denote it by Q and call it the *quantum D–module*.

The *mirror symmetry conjecture* states that the solution to the quantum D–module, convoluted with the canonical exponent, can be represented as a period in some family of varieties, called a (parametric) Landau-Ginzburg model. Let us make this more precise.

Definition 1.3 (The exponent object on a one-dimensional torus). Let $\mathbb{A}^1 = \operatorname{Spec} \mathbb{C}[t]$, $\mathbb{G}_m = \operatorname{Spec} \mathbb{C}[t, t^{-1}]$, and let $j \colon \mathbb{G}_m \longrightarrow \mathbb{A}^1$ denote the corresponding open immersion. Let $\partial = \frac{\partial}{\partial t}$ and $D = D_t = t\frac{\partial}{\partial t}$ be the (invariant) derivations on $\mathbb{A}^1$ and $\mathbb{G}_m$ respectively.

The D–module $E = \mathcal{D}_{\mathbb{A}^1}/\mathcal{D}_{\mathbb{A}^1}(\partial - 1)$, and its restriction to $\mathbb{G}_m$ $j^*E = \mathcal{D}_{\mathbb{G}_m}/\mathcal{D}_{\mathbb{G}_m}(D - t)$, will be called the *exponent object*.

In general, the quantum D–module Q is irregular. As such, it cannot possibly be of geometric origin, that is, arise from a Gauss-Manin connection of an algebraic family: Gauss-Manin connections are known to be regular [7]. In order to make a suitable geometricity assertion one should pass to a regular object first.

Consider the inclusion $\mathbb{Z}K_X \to NS_X$. Dualizing twice, we have a morphism of tori $\iota \colon \mathbb{G}_m \to T_{NS^\vee}$ (the canonical torus map). Consider the exponent object j^*E on $\mathbb{G}_m$, and the pushforward $\iota_*(j^*E)$. Define the *regularized quantum object* as follows

$$Q^{\text{reg}} = Q * \iota_*(j^*E).$$

The mirror symmetry conjecture. The object Q^{reg} is of geometric origin.

This assertion in its strong interpretation means that for any irreducible constituent of Q^{reg} there exists a family of varieties $\pi\colon \mathcal{E} \longrightarrow T_{\mathrm{NS}^{\vee}}$ such that the restriction of that constituent to some open subset U is isomorphic to a constituent of $R^{j}\pi_{*}(\mathcal{O})$.

Remark 1.4. In practice (e.g. Proposition 2.7 below) we will forget about the trivial constituents that may arise as a by-product of the convolution construction, and deal only with the essential subquotient of a single cohomology D–module of the regularized quantum object. We will call it the *regularized quantum D–module*.

Let $\iota_{x}\colon \mathbb{G}_{m} \longrightarrow \iota(\mathbb{G}_{m})x$ be an orbit of $\mathbb{G}_{m}$ in $T_{\mathrm{NS}^{\vee}}$. The Gauss-Manin connection in the Landau-Ginzburg model with parameter x is then essentially the pullback to $\mathbb{G}_{m}$ of the regularized quantum D–module with respect to ι_{x}.

Classification strategy. Let us lay out broadly our classification strategy. It is logical to start with the Picard rank 1 case, as in this case $T_{\mathrm{NS}^{\vee}}$ is one-dimensional and the regularized quantum D–module is essentially a linear ordinary differential equation with polynomial coefficients.

Assume we are interested in finding all families of Fano varieties in a given class. (From our point of view, a class comprises varieties with similar cohomology structure. For instance, an interesting, if too narrow, class is that of *minimal Fanos* of a given dimension, i.e. those whose non-trivial cohomology groups are just $\mathbb{Z}$ in every even dimension. In the class of *almost minimal* odd-dimensional Fanos we allow nontrivial cohomology in the middle dimension.) Assume that a variety X in the class is known, together with the values $A_{X} = \{a_{ij}(X)\}$ of the three-point correlators between two arbitrary cohomology classes and the divisor class. Compute the regularized quantum D–module and represent it as $\mathcal{D}_{\mathbb{G}_{m}}/\mathcal{D}_{\mathbb{G}_{m}}L_{A_{X}}$ for some $L_{A_{X}} \in \mathcal{D}_{\mathbb{G}_{m}}$. We will say that $L_{A_{X}}$ is the *counting differential operator* for X. Doing the same construction starting with a matrix variable $A = \{a_{ij}\}$, we obtain a differential operator L_{A} depending on the set of parameters $\{a_{ij}\}$. (We will do this in detail for almost minimal Fanos in 2.8, getting what we call a differential operator of *type DN*.) Thus, we can restate the original classification problem as follows: determine which L_{A} can be counting differential operators $L_{A_{X}}$ of some Fano variety X.

What are the properties that distinguish $L_{A_{X}}$s as points in the affine space of all L_{A}s? As we have seen, the mirror symmetry conjecture asserts that the $L_{A_{X}}$s are of Picard-Fuchs type: we expect that there exist a pencil $\pi \colon \mathcal{E} \longrightarrow \mathbb{G}_m$ defined over $\mathbb{Q}$ and ω a meromorphic section of a sheaf of relative differential forms, such that a period Φ of ω satisfies $L_{A_X}\Phi = 0$. A believer in the mirror symmetry conjecture would therefore approach the problem of identifying the possible L_{A_X}'s by first telling which among all L_A's are Picard-Fuchs. This will significantly narrow one's search, as being Picard-Fuchs is a very strong condition.

Identifying Picard-Fuchs operators among all L_A's apparently is not an algorithmic problem. The very first idea is to translate (and this can be done algorithmically) the basic properties that an (irreducible) variation of Hodge structures must have – regularity, polarizability, quasiunipotence of local monodromies – into algebraic conditions on the coefficients of the operator that represents it. One might hope that these conditions cut out a variety of positive codimension from the affine space of all L_A's, thereby facilitating further search. However, the hope is vain: a theorem proved recently by J. Stienstra and myself asserts that a generic DN equation is regular, polarizable and has quasiunipotent local monodromies everywhere (see Theorem 2.12).

Algebraic requirements being met by virtue of the construction, we have to shift the emphasis toward non-algebraicizable conditions of analytic or arithmetic nature imposed by the PF property.

It is known that if a differential equation $L_A\Phi = 0$ with coefficients in $\mathbb{Q}$ is of Picard-Fuchs type, then it is also

H) Hodge (that is, it describes an abstract variation of $\mathbb{Q}$-Hodge structures);

N) globally nilpotent† in the sense of Dwork-Katz, see [9, 15].

It is expected that, at least for small order r and degree d, both (**H**) and (**GN**) are also sufficient conditions. Unfortunately, there is no algorithmic way, given a_{ij}, to verify that (**H**) or (**GN**) holds: in the former case, because of the fact that (**H**) is, in particular, a condition on the global monodromy which depends transcendentally on the coefficients of the equation; in the latter case, because one does not know, given a_{ij}, how to estimate the number of places (p) of $\mathbb{Q}$ where the nilpotence

---

† We briefly recall what global nilpotence is. Let $\partial \xi = \xi M$ be an algebraic differential equation over $\mathbf{F}_p$. Consider $\partial \partial \xi = \partial \xi M = \xi(M^2 + M')$, $\partial \partial \partial \xi$, etc. Then

$$\underbrace{\partial \partial \ldots \partial}_{p \text{ times}} \xi = C_p \xi, \qquad \text{for some matrix } C_p = C_p(M)$$

$C_p(M)$ is called the p-*curvature matrix*. A differential equation $\partial \xi = \xi M$ over $\mathbb{Q}$ with M having p-integral entries is said to be p-*nilpotent* if $C_p(M \mod p)$ is nilpotent. It is *globally nilpotent* if it is p-nilpotent for almost all p.

of the p-curvature operator must be verified in order to conclude that global nilpotence holds.

The hypergeometric pullback conjecture. In order to state this conjecture, we will need some basic facts about hypergeometric D–modules. Roughly, a D–module is hypergeometric if the coefficients of the series expansion of its solution are products/quotients of the gamma function applied to values of nonhomogeneous linear forms in the degrees:

$$\Phi = \sum u(n_1, \ldots, n_p) t_1^{n_1} \ldots t_p^{n_p}$$

with $u(n_1, \ldots, n_p) = \prod c_i^{n_i} \prod_j \Gamma(l_i^{(j)} n_i - \sigma^{(j)})^{\gamma_j}$. To put it precisely, one might use the language of *algebraic Mellin transforms*, introduced by Loeser and Sabbah [18].

Let $\mathbb{C}[s] = \mathbb{C}[s_1, \ldots, s_p]$ be the ring of polynomials in p variables and let $\mathbb{C}(s)$ be the corresponding fraction field.

Definition 1.5. A *rational system of finite difference equations* (FDE) is a finite-dimensional $\mathbb{C}(s)$-vector space together with $\mathbb{C}$-linear automorphisms $\tau_1, \ldots, \tau_p$ that commute with each other and satisfy the relations

$$\tau_i s_j = s_j \tau_i \quad \text{if } i \neq j$$
$$\tau_i s_i = (s_i + 1) \tau_i \quad \text{for all } i = 1, \ldots, p.$$

If $\mathfrak{M}(s)$ and $\mathfrak{M}'(s)$ are rational systems of FDE, then so are $\mathfrak{M}(s) \otimes_{\mathbb{C}(s)} \mathfrak{M}'(s)$, $\mathrm{Hom}_{\mathbb{C}(s)}(\mathfrak{M}(s), \mathfrak{M}'(s))$. Therefore, the set of isomorphism classes of one-dimensional systems forms a group, which Sabbah and Loeser call the *hypergeometric group* and denote $\mathcal{HG}(p)$.

Denote by $\mathcal{L}$ a subset of non-zero linear forms on $\mathbb{Q}^p$ with coprime integer coefficients such that for all such forms L either L or $-L$ is in $\mathcal{L}$. Let $\mathbb{Z}^{[\mathcal{L} \times \mathbb{C}/\mathbb{Z}]}$ be the subset of finitely supported functions $\mathcal{L} \times \mathbb{C}/\mathbb{Z} \longrightarrow \mathbb{Z}$ with the natural group structure.

Proposition 1.6 ([18, 1.1.4]). *Let σ be a section of the projection $\mathbb{C} \longrightarrow \mathbb{C}/\mathbb{Z}$. Then, the map*

$$(\mathbb{C}^*)^p \times \mathbb{Z}^{[\mathcal{L} \times \mathbb{C}/\mathbb{Z}]} \longrightarrow \mathcal{HG}(p)$$

that attaches to $[(c_1, \ldots, c_p), \gamma]$ the isomorphism class of the system satisfied by

$$(c_1)^{s_1} \ldots (c_p)^{s_p} \prod_{L \in \mathcal{L}} \prod_{\alpha \in \mathbb{C}/\mathbb{Z}} \Gamma(L(s) - \sigma(\alpha))^{\gamma_{L, \alpha}}$$

is a group isomorphism which does not depend on the choice of σ.

Let $T^p \simeq \mathbb{G}_m{}^p$ be a complex torus of dimension p. Put $D_i = t_i \frac{\partial}{\partial t_i}$. Let $\mathbb{C}[t, t^{-1}]\langle D \rangle$ denote the algebra of algebraic differential operators on the torus (here t stands for $(t_1, \ldots, t_p)$ and D for $(D_1, \ldots, D_p)$).

The correspondence $\tau_i = t_i$ and $s_i = -D_i$ identifies this algebra with the algebra $\mathbb{C}[s]\langle \tau, \tau^{-1} \rangle$ of finite difference operators (which is the quotient of the algebra freely generated by $\mathbb{C}[s]$ and $\mathbb{C}[\tau, \tau^{-1}]$ by the relations in Definition 1.5).

If $\mathcal{M}$ is a holonomic D–module on T^p, then its global sections form a $\mathbb{C}[t, t^{-1}]\langle D \rangle$-module. The *algebraic Mellin transform* $\mathfrak{M}(\mathcal{M})$ of the D–module $\mathcal{M}$ is this module of global sections considered as $\mathbb{C}[s]\langle \tau, \tau^{-1} \rangle$-module. We say that $\mathfrak{M}(M)$ is a holonomic algebraic system of FDE if M is holonomic.

Theorem 1.7 (The algebraic Mellin transform theorem [18, 1.2.1]). *Let $\mathfrak{M}$ be a holonomic algebraic system of FDE. Then $\mathfrak{M}(s) = \mathbb{C}(s) \otimes_{\mathbb{C}[s]} \mathfrak{M}$ is a rational holonomic system of FDE. Conversely, if $\mathfrak{M}(s)$ is a rational holonomic system of FDE, then for any $\mathbb{C}[s]\langle \tau, \tau^{-1} \rangle$-submodule $\mathfrak{M} \subset \mathfrak{M}(s)$ such that $\mathfrak{M}(s) = \mathbb{C}(s) \otimes_{\mathbb{C}[s]} \mathfrak{M}$ there exists a holonomic algebraic system $\mathfrak{M}' \subset \mathfrak{M}$ such that $\mathfrak{M}(s) = \mathbb{C}(s) \otimes_{\mathbb{C}[s]} \mathfrak{M}'$.*

Proposition 1.8 ([17]). *One has $\chi((\mathbb{G}_m)^p, \mathcal{M}) = \dim_{\mathbb{C}(s)} \mathfrak{M}(\mathcal{M})(s)$.*

Definition 1.9. A D–module $\mathcal{M}$ on T^p is said to be *hypergeometric* if $\mathfrak{M}(\mathcal{M})(s)$ has rank 1.

Every one-dimensional $\mathbb{C}(s)$ vector space with invertible τ-action contains a unique irreducible holonomic $\mathbb{C}[s]\langle \tau, \tau^{-1} \rangle$-module and every such module of generic rank one is obtained in this way.

Passing back to the subject of quantum D–modules, we are finally set to state the following conjecture.

The hypergeometric pullback conjecture. Let X be a Fano variety. We conjecture that for any constituent $\mathcal{C}$ of the quantum D–module Q there exists a torus $T_\mathcal{C}$, a morphism of tori $h_\mathcal{C} \colon T_{\mathrm{NS}^\vee} \longrightarrow T_\mathcal{C}$ and a hypergeometric D–module $\mathcal{H}_\mathcal{C}$ on $T_\mathcal{C}$ such that $\mathcal{C}$ is isomorphic to a constituent of the pullback $h^! \mathcal{H}_\mathcal{C}$ on some open subset U of $T_{\mathrm{NS}^\vee}$.

Remark 1.10. One can show that the D–module Q is essentially the restriction of the 'extended first structural connection' onto the divisorial direction the Frobenius manifold associated to X while Q^{reg} corresponds to the 'second structural connection'; see Chapter 2 of [19].

2 The Iskovskikh classification and $D3$ equations.

Let X be a Fano 3-fold with one-dimensional Picard lattice, and let $H = -K_X$ be the anticanonical divisor. V. A. Iskovskikh classified all deformation families of these varieties (see [14]). Recall that if X is a smooth rank 1 Fano variety and $G \in H^2(X, \mathbb{Z})$ is the positive generator then the *index* of X is defined by $H = (\mathrm{ind}\, X)\, G$.

Theorem 2.1. *The possible pairs of invariants* $\left(\dfrac{H^3}{2\, \mathrm{ind}^2 X},\ \mathrm{ind}\, X \right)$ *are*

$$(1,1), (2,1), (3,1), (4,1), (5,1), (6,1), (7,1), (8,1), (9,1), (11,1),$$
$$(1,2), (2,2), (3,2), (4,2), (5,2), (3,3), (2,4).$$

To realize our strategy (p. 92) for rank one Fano 3-folds, one must first compute the quantum D–module Q.

Proposition 2.2. *The subspace of algebraic classes in the total cohomology $H^\bullet(X)$ is stable under quantum multiplication by H. Therefore, the connection ∇ restricts to the rank 4 subbundle of $\mathcal{H}(X)$ generated by the algebraic classes.*

Proof. This follows easily from the 'dimension axiom' (see the formula in the footnote on page 90). $\qquad\square$

We compute this divisorial submodule explicitly, according to the definition. Let us normalize a_{ij} so that

$$a_{ij} = \frac{1}{\deg X}(j - i + 1) \cdot \left(\begin{array}{c} \text{the expected number of maps } \mathbb{P}^1 \mapsto X \\ \text{of degree } j - i + 1 \text{ that send 0 to the class} \\ \text{of } H^{3-i} \text{ and send } \infty \text{ to the class of } H^j. \end{array} \right)$$

The degrees of the variety and of curves on it are considered with respect to H. Assume now for simplicity that X has index 1.

As always, $\mathbb{G}_m = \mathrm{Spec}\,\mathbb{C}[t, t^{-1}]$ and $D = t\frac{\partial}{\partial t}$. Let h^i be the constant sections of $\mathcal{H}(X)$ that correspond to the classes H^i.

Proposition 2.3. *The connection ∇ is given by*

$$D(h^0, h^1, h^2, h^3) = (h^0, h^1, h^2, h^3) \begin{pmatrix} a_{00}t & a_{01}t^2 & a_{02}t^3 & a_{03}t^4 \\ 1 & a_{11}t & a_{12}t^2 & a_{13}t^3 \\ 0 & 1 & a_{22}t & a_{23}t^2 \\ 0 & 0 & 1 & a_{33}t \end{pmatrix}$$

Proof. This follows from the definition. $\qquad\square$

Corollary 2.4. *Put*

$$
\widehat{L}_A = \det{}_{right}\left(D - \begin{pmatrix} a_{00}t & a_{01}t^2 & a_{02}t^3 & a_{03}t^4 \\ 1 & a_{11}t & a_{12}t^2 & a_{13}t^3 \\ 0 & 1 & a_{22}t & a_{23}t^2 \\ 0 & 0 & 1 & a_{33}t \end{pmatrix}\right).
$$

where $\det_{right}$ *means the 'right determinant', i.e. the one that expands as*

$$
\sum element \cdot its\ cofactor,
$$

the summation being over the rightmost *column, and the cofactors being themselves right determinants. Then* h^0 *is annihilated by* $\widehat{L}$.

Proof. This is a non-commutative version of Cayley-Hamilton. □

Corollary 2.5. *The quantum D-module Q is isomorphic to (a subquotient of) $\mathcal{D}/\mathcal{D}\widehat{L}$.*

Having thus computed Q, we proceed with regularization. We must convolute Q with the pushforward under the morphism inv: $x \mapsto 1/x$ of the exponent object $\mathcal{D}/(z\partial - z)\mathcal{D}$. Convolution with the exponent of the inverse argument on a torus is essentially† the Fourier(-Laplace) transform, as the following formula suggests:

$$
\left(F(x) * \left(\frac{1}{x}e^{1/x}\right)\right)(t) = \int F(y)\frac{y}{t}e^{y/t}\frac{dy}{y} = \frac{1}{t}(\mathrm{FT}(F))\left(\frac{1}{t}\right)
$$

More precisely, one has the following definition.

Definition 2.6 (The Fourier transform on $\mathbb{A}^1$ [16, 2.10.0]). The *Fourier transform* of a differential operator $L = \sum f_i(t)\partial^i \in \mathcal{D}_{\mathbb{A}^1}$ is defined by $\mathrm{FT}(L) = \sum f_i(\partial)(-t)^i$. The Fourier transform of the left D-module $M = \mathcal{D}_{\mathbb{A}^1}/\mathcal{D}_{\mathbb{A}^1}L$ is $\mathrm{FT}(M) = \mathcal{D}_{\mathbb{A}^1}/\mathcal{D}_{\mathbb{A}^1}\mathrm{FT}(L)$.

Proposition 2.7 ([16, 5.2.3, 5.2.3.1]). *Retain the notation of Example 1.3. Then for any holonomic D-module M on $\mathbb{G}_m$ we have*

$$
j^* \mathrm{FT}(j_* \mathrm{inv}_*(M)) \approx M * j^*E \quad and \quad \mathrm{inv}_* j^* \mathrm{FT}(j_* M) \approx M * (\mathrm{inv}_* j^*E).
$$

The second formula shows that the convolution is in fact a single D-module, though not in general an irreducible one. We need to isolate the essential subquotient, combing out the parasitic ones.

---

† Since everything is considered on $\mathbb{G}_m$, the functions Φ and $t\Phi$ are solutions to isomorphic D-modules.

Note that the operator $\widehat{L}_A$ is divisible in $\mathbb{C}[t, \partial]$ by t on the left (because the rightmost column of the matrix is divisible by t on the left). Extend the D–module $\mathcal{D}_{\mathbb{G}_m}/\mathcal{D}_{\mathbb{G}_m}t^{-1}\widehat{L}_A \approx \mathcal{D}_{\mathbb{G}_m}/\mathcal{D}_{\mathbb{G}_m}\widehat{L}_A$ naively to $\mathbb{A}^1$ as $\mathcal{D}_{\mathbb{A}^1}/\mathcal{D}_{\mathbb{A}^1}t^{-1}\widehat{L}_A$. Do the Fourier transform. We get a D–module that corresponds to the differential operator

$$\partial^{-1}\det{}_{\text{right}}\left(-D-1-\begin{pmatrix} a_{00}\partial & a_{01}\partial^2 & a_{02}\partial^3 & a_{03}\partial^4 \\ 1 & a_{11}\partial & a_{12}\partial^2 & a_{13}\partial^3 \\ 0 & 1 & a_{22}\partial & a_{23}\partial^2 \\ 0 & 0 & 1 & a_{33}\partial \end{pmatrix}\right). \qquad \text{(FT)}$$

Pass to the inverse: under inv, D is sent to $-D$ and ∂ to $-t^2\partial$. For further convenience we do two more things: shift the differential operator by -1 on the torus (D goes to D and ∂ to $-\partial$) and multiply it by t on the right. The result is then what we call a *counting differential operator of type D3*.

Abstracting our situation to any dimension and arbitrary $\{a_{ij}\}$, we introduce the following.

Definition 2.8. Let N be a positive integer. Let $a_{ij} \in \mathbb{Q}$, $0 \leq i \leq j \leq N$. Let M be an $(N+1) \times (N+1)$ matrix such that for $0 \leq k, l \leq N$:

$$M_{kl} = \begin{cases} 0, & \text{if } k > l+1, \\ 1, & \text{if } k = l+1, \\ a_{kl} \cdot (Dt)^{l-k+1}, & \text{if } k < l+1. \end{cases}$$

We will also assume that the set a_{ij} is symmetric with respect to the SW-NE diagonal: $a_{ij} = a_{N-j,N-i}$.

Put

$$\widetilde{L} = \det{}_{\text{right}}(D - M).$$

Since the rightmost column is divisible by D on the left, the resulting operator $\widetilde{L}$ is divisible by D on the left. Put

$$\widetilde{L} = DL.$$

The differential equation $L\,\Phi(t) = 0$ will be called a *determinantal equation of order N*, or just a *DN equation*.

Sometimes we write $DN_{0,0}$ to signify that 0 is a point of maximally unipotent monodromy, and that the local expansion $\Phi = c_0 + c_1 t + \cdots$ of an analytic solution Φ at 0 starts with a nonzero constant term. (One may have made other choices; for instance, the differential operator marked (FT) above is of type $D3_{\infty,1}$ in this language.)

Example 2.9. A $D3$ equation expands as

$$
\left[
\begin{array}{l}
D^3 - t(2D+1)\left(a_{00}D^2 + a_{11}D^2 + a_{00}D + a_{11}D + a_{00}\right) \\[2mm]
+ t^2(D+1)\left(
\begin{array}{c}
(a_{11}{}^2 + a_{00}{}^2 + 4a_{11}a_{00} - a_{12} - 2a_{01})D^2 \\[1mm]
+ \left(\begin{array}{c} 8a_{11}a_{00} - 2a_{12} + 2a_{00}{}^2 \\ -4a_{01}D + 2a_{11}{}^2 \end{array}\right) D \\[1mm]
+ 6a_{11}a_{00} + a_{00}{}^2 - 4a_{01}
\end{array}
\right) \\[6mm]
- t^3(2D+3)(D+2)(D+1)\left(
\begin{array}{c}
a_{00}{}^2 a_{11} + a_{11}{}^2 a_{00} - a_{12}a_{00} \\
+ a_{02} - a_{11}a_{01} - a_{01}a_{00}
\end{array}
\right) \\[5mm]
+ t^4(D+3)(D+2)(D+1)\left(
\begin{array}{c}
-a_{00}{}^2 a_{12} + 2a_{02}a_{00} \\
+ a_{00}{}^2 a_{11}{}^2 - a_{03} \\
+ a_{01}{}^2 - 2a_{01}a_{11}a_{00}
\end{array}
\right)
\end{array}
\right]
\Phi(t) = 0
$$

Definition 2.10. We say that two DN equations defined by sets a_{ij} and $a_{ij}{}'$ are *in the same class* if there exists an a such that $a_{ii} = a_{ii}{}' + a$ for $i = 0, \ldots, N$ and $a_{ij} = a_{ij}{}'$ for $i \neq j$, i.e. if the matrices defined by a_{ij} and $a_{ij}{}'$ differ by a scalar matrix.

Shifting the Fourier transformed differential operator FT on $\mathbb{A}^1$ corresponds exactly to shifting the DN matrix in its class.

Definition 2.11. We say that:

(i) a holonomic D-module M is a *variation of type DN* if there exists a set of parameters $A - \{a_{ij}\}$ such that $\mathcal{D}/\mathcal{D}L_A \approx M$. Here $\approx$ denotes equivalence in the category of $\mathcal{D}$-modules up to modules with punctual support;

ii) a constructible sheaf S is a *variation of type DN* if there exists a D-module M of type DN with regular singularities, such that

$$H^{-1}(DR(M)) \approx S.$$

Here $\approx$ denotes equivalence in the category of constructible sheaves up to sheaves with punctual support; DR is the Riemann-Hilbert correspondence functor.

Theorem 2.12. *A D–module $\mathcal{D}/\mathcal{D}L$ of type DN has the following properties:*

(i) *it is holonomic with a regular singularity at 0;*
ii) *it is self-adjoint;*

(iii) *the local monodromy around zero is maximally unipotent (i.e. is conjugate to a Jordan block of size N);*

(iv) *for a generic set $A = \{a_{ij}\}$, the D-module $\mathcal{D}/\mathcal{D}L_A$ has $N+1$ non-zero singularities, all of which are regular; the local monodromies at those singularities are symplectic (for N even) or orthogonal (for N odd) reflections, and the global monodromy is irreducible;*

(v) *the set $A = \{a_{ij}\}$ can be recovered from the respective L_A: if $A \neq A'$, then $L_A \neq L_{A'}$.*

A proof can be found in a forthcoming paper by Jan Stienstra and myself.

Definition 2.13. We say that a DN variation M (resp. local system S) is of *geometric origin* if there exists a flat morphism $\pi \colon \mathcal{E} \longrightarrow \mathbb{G}_m$ of relative dimension d such that M (resp. S) is isomorphic to a subquotient of the variation arising in its middle relative cohomology ($R^0\pi_*(\mathcal{O})$, resp. $R^d\pi_*(\mathbb{C})$) up to a D-module (resp. a sheaf) with punctual support.

Recall that we had assumed (p. 96) that the variety in question had index 1 before proceeding with the construction of the counting differential operator. What happens in the higher index cases? It turns out that Definition 2.8 with the values of a_{ij} as defined earlier is still valid, in the sense that it yields a counting operator that corresponds to the pullback of the regularized quantum D-module with respect to the anticanonical isogeny $\mathbb{G}_m \overset{\text{ind}X}{\longrightarrow} \mathbb{G}_m$ (see Remark 1.4.) We leave the proof to the reader. Use, for instance, the following property:

Proposition 2.14 ([16, 5.1.9 1b])**.** *Let G be a smooth separated group scheme of finite type, $\varphi \colon G \longrightarrow G$ a homomorphism. Then for any two objects K, L of $D^{b,holo}(G)$ one has*

$$\varphi^!((\varphi_* K) * L) \approx K * (\varphi^! L).$$

Remark 2.15. In this language, the mirror symmetry conjecture for Fanos states: the counting DN equations of almost minimal Fano N-folds† are of geometric origin. In order to recover all counting DN equations one should pose and then solve a mirror dual problem: find all geometric DN equations that possess some special property. In general, we do not know what that property is. However, in the $D3$ case we have an additional insight: a counting $D3$ should come from an (N, d)-modular family.

---

† One expects that the analogue of Proposition 2.2 holds in even dimensions as well, so that all almost minimal N-folds are controlled by DN's.

Definition 2.16. A non-zero singularity of a D–module of type $D3$ with regular singularities is said to be:

(i) *simple*, if the local monodromy around that singularity is a reflection (i.e. conjugate to the operator $diag(-1, 1, 1)$);

(ii) *complex*, if it is not simple and is of determinant 1;

(iii) *very complex*, if it is not simple and is of determinant -1.

3 (N, d)-modular variation.

Warning. In this section N stands for level. This is not the N of the previous section, which denoted the order of a differential operator.

The quantum weak Lefschetz principle implies that the fibres of the Landau-Ginzburg model of a Fano variety are mirror dual to the sections of the anticanonical line bundle on it. For rank 1 Fano 3-folds, these sections are rank 1 $K3$ surfaces.

The first picture of mirror symmetry for families of $K3$ surfaces arose as an attempt to explain Arnold's strange duality. Let $L = 3U \oplus -2E_8$ be the $K3$ lattice. For a wide class of primitive sublattices M of L there is a unique decomposition

$$M^\perp = U \oplus M^D,$$

so that there is a duality between M and M^D :

$$(M^D)^\perp = U \oplus M.$$

The Picard lattices of mirror dual families of $K3$ surfaces are dual in this sense. Therefore, it is natural to expect that the dual Landau-Ginzburg model of a Fano 3-fold is a family of $K3$ surfaces of Picard rank 19. We recall that a Kummer $K3$ is the minimal resolution of the quotient of an abelian surface by the canonical involution which sends x to $-x$ in the group law.

The following construction was described in [21, 12].

Consider the modular curve $X_0(N)$, and the 'universal elliptic curve' over it. Strictly speaking, the universal elliptic curve is a fibration not over $X_0(N)$ but over a Galois cover with group Γ, e.g. $X(3N) - \{\text{cusps}\}$, such that one can choose a Γ-form of the universal elliptic curve; call it 'the' universal elliptic curve and denote it by E_t. Denote by W the Atkin-Lehner involution of $X_0(N)$. Consider the fibred product of E_t

with the N-isogenous universal elliptic curve E_t^W over $X_0(N)$. We quotient out this relative abelian surface V_t by the canonical involution $x \mapsto -x$ and then resolve to get a family of Kummer $K3$ surfaces. Let $X_0(N)^\circ$ stand for $X_0(N) - \{\text{cusps}\} - \{\text{elliptic points}\}$. Denote by $H(V_{\bar{t}_0})$ the cohomology of the generic fibre of V_t, that is, of the pullback of the family V_t to the universal cover of the base.

The monodromy representation

$$\psi \colon \pi_1(X_0(N)^\circ) \longrightarrow H^2(V_{\bar{t}_0})$$

is well defined. We are going to compute ψ in terms of the tautological projective representation

$$\varphi \colon \pi_1(X_0(N)^\circ) \longrightarrow PGL(H^1(E_{\bar{t}_0})) = PSL_2(\mathbb{Z}).$$

The monodromy that acts on H^1 of the fibre of the universal elliptic curve is given by a lift of φ to a linear representation

$$\bar{\varphi} \colon \gamma \mapsto \begin{pmatrix} a & b \\ c & d \end{pmatrix}, \quad c = 0 \mod N.$$

Then, the monodromy that acts on H^1 of the fibre of the isogenous curve is

$$\bar{\varphi}_N \colon \gamma \mapsto \begin{pmatrix} d & -\frac{c}{N} \\ -bN & a \end{pmatrix} = \begin{pmatrix} 0 & -\frac{1}{N} \\ 1 & 0 \end{pmatrix} \begin{pmatrix} a & b \\ c & d \end{pmatrix} \begin{pmatrix} 0 & 1 \\ -N & 0 \end{pmatrix}$$

where we have chosen symplectic bases $\langle e_1, e_2 \rangle, \langle f_1, f_2 \rangle$ of $H^1(E_{\bar{t}_0})$, $H^1(E_{\bar{t}_0}^W)$ such that the matrix of the isogeny W in these bases is

$$\begin{pmatrix} 0 & 1 \\ -N & 0 \end{pmatrix}.$$

The cohomology ring of the generic fibre $V_{\bar{t}_0}$ of our relative abelian surface is $H(E_{\bar{t}_0}) \otimes H(E_{\bar{t}_0}^W)$. The vector subspace of algebraic classes in $H^2(V_{\bar{t}_0})$ is generated by the pullbacks from the factors and the graph of the isogeny:

$$e_1 \wedge e_2 \otimes 1, \quad 1 \otimes f_1 \wedge f_2, \quad -e_1 \otimes f_1 - Ne_2 \otimes f_2.$$

These classes are invariant under monodromy. The orthogonal lattice of transcendental classes is generated by

$$e_2 \otimes f_1, \quad e_1 \otimes f_1 - Ne_2 \otimes f_2, \quad e_1 \otimes f_2.$$

Identifying the e's and f's with their pullbacks to the product, we write, abusing notation,

$$e_2 \wedge f_1, \quad e_1 \wedge f_1 - Ne_2 \wedge f_2, \quad e_1 \wedge f_2.$$

In this basis the monodromy representation is (cf. [21, 8])

$$\psi \colon \gamma \mapsto \mathrm{Sym}_N^2 \, \varphi(\gamma) = \begin{pmatrix} d^2 & 2cd & -c^2/N \\ bd & bc+ad & -ac/N \\ -Nb^2 & -2Nab & a^2 \end{pmatrix}.$$

Let $\bar{\omega}$ be a meromorphic section of the sheaf of relative holomorphic differential forms on the universal elliptic curve. Identify e_1, e_2 and f_1, f_2 with cohomology classes in the pullback of the universal elliptic curve to the universal cover of the base. Denote by ω the pullback of $\bar{\omega}$. Introduce a coordinate τ on the universal cover by writing:

$$[\omega] = \tau F e_1 + F e_2$$

(where F is a function on the universal cover) identifying it with the upper halfplane. The class ω^W is then

$$[\omega^W] = F f_1 - N\tau F f_2.$$

Let ω and ω^W also denote, by abuse of notation, the pullbacks of the respective forms to $V_{\bar{t}_0}$. Clearly

$$[\omega \wedge \omega^W] = F^2 e_2 \wedge f_1 + \tau F^2(e_1 \wedge f_1 - Ne_2 \wedge f_2) - \tau^2 N F^2 e_1 \wedge f_2.$$

Now, as ω is $\Gamma_0(N)$-equivariant,

$$\psi(\gamma) \begin{pmatrix} F^2(\tau) \\ \tau F^2(\tau) \\ -N\tau^2(F^2(\tau)) \end{pmatrix} = \begin{pmatrix} F^2(\gamma(\tau)) \\ \gamma(\tau)) F^2(\gamma(\tau)) \\ -N\gamma(\tau))^2 (F^2(\gamma(\tau))) \end{pmatrix}$$

where

$$\gamma(\tau) = \frac{a\tau + b}{c\tau + d}.$$

This is equivalent to the identity

$$F^2 \left(\frac{a\tau + b}{c\tau + d} \right) = (c\tau + d)^2 F^2(\tau).$$

Therefore, the period F^2 in our family of abelian surfaces, as a function of τ, is a $\Gamma_0(N)$-automorphic function of weight 2 on the upper halfplane. Now for any $\Gamma_0(N)$-automorphic function G of weight 2, the quotient $\frac{G}{F^2}$ is $\Gamma_0(N)$-invariant on the upper halfplane, hence a rational function on $X_0(N)$. This identifies G with a (meromorphic) section of the sheaf of relative holomorphic 2-forms in our family.

Finally, delete the W-invariant points from $X_0(N)^\circ$ and let $X_0(N)^{W\circ}$ be the quotient of the resulting curve by W. The involution W extends to the fibration V_t in an obvious way, and yields a family V_t^W over $X_0(N)^{W\circ}$. The fundamental group $X_0(N)^{W\circ}$ is generated by $\pi_1(X_0(N)^\circ)$ and a loop ι around the point that is the image of a point s on the upper halfplane stabilized by $\begin{pmatrix} 0 & 1 \\ -N & 0 \end{pmatrix}$. Extend ψ to ι, by setting $\psi(\iota) = \begin{pmatrix} 0 & 0 & 1 \\ 0 & 1 & 0 \\ 1 & 0 & 0 \end{pmatrix}$. The resulting representation is the monodromy representation of the family V_t^W over $X_0(N)^{W\circ}$.

If a relative holomorphic form in the family V_t is a pullback from V_t^W, then, denoting its first period by G, one has

$$\psi(\iota) \begin{pmatrix} G(\tau) \\ \tau G(\tau) \\ -N\tau^2 G(\tau) \end{pmatrix} = \begin{pmatrix} G(\frac{-1}{N\tau}) \\ \tau G(\frac{-1}{N\tau}) \\ -N\tau^2 G(\frac{-1}{N\tau}) \end{pmatrix}.$$

G is odd Atkin-Lehner as, by definition,

$$G^W(\tau) = G(\frac{-1}{N\tau})N^{-1}\tau^{-2}.$$

Now let N be a level such that the curve $X_0(N)^W$ is rational. We choose a coordinate T on it such that $T = 0$ at the image of the cusp $(i\infty)$ (the inverse of a Conway-Norton uniformizer, see Table 1 below); this defines an immersion of the torus $\iota: X_0(N)^W \hookleftarrow \mathbb{G}_m' = \operatorname{Spec} \mathbb{C}[T, T^{-1}]$. Let $\operatorname{Spec} \mathbb{C}[t, t^{-1}] = \mathbb{G}_m \longrightarrow \mathbb{G}_m'$ be the Kummer covering of degree d, given by the homomorphism $T \mapsto t^d$.

The pullback of the variation described above (that is, the pullback of the family itself, or the monodromy representation, or the D-module, depending on the context) to $\mathbb{G}_m$ will be called the (N, d)-*modular variation*. Let us emphasize: (N, d)-modular variations are variations on *tori*, even if we speak of them as of variations on $\mathbb{P}^1$, as in the proof of Theorem 4.1 below.

In the case $N = 1$ the construction is modified, since the 'Atkin-Lehner involution' $\begin{pmatrix} 0 & 1 \\ -N & 0 \end{pmatrix}$ acts trivially on $X_0(N)$. In this case we work with the fibred product of the 'universal elliptic curve' over $X_0(1)$ with its quadratic twist with respect to the degree two branched covering ramified at the two elliptic points. In this case the relative 2-form can no longer be identified with a weight 2 level 1 modular function because of the sign multiplier. However, squaring the corresponding period, we get a bona fide modular function of weight 4 and level 1.

4 (N, d)-modular $D3$ equations: the necessary condition.

Problem: Find all pairs N, d such that the (N, d)-modular variation described in the previous section is of type $D3$.

Theorem 4.1 (Necessary condition). *If an (N, d)-modular variation is of type $D3$, then the pair (N, d) belongs to the set*

$$\mathcal{M} = \left\{ \begin{array}{c} (1, 1), (2, 1), (3, 1), (4, 1), (5, 1), (6, 1), (7, 1), (8, 1), (9, 1), (11, 1), \\ (1, 2), (2, 2), (3, 2), (4, 2), (5, 2), (3, 3), (2, 4) \end{array} \right\}$$

Proof. We begin by noticing that no case with $d > 5$ is possible as there would have to be at least 6 singularities.

Case d = 1. Assume $N \neq 1$. We make the following remarks:

(1) *All ramification points of the quotient map*

$$\sigma \colon X_0(N) \longrightarrow X_0(N)^W$$

map to singularities of the $(N, 1)$-modular variation. The corresponding local monodromy is projectively (dual to) the symmetric square of the element in $\Gamma_0(N) + N$ that stabilizes this ramification point. This element is elliptic or cuspidal, therefore its symmetric square cannot be a scalar.

(2) *Every elliptic point or a cusp point p on $X_0(N)$ maps to a complex or very complex point $\sigma(p)$ on $X_0(N)^W$.* If $\sigma(p)$ were an apparent singularity or a simple singularity, then the local monodromy around p would vanish, which is precluded by the reason given above in (1).

(3) *The point s on the upper halfplane is neither elliptic nor a cusp. It goes to a simple point on $X_0(N)^W$.* We defined the monodromy ι(image of s) in the previous section to be a reflection.

(4) *If a $D3$ equation is $(N, 1)$-modular, then its set of non-zero singularities consists of either 4 simple points, or of 1 complex and 2 simple points, or of 1 very complex and 1 simple point.* The non-zero singularities of a $D3$ equation are inverse to roots of a polynomial of degree 4, as can be seen from the expansion in Example 2.9. It has one simple singularity, according to (3). Any singularity of multiplicity 1 is simple. A singularity of multiplicity 3 is very complex because the determinant

must be -1 and it cannot be simple (otherwise the global monodromy would be generated by two reflections and therefore would be reducible).

(5) *The genus g of $X_0(N)$ is related to the numbers of elliptic points ν_2 and ν_3 of order 2 and 3 on $X_0(N)$ by the formula*

$$g = 1 + \frac{N}{12} \prod_{p \mid N} (1 + p^{-1}) - \frac{\nu_2}{4} - \frac{\nu_3}{3} - \frac{\nu_\infty}{2}.$$

This is Proposition 1.40 from [23].

These remarks show that $g \leq 1$, (otherwise the variation would have at least 6 singularities according to (1)); that if $g = 1$, then all of the singularities are simple (this is from (1) and (4)) and $\nu_2 = 0$, $\nu_3 = 0$, $\nu_\infty = 2$ so $N = 11$; and that if $g = 0$, then $N < 12$ (otherwise there would be too many singularities, which would contradict (2) and (4)). The last argument also shows that $N \neq 10$, as in this case $\nu_2 = 2$ and $\nu_\infty = 4$.

Case d $= 2$. Again, assume $N \neq 1$.

(1) *The genus g of $X_0(N)$ is zero.* If it were greater than zero, there would be at least four singularities besides the one at 0. Therefore, the $(N, 2)$-modular variation would have at least 7 singularities.

(2) *There can be no more than 3 cusps on $X_0(N)$.* Assume there are at least 4 cusps on $X_0(N)$. Consider the ramification points of the Atkin-Lehner involution. One of them being s, the other is either a cusp or not a cusp. In the former case we get at least three cusps on $X_0(N)^W$. Pulling them back we get at least 4 singularities of a $D3$ variation that are not simple, a contradiction. In the latter case, we get at least two cusps and at least two other singularities of the $(N, 1)$-modular variation. Pulling them back to the $(N, 2)$-modular variation we get either:

- at least four simple points and two non-simple points, or:
- at least two simple points and three non-simple points,

and in neither case can the resulting variation be of type $D3$.

(3) *There can be no more than 1 order 3 elliptic point on $X_0(N)$.* The proof is the same as above.

(4) *There can be no more than 7 order 2 elliptic points on $X_0(N)$.* These would give at least 5 singularities on $X_0(N)^W$ and therefore at least 7 singularities on the pullback.

(5) *The level N is smaller than 48.* This bad but easy estimate follows from the genus formula and the above remarks.

Having made these remarks, one proceeds (for instance) by inspecting the values $g, \nu_2, \nu_3, \nu_\infty$ for all levels $N < 48$. One uses the formulas of [23, Proposition 1.43]:

$$\nu_2 = \begin{cases} 0, & \text{if } N = 4k, \\ \frac{1}{2}\prod_{p|N}\left(1 + \left(\frac{-1}{p}\right)\right), & \text{if } N = 4k + 2, \\ \prod_{p|N}\left(1 + \left(\frac{-1}{p}\right)\right), & \text{if } 2 \nmid N; \end{cases}$$

$$\nu_3 = \begin{cases} 0, & \text{if } N = 9k, \\ \prod_{p|N}\left(1 + \left(\frac{-3}{p}\right)\right), & \text{if } 9 \nmid N; \end{cases}$$

$$\nu_\infty = \sum_{d|N} \varphi(\gcd(d, N/d))$$

where $\varphi(n)$ is as usual the number of positive integers not exceeding n and relatively prime to n.

One thus finds that the only levels that satisfy the requirements above are $N = 2, 3, 4, 5$.

Case d = 3. Pulling back under the degree 3 map dramatically multiplies singularities; the analysis, which goes along the same rails, is this time much easier and leaves one with the only possibility of a curve of genus 0 that has just 2 cusps, 1 order 3 elliptic point and no order 2 points, which corresponds to level 3.

Case d = 4. Yet easier. The curve must be of genus 0 and have 2 cusps, 1 order 2 elliptic point and no order 3 points. The level is 2. $\square$

5 (N, d)-modular $D3$ equations: a sufficient condition.

Theorem 5.1. † *For all pairs (N, d) in $\mathcal{M}$ the corresponding (N, d)-modular variation is of type $D3$.*

Proof. Assume for simplicity that $d = 1$. Reshape the assertion this way: for any pair $(N, 1)$ in $\mathcal{M}$ there is a period Φ of a section of the line bundle $\pi_* \Omega^2_{V_t/X_0(N)^W}$ in our $(N, 1)$-modular variation that satisfies, as a multivalued function on $X_0(N)^W$, a $D3$ equation with respect to a

---

† Rather, a 'fact', as its proof requires computations too tedious to be done by hand. Note however that, in the cases of complete intersections in projective spaces, the respective $(N, 1)$-modular local systems are rigid and can be identified with the global monodromies of given $D3$ equations by comparison of local monodromies, see [12].

coordinate t on $X_0(N)^W$. Given an expansion of $\Phi(t)$ as a series in t, it is easy to find the differential equation that it satisfies.

To be more specific, recall that in Section 3 we chose a coordinate T on $X_0(N)^W$ such that $T = 0$ at the image of the cusp $(i\infty)$ (see page 104). The local monodromy at $T = 0$ of the cycles against which our fibrewise 2-form is integrated is conjugate to a unipotent Jordan block of size 3. Therefore, the analytic period $\Phi = \Phi_0$ is well defined as the integral against the monodromy-invariant cycle. In the same way, the logarithmic period Φ_1, being the integral against a cycle in the second step of the monodromy filtration, is well defined up to an integral multiple of the analytic period. This defines τ locally as $\frac{\Phi_1}{\Phi_0}$, and q as $\exp(2\pi i \frac{\Phi_1}{\Phi_0})$.

Now q being a local coordinate around 0, one can expand both Φ and T as q-series. Note that the expansion of T^{-1} is a q-series that is uniquely defined up to a constant term. The q-expansions of coordinates on $X_0(N)^W$ appeared in a paper by Conway and Norton [5] and are called *Conway-Norton uniformizers*. Table 1, which gives the uniformizers for the levels that we need, is taken from [5].

Recall also that we have identified periods Φ with odd Atkin-Lehner weight 2 level N modular functions in Section 3. Therefore, to prove our theorem explicitly one may: (1) produce a q-expansion of such a modular function Φ; (2) fix the constant term in the uniformizer T^{-1}; (3) express q in T; (4) expand Φ in T; (5) recover the differential equation that Φ satisfies with respect to T. If it is a $D3$ equation, we are done.

The same essentially goes for the cases $d = 2, 3, 4$, except that the coordinate on the Kummer covering is $t = T^{1/d}$ and the local parameter is $Q = q^{1/d}$. Tables 2 contain the Q-expansions of Φ, the recovered $D3$ matrices and the eta-expansions of the I-function that we introduce below. The uniformizers, as we said, are in Table 1. For level N and index d one should set the constant term of the uniformizer to a_{11} in the $(N, 1)$ matrix in Tables 2 (e.g. take $c = 744$ for level 1 and index 2). $\quad\square$

Table 1: Conway-Norton uniformizers.

The constant term is denoted indiscriminately by c below. We put $\mathbf{i} = q^{i/24} \prod(1 - q^{in})$ in this table.

$N=1$	$N=2$	$N=3$
$j + c$	$\frac{1^{24}}{2^{24}} + 4096 \frac{2^{24}}{1^{24}} + c$	$\frac{1^{12}}{3^{12}} + 729 \frac{3^{12}}{1^{12}} + c$
$N=4$	$N=5$	$N=6$
$\frac{1^8}{4^8} + 256 \frac{4^8}{1^8} + c$	$\frac{1^6}{5^6} + 125 \frac{5^6}{1^6} + c$	$\frac{1^5 3^1}{2^1 6^5} + 72 \frac{2^1 6^5}{1^5 3^1} + c$
$N=7$	$N=8$	$N=9$
$\frac{1^4}{7^4} + 49 \frac{7^4}{1^4} + c$	$\frac{1^4 4^2}{2^2 8^4} + 32 \frac{2^2 8^4}{1^4 4^2} + c$	$\frac{1^3}{9^3} + 27 \frac{9^3}{1^3} + c$
$N=11$		
$\frac{1^2 11^2}{2^2 22^2} + 16 \frac{2^2 22^2}{1^2 11^2} + 16 \frac{2^4 22^4}{1^4 11^4} + c$		

In most of the cases, the form Φ will be expressed as a finite linear combination of 'elementary Eisenstein series'

$$E_{2,i}(Q) \overset{\text{def}}{=} -\frac{1}{24} i \left(1 - 24 \sum_{n=1}^{\infty} \sigma(n) Q^{in}\right).$$

A sequence $e_1, e_2, e_3, \ldots$ determines the Eisenstein series

$$\sum e_j E_{2,j}(Q).$$

We use notation $\Phi = e_1 \cdot [\mathbf{1}] + e_2 \cdot [\mathbf{2}] + \ldots$ in the third column of Tables 2.

Remark . We have proved Theorem 5.1 by producing *some* modular function Φ and *some* Conway-Norton uniformizer T^{-1} of level N such that Φ expanded in T satisfies a $D3$ equation. Is the pair Φ, T^{-1} that we have produced determined by this condition uniquely? The answer is in general no, even if Φ is known to be an Eisenstein series: at certain composite levels the space spanned by Eisenstein series has dimension higher than 1, and it is possible to find two different Eisenstein series and two uniformizers (that differ by a constant term) such that the respective expansions give rise to different $D3$ matrices.

The extra piece that we use to characterize the matrices and the solutions Φ in Tables 2 uniquely is the following.

The miraculous eta-product formula. Define $I = \Phi \cdot t^{d\frac{N+1}{12}}$. Let $H_j(Q) = Q^{j/24} \prod(1 - Q^{jn})$. It turns out that I expands as a finite product of series of the form $\prod H_j^{h_j}(Q)$ in a remarkably uniform way:

$$I = H_d(Q)^2 H_{Nd}(Q)^2.$$

We reflect this phenomenon in the right-hand column of Tables 2. The notation used is $I = \mathbf{1}^{h_1} \mathbf{2}^{h_2} \cdot \ldots$. No intrinsic explanation of the eta-product formula is known to the author.

Table 2: Level, matrix, solution, I-function.

Key to notation: in the Φ column, $[\mathbf{j}] = E_{2,j}(Q)$; in the I column, $\mathbf{k} = H_k(Q)$.

$d = 1$

N	a_{ij}				Φ	I
1	120	137520	119681280	21690374400	$\sqrt{E_4(q)}$	$1^2 1^2$
	0	744	650016	119681280		
	0	0	744	137520		
	0	0	0	120		
2	24	3888	504576	18323712	$+24 \cdot [\mathbf{1}] - 24 \cdot [\mathbf{2}]$	$1^2 2^2$
	0	104	13600	504576		
	0	0	104	3888		
	0	0	0	24		
3	12	792	43632	793152	$+12 \cdot [\mathbf{1}] - 12 \cdot [\mathbf{3}]$	$1^2 3^2$
	0	42	2340	43632		
	0	0	42	792		
	0	0	0	12		
4	8	304	9984	121088	$+8 \cdot [\mathbf{1}] - 8 \cdot [\mathbf{4}]$	$1^2 4^2$
	0	24	800	9984		
	0	0	24	304		
	0	0	0	8		
5	6	156	3600	33120	$+6 \cdot [\mathbf{1}] - 6 \cdot [\mathbf{5}]$	$1^2 5^2$
	0	16	380	3600		
	0	0	16	156		
	0	0	0	6		
6	5	96	1692	12816	$+5 \cdot [\mathbf{1}] \quad 1 \cdot [\mathbf{2}]$ $+1 \cdot [\mathbf{3}] - 5 \cdot [\mathbf{6}]$	$1^2 6^2$
	0	12	216	1692		
	0	0	12	96		
	0	0	0	5		
7	4	64	924	5936	$+4 \cdot [\mathbf{1}] - 4 \cdot [\mathbf{7}]$	$1^2 7^2$
	0	9	140	924		
	0	0	9	64		
	0	0	0	4		
8	4	48	576	3328	$+4 \cdot [\mathbf{1}] - 2 \cdot [\mathbf{2}]$ $+2 \cdot [\mathbf{4}] - 4 \cdot [\mathbf{8}]$	$1^2 8^2$
	0	8	96	576		
	0	0	8	48		
	0	0	0	4		
9	3	36	378	1944	$+3 \cdot [\mathbf{1}] - 3 \cdot [\mathbf{9}]$	$1^2 9^2$
	0	6	72	378		
	0	0	6	36		
	0	0	0	3		
11	12/5	24	198	880	$+12/5 \cdot [\mathbf{1}]$ $-12/5 \cdot [\mathbf{11}]$	$1^2 11^2$
	0	22/5	44	198		
	0	0	22/5	24		
	0	0	0	12/5		

$d = 2$	N	a_{ij}				Φ	I
	1	0	240	0	57600	$\sqrt{E_4(Q^2)}$	$\mathbf{2^2 2^2}$
		0	0	1248	0		
		0	0	0	240		
		0	0	0	0		
	2	0	48	0	2304	$+12 \cdot [\mathbf{2}] - 12 \cdot [\mathbf{4}]$	$\mathbf{2^2 4^2}$
		0	0	160	0		
		0	0	0	48		
		0	0	0	0		
	3	0	24	0	576	$+6 \cdot [\mathbf{2}] - 6 \cdot [\mathbf{6}]$	$\mathbf{2^2 6^2}$
		0	0	60	0		
		0	0	0	24		
		0	0	0	0		
	4	0	16	0	256	$+4 \cdot [\mathbf{2}] - 4 \cdot [\mathbf{8}]$	$\mathbf{2^2 8^2}$
		0	0	32	0		
		0	0	0	16		
		0	0	0	0		
	5	0	12	0	160	$+3 \cdot [\mathbf{2}] - 3 \cdot [\mathbf{10}]$	$\mathbf{2^2 10^2}$
		0	0	20	0		
		0	0	0	12		
		0	0	0	0		
$d = 3$	N	a_{ij}				Φ	I
	3	0	0	54	0	$+4 \cdot [\mathbf{3}] - 4 \cdot [\mathbf{9}]$	$\mathbf{3^2 9^2}$
		0	0	0	54		
		0	0	0	0		
		0	0	0	0		
$d = 4$	N	a_{ij}				Φ	I
	2	0	0	0	256	$+6 \cdot [\mathbf{4}] - 6 \cdot [\mathbf{8}]$	$\mathbf{4^2 8^2}$
		0	0	0	0		
		0	0	0	0		
		0	0	0	0		

Remark 5.2. The functions Φ and I that describe in these tables the cases with the same level N, but with different d, are equal.

Table 3: Respective differential operators.

$d = 1$

1	$D^3 - 24t(1 + 2D)(6D + 5)(6D + 1)$
2	$D^3 - 8t(1 + 2D)(4D + 3)(4D + 1)$
3	$D^3 - 6t(1 + 2D)(3D + 2)(3D + 1)$
4	$D^3 - 8t(1 + 2D)^3$
5	$D^3 - 2t(1 + 2D)(11D^2 + 11D + 3)$ $\quad - 4t^2(D + 1)(2D + 3)(1 + 2D)$
6	$D^3 - t(1 + 2D)(17D^2 + 17D + 5) + t^2(D + 1)^3$
7	$D^3 - t(1 + 2D)(13D^2 + 13D + 4)$ $\quad - 3t^2(D + 1)(3D + 4)(3D + 2)$
8	$D^3 - 4t(1 + 2D)(3D^2 + 3D + 1) + 16t^2(D + 1)^3$
9	$D^3 - 3t(1 + 2D)(3D^2 + 3D + 1) - 27t^2(D + 1)^3$
11	$D^3 - \dfrac{2}{5}t(2D + 1)(17D^2 + 17D + 6)$ $\quad - \dfrac{56}{25}t^2(D + 1)(11D^2 + 22D + 12)$ $\quad - \dfrac{126}{125}t^3(2D + 3)(D + 2)(D + 1)$ $\quad - \dfrac{1504}{625}t^4(D + 3)(D + 2)(D + 1)$

$d = 2$	$N = 1$	$D^3 - 192t^2(3D + 5)(3D + 1)(D + 1)$
	$N = 2$	$D^3 - 64t^2(2D + 3)(2D + 1)(D + 1)$
	$N = 3$	$D^3 - 12t^2(3D + 2)(3D + 4)(D + 1)$
	$N = 4$	$D^3 - 64t^2(D + 1)^3$
	$N = 5$	$D^3 - 4t^2(D + 1)(11D^2 + 22D + 12)$ $-16t^4(D + 3)(D + 2)(D + 1)$
$d = 3$	$N = 3$	$D^3 - 54t^3(2D + 3)(D + 2)(D + 1)$
$d = 4$	$N = 2$	$D^3 - 256t^4(D + 3)(D + 2)(D + 1)$

6 A conjecture on counting matrices. The Iskovskikh classification revisited.

Corollary 6.1 (of Theorems 2.1 and 4.1). *The d-Kummer pullback of the Picard-Fuchs equation of the twisted symmetric square of the universal elliptic curve over $X_0(N)^W$ is of type D3 if and only if there exists a family of rank 1 Fano 3-folds of index d and anticanonical degree $2d^2N$.*

Modularity conjecture. The counting matrix of a generic Fano 3-fold in the Iskovskikh family with parameters (N, d) is in the same class as the corresponding matrix in Tables 2.

More concretely, the conjecture states that the matrix a_{ij} of normalized Gromov-Witten invariants of a Fano 3-fold with invariants (N, d) can be obtained in the following uniform way. Let $T = T(q)$ be the inverse of the suitable Conway-Norton uniformizer on $X_0(N)$ (that is, the one with the 'right' constant term). Consider

$$\Phi = (q^{1/24} \prod (1 - q^n) q^{N/24} \prod (1 - q^{Nn}))^2 T^{-\frac{N+1}{12}}.$$

Then Φ satisfies a $D3$ equation with respect to $t = T^{\frac{1}{d}}$. Recover the matrix of a_{ij} that corresponds to this equation (e.g. by looking at the expansion in Example 2.9), and normalize it by subtracting $a_{00}I$.

This uniform description is somewhat unexpected, since it does not have an obvious translation in terms of the geometry of Fano 3-folds. Let us now take a more detailed view at the Iskovskikh classification, according to the index and the degree.

Table 4: The Iskovskikh classification revisited.

$d = 1$

1	hypersurface of degree 6 in $\mathbb{P}(1,1,1,1,3)$
2	quartic in $\mathbb{P}^4$
3	complete intersection of a quadric and a cubic in $\mathbb{P}^5$
4	complete intersection of 3 quadrics in $\mathbb{P}^6$
5	a section of the Grassmannian $G(2,5)$ by a quadric and a codimension 2 plane
6	a section of the orthogonal Grassmannian $O(5,10)$ by a codimension 7 plane
7	a section of the Grassmannian $G(2,6)$ by a codimension 5 plane
8	a section of the Lagrangian Grassmannian $L(3,6)$ by a codimension 3 plane
9	a section of G_2/P by a codimension 2 plane
11	variety V_{22}

$d = 2$

1	hypersurface of degree 6 in $\mathbb{P}(1,1,1,2,3)$
2	hypersurface of degree 4 in $\mathbb{P}(1,1,1,1,2)$
3	a cubic in $\mathbb{P}^4$
4	complete intersection of 2 quadrics in $\mathbb{P}^6$
5	a section of the Grassmannian $G(2,5)$ by a codimension 3 plane

$d = 3$

3	quadric in $\mathbb{P}^4$

$d = 4$

2	$\mathbb{P}^3$

The description of families $(6,1), (8,1), (9,1)$ as hyperplane sections in Grassmannians is due to Sh. Mukai [20].

How can one prove the modularity conjecture? The uniformity of the assertion calls for a uniform proof, but I do not know how such a proof might work. The only way I know how to prove the conjecture is to explicitly calculate the quantum cohomology of Fano 3-folds on a case by case basis.

Kuznetsov calculated the quantum cohomology of V_{22}. All other cases are complete intersections in weighted projective spaces or Grassmannians of simple Lie groups.

For complete intersections in usual projective space, Givental's result allows to compute the $D3$ equations and the result agrees with the conjecture. Przyjalkowski [22] has recently extended Givental's result to the cases of smooth complete intersections in weighted projective spaces and established the predictions in the cases $(N, d) \in \{(1, 1), (1, 2), (2, 2)\}$.

In the remaining cases we use the quantum Lefschetz principle to reduce the computation of the quantum D–module of a hyperplane section to that of the ambient variety. The following is due to Coates-Givental-Lee-Gathmann, see e.g. [11].

Theorem 6.2 (Quantum Lefschetz hyperplane section theorem). *Let* Y *be a section of a very ample line bundle* $\mathcal{L}$ *on* X. *We assume that both varieties are of Picard rank 1. Let* $\iota_{\widehat{\mathcal{L}}} \colon \mathbb{G}_m \to T_{NS^\vee}$ *be the morphism of tori double dual to the map* $\mathbb{Z}[\widehat{\mathcal{L}}] \longrightarrow NS_X$.

For $\lambda \in \mathbb{C}^*$ *let* $[\lambda] \colon \mathbb{G}_m \longrightarrow \mathbb{G}_m$ *be the corresponding translationi, and for* $\alpha \in \mathbb{C}^*$ *let* $[\alpha] \colon \mathbb{A}^1 \longrightarrow \mathbb{A}^1$ *be the corresponding multiplication morphism. Then the quantum D–modules are related as follows:*

(i) *if the index of* $Y > 1$, *there exists* λ *in* $\mathbb{C}^*$ *such that*

$$Q_Y \text{ is a subquotient of } [\lambda]^*(Q_X * \iota_{\widehat{\mathcal{L}}*}(j^*E)) \; ;$$

(ii) *if the index of* $Y = 1$, *there exist* λ *and* α *in* $\mathbb{C}^*$ *such that*

$$Q_Y \text{ is a subquotient of } [\lambda]^*(Q_X * \iota_{\widehat{\mathcal{L}}*}(j^*E)) \otimes j^*([\alpha]_*E) \; .$$

The quantum cohomology of ordinary, orthogonal and Lagrangian Grassmannians is known (Givental-Kim-Siebert-Tian-Peterson-Kresch-Tamvakis). Przyjalkowski calculated the quantum Lefschetz reduction for the cases $(5, 1)$ and $(7, 1)$, confirming the conjecture.

Note that we do not need the whole cohomology structure: we just need to know quantum multiplication by the divisor classes, and this can be computed using Peterson's quantum Chevalley formula [10]. I calculated the quantum Lefschetz reduction for the cases $(6,1)$, $(8,1)$ and $(9,1)$ and the results again agreed with the ones predicted by the conjecture.

To our knowledge, quantum multiplication by the divisor class on V_5 (case $(5,2)$) was first computed by Beauville [3].

To summarize, we have checked the conjecture in all 17 cases by a case by case analysis. This proof, however, does not explain why the conjecture is true. A more uniform approach, yet to be discovered, would presumably start from the embedded $K3$ rather than the ambient space.

Remark . If the conjecture is true, then there is a mysterious relation between varieties of different index, as implied by Remark 5.2.

7 What next?

Classification of smooth rank 1 Fano 4-folds. This is an open question. For a variety of index ≥ 2 one can pass to the hyperplane section (which has to be a Fano 3-fold) and thus reduce the problem to lower dimension. On the other hand, the classification of index one Fano 4-folds seems to be beyond reach of today's geometric methods. Our program, if carried out in this case, would suggest a blueprint of a future classification.

The first step is to classify counting $D4$ equations. Unlike $D2$ and $D3$ variations, whose differential Galois group is $SL_2 - Sp_2 \rightarrow SO_3$, a variation of type $D4$ is controlled by Sp_4 and has in general no chance of being modular. Thus, as we remarked in 2.15, in the $D4$ case we lack the consequences of modularity that enabled us in the $D3$ case first to state the correct mirror dual problem and then to effectively handle it.

With no idea of what the mirror dual problem might be, one can still rely on the basic conjectures of Section 1 to compose a list of candidate $D4$ equations. If the list is not too long and it contains all $D4$ equations, the problem is reduced to weeding out the extra non-counting $D4$ equations that have sneaked into the list.

Which $D4$ equations are of Picard-Fuchs type? Of the approaches that we discuss in Section 1 (p. 93), establishing the $\mathbb{Q}$-Hodge or even the $\mathbb{R}$-Hodge property of a differential equation, given its coefficients, seems hopeless. On the other hand, a necessary, though not sufficient, condition for global nilpotence is that the p-curvatures are nilpotent for sufficiently many prime p. In principle †, one needs to guess the upper bound $h_{\max}$ of the height ($h = \max(p, q)$ for $p/q \in \mathbb{Q}$ in lowest terms) of possible Gromov-Witten invariants a_{ij}, and then run the above search over the corresponding box.

A non-systematic search for $D4$ equations whose analytic solution expands as a series in $\mathbb{Z}[[t]]$ was pioneered by Almkvist, van Enckevort, van Straten and Zudilin [1, 24]. See [24] for a systematic approach to recognizing a given globally nilpotent $D4$ equation as the mirror DE of a Calabi-Yau family by computing invariants of its global monodromy.

The hypergeometric pullback conjecture suggests a (presumably) more restrictive candidate list, but it is not clear how one can identify these among all $D4$ equations, without further assumptions.

Del Pezzo surfaces and $D2$s.

The only rank 1 del Pezzo is $\mathbb{P}^2$, so it might seem that our program is just not applicable here. However, it turns out (Orlov and Golyshev, unpublished) that the three-dimensional subspace of the total cohomology generated by the classical powers of the anticanonical class is stable under quantum multiplication by the anticanonical class, and it gives rise to $D2$ equations for del Pezzo surfaces of degrees $9, 6, 5, 4, 3, 2, 1$. In [12] the parametric $D2$ equation was identified with a particular case of the classical Heun equation that had been studied by Beukers [4] and Zagier [25]. Zagier had run a search over a large box for $D2$ equations with analytic solution in $\mathbb{Z}[[t]]$, see the list in [25]. Our counting $D2$ equations are hypergeometric in degrees $4, 3, 2, 1$ and are hypergeometric pullbacks in degrees $9, 6, 5$.

The classification of del Pezzo surfaces is of course well known; however, it might be interesting to understand the significance of the non-$D2$ equations arising as canonical pullbacks in degrees 8 and 7.

Singular Fano 3-folds.

The classification of singular Fano 3-folds of Picard rank 1 is of interest in birational geometry, see [6]. Corti has suggested extending the mirror approach to the classification of $\mathbb{Q}$-Fano 3-folds with prescribed (say terminal, or canonical) singularities. One expects that to a $\mathbb{Q}$-Fano 3-fold one can associate a differential equation that reflects its properties in much the same way as $D3$ equations do for smooth 3-folds. In order to construct it as a counting DE one would have to rely on a theory of Gromov-Witten invariants of singular varieties, which is not yet sufficiently developed. A provisional solution is to model

---

† But not in practice. The generic $D4$ depends on 9 parameters, and the computation involved needs unrealistic resources.

the construction of such a DE on the known smooth examples, formally generalizing them in the simplest cases such as complete intersections in weighted projective spaces.

An instance of a pair of mirror dual problems in this setup is due to Corti and myself. Let $\mathbb{P}(w_0, w_1, w_2, w_3)$ be a weighted projective space, $d = \sum w_j$. The operator

$$\prod_{i=0}^{3} \left(w_i^{w_i} \left(D - \frac{w_i - 1}{w_i} \right) \left(D - \frac{w_i - 2}{w_i} \right) \cdots D \right)$$
$$-d^d t \left(D + \frac{1}{d} \right) \left(D + \frac{2}{d} \right) \cdots \left(D + \frac{d-1}{d} \right) (D + 1)$$

gives rise to a hypergeometric D–module, whose essential constituent we call the *anticanonical Riemann-Roch D–module*. It is easy to show that the monodromy of this D–module respects a real orthogonal form. The problem of classification of weighted $\mathbb{P}^3$ with canonical singularities happens to admit a mirror dual problem: to classify the anticanonical Riemann-Roch D–modules such that the form above is of signature $(2, n - 2)$.

Acknowledgements

I am obliged to Helena Verrill who checked the modular formulas and made a number of suggestions.

I thank Alexander Givental for references to quantum Lefschetz, Victor Przyjalkowski for explanations, Constantin Shramov and Jan Stienstra for comments. I thank Yuri Manin for his interest in my work.

Special thanks go to Alessio Corti who organized the lecture series at Cambridge and edited these notes.

Bibliography

[1] G. Almkvist and W. Zudilin. Differential equations, mirror maps and zeta values. Preprint `math.AG/0507430`.
[2] A. Bayer and Yu. I. Manin. (Semi)simple exercises in quantum cohomology. In *The Fano Conference*, pages 143–173. Univ. Torino, Turin, 2004.
[3] A. Beauville. Quantum cohomology of complete intersections. *Mat. Fiz. Anal. Geom.*, 2(3–4):384–398, 1995.
[4] F. Beukers. On Dwork's accessory parameter problem. *Math. Z.*, 241(2):425–444, 2002.
[5] J. H. Conway and S. Norton. Monstrous moonshine. *Bull. Lond. Math. Soc.*, 11:308–339, 1979.

[6] A. Corti, A. Pukhlikov, and M. Reid. Fano 3-fold hypersurfaces. In *Explicit birational geometry of 3-folds*, volume 281 of *London Math. Soc. Lecture Note Ser.*, pages 175–258. Cambridge Univ. Press, Cambridge, 2000.

[7] P. Deligne. *Équations différentielles à points singuliers réguliers.* Springer-Verlag, Berlin, 1970. Lecture Notes in Mathematics, Vol. 163.

[8] I. Dolgachev. Mirror symmetry for lattice polarized $K3$ surfaces. *J. Math. Sci. (New York)*, 81(3):2599–2630, 1996.

[9] B. Dwork. Differential operators with nilpotent p-curvature. *Amer. J. Math.*, 112(5):749–786, 1990.

[10] W. Fulton and C. Woodward. On the quantum product of Schubert classes. *J. Algebr. Geom.*, 13(4):641–661, 2004.

[11] A. Gathmann. Relative Gromov-Witten invariants and the mirror formula. *Math. Ann.*, 325(2):393–412, 2003.

[12] V. Golyshev. Geometricity problem and modularity of certain Riemann-Roch variations. *Doklady Akad. Nauk*, 386(5):583–588, 2002.

[13] V. Golyshev. Modularity of equations $D3$ and the Iskovskikh classification. *Doklady Akad. Nauk*, 396(6):733–739, 2004.

[14] V. A. Iskovskikh and Yu. G. Prokhorov. Fano varieties. In *Algebraic geometry, V*, volume 47 of *Encyclopaedia Math. Sci.*, pages 1–247. Springer, Berlin, 1999.

[15] N. Katz. Nilpotent connections and the monodromy theorem – applications of a result of Turritin. *Publ. Math. IHES*, 39:355–412, 1970.

[16] N. M. Katz. *Exponential sums and differential equations*, volume 124 of *Annals of Mathematics Studies*. Princeton University Press, Princeton, NJ, 1990.

[17] F. Loeser and C. Sabbah. Caractérisation des $\mathcal{D}$-modules hypergéométriques irréductibles sur le tore. *C. R. Acad. Sci., Paris, Sér. I*, 312(10):735–738, 1991.

[18] F. Loeser and C. Sabbah. Équations aux différences finies et déterminants d'intégrales de fonctions multiformes. *Comment. Math. Helv.*, 66(3):458–503, 1991.

[19] Yu. I. Manin. *Frobenius manifolds, quantum cohomology, and moduli spaces*, volume 47 of *American Mathematical Society Colloquium Publications*. American Mathematical Society, Providence, RI, 1999.

[20] S. Mukai. Fano 3-folds. In *Complex projective geometry (Trieste, 1989/Bergen, 1989)*, volume 179 of *London Math. Soc. Lecture Note Ser.*, pages 255–263. Cambridge Univ. Press, Cambridge, 1992.

[21] C. Peters and J. Stienstra. A pencil of $K3$-surfaces related to Apéry's recurrence for $\zeta(3)$ and Fermi surfaces for potential zero. In *Arithmetic of complex manifolds (Erlangen, 1988)*, volume 1399 of *Lecture Notes in Math.*, pages 110–127. Springer, Berlin, 1989.

[22] V. Przyjalkowski. Quantum cohomology of smooth complete intersections in weighted projective spaces and singular toric varieties. Preprint math.AG/0507232.

[23] G. Shimura. *Introduction to the arithmetic theory of automorphic functions.* Publications of the Mathematical Society of Japan, No. 11. Iwanami Shoten, Publishers, Tokyo, 1971.

[24] C. van Enckevort and D. van Straten. Monodromy calculatons of fourth order equations of Calabi-Yau type. Preprint math.AG/0412539.

[25] D. Zagier. Integral solutions of Apéry-like recurrence equations. Preprint.

Number Theory Section
Steklov Mathematical Institute
Gubkina str. 8
119991 Moscow
Russia
golyshev@mccme.ru

Birational models of del Pezzo fibrations

Mikhail M. Grinenko

Preliminaries

Acknowledgment. This review is a printed version of the talks given by the author in the University of Liverpool and the Isaac Newton Institute for Mathematical Sciences in spring of 2002, and organized by the London Mathematical Society. I express my gratitude to the Society for the kind support. I am especially grateful to Professor Vyacheslav Nikulin and Professor Thomas Berry for their close reading of the text. They helped me very much to prepare the final version of the paper.

Given an algebraic variety X, we can naturally attach some objects to it, such as its field of functions $k(X)$, the essential object in birational geometry. Assuming classification to be one of the most important problems in algebraic geometry, we may be asked to describe all algebraic varieties with the same field of functions, that is, all varieties that are birationally isomorphic to X. Of course, 'all' is far too large a class, and usually we restrict to projective and normal varieties (though non-projective or non-normal varieties may naturally arise in some questions). Typically there are two main tasks:

A. Given a variety V, determine whether it is birational to another variety W.

B. Given that V and W are birational to each other, determine a decomposition of a birational map between them into 'elementary links', that is, birational maps that are, in an appropriate sense, particularly simple.

The rationality problem, i.e. to determine if a given variety is rational, is an essential example of task A. (Recall that a variety is said to be rational if it is birational to $\mathbb{P}^n$ or, equivalently, if its field of functions is

$k(x_1, \ldots, x_n))$. As to task B, examples will be given below. Let us only note that varieties joined by 'elementary links' should belong to a more or less restrictive category (otherwise the task becomes meaningless), and such a category has to meet the following requirement: any variety can be birationally transformed to one lying in it. In dimension 3 minimal models and Mori fibrations give examples of such categories. It often happens however that the indicated category is too large to be convenient, and we introduce a subcategory of 'good models' (for example, relatively minimal surfaces). Here 'good' means a class of varieties that are simple enough for describing, handling, classifying, and so on. Now let us look what we have in the first three dimensions.

Curves

Normal algebraic curves are exactly the smooth ones. A birational map between projective smooth curves is an isomorphism, so the birational and biregular classifications coincide in dimension 1. Projective spaces have no moduli, so $\mathbb{P}^1$ is a unique representative of rational curves.

Surfaces

It is well known that any birational map between smooth surfaces can be decomposed into a chain of blow-ups of points and contractions of (-1)-curves without loss of the smoothness of the intermediate varieties. It is clear that we can successively contract (in any order) all (-1)-curves to get the so-called (relatively) minimal model (which has no (-1)-curves, thus nothing to contract). It is therefore very convenient to use minimal models as the class of 'good' models and the indicated blow-ups and contractions as elementary links. As to the rationality problem, the famous theorem of Castelnuovo says that a smooth surface X is rational if and only if $H^1(X, \mathcal{O}_X) = H^0(X, 2K_X) = 0$. This is one of the most outstanding achievements of classical algebraic geometry.

Summarizing the results in dimensions 1 and 2, we can formulate the rationality criterion as follows: X is rational if and only if all the essential differential-geometric invariants $H^0\left(X, (\Omega_X^p)^{\otimes m}\right)$ vanish.

Threefolds

As soon as we get to dimension 3, the situation becomes much harder. Now it is not at all obvious at all what is 'a good model'. We may proceed in the same way as in dimension 2, by contracting everything that can be contracted. This is the viewpoint of Mori theory. But then, starting from a smooth variety we may lose smoothness very quickly, and even get a 'very bad' variety with two Weil divisors intersecting in a point (this is the case of 'a small contraction', that is, a birational morphism which is an isomorphism in codimension 1).

Nevertheless, Mori theory states that there is a smallest category of varieties which is stable under divisorial contractions and flips (a flip is exactly a tool which allows us to 'correct' small contractions). In what follows we shall not need the details of this theory; the reader can find them in many monographs (e.g. [23]). We only point out that X belongs the Mori category if it is a projective normal variety with at most $\mathbb{Q}$-factorial terminal singularities. This means that every Weil divisor is $\mathbb{Q}$-Cartier, that is it becomes Cartier if we take it with some multiplicity. Moreover, for every resolution of singularities $\varphi : Y \to X$ we have

$$K_Y = \varphi^*(K_X) + \sum a_i E_i,$$

where the E_i are exceptional divisors and all the discrepancies a_i are positive rational numbers; the equality '=' means 'equal as $\mathbb{Q}$-divisors', that is, after multiplying by a suitable integer we get linear equivalence. There exists an intersection theory on such varieties, which is very similar to the usual one; the difference is mainly that we must accept rational numbers as intersection indices.

The Mori category has some nice properties. In particular, the Kodaira dimension is a birational invariant under maps in this category. We recall that the *Kodaira dimension* $\mathrm{kod}(X)$ of a variety X is the largest dimension of images of X under (rational) maps defined by linear systems $|mK_X|$. We are mostly interested in studying varieties of negative Kodaira dimension, that is, varieties such that $H^0(X, mK_X) = 0$ for all $m > 0$, because their birational geometry is especially non-trivial.

From now on, we will only consider varieties of negative Kodaira dimension. What are 'minimal models' for such varieties in the Mori theory? They are the Mori fibre spaces. By definition, a triple $\mu : X \to S$ is *a Mori fibre space* if X is projective, $\mathbb{Q}$-factorial and terminal, S is a projective normal variety with $\dim S < \dim X$, and μ is an extremal contraction of fibring type, i.e. the relative Picard number $\rho(X/S) = \mathrm{rk}\,\mathrm{Pic}(X) - \mathrm{rk}\,\mathrm{Pic}(S)$ is equal to 1 and $(-K_X)$ is μ-ample.

In dimension 3 (the highest dimension where Mori theory has been established) we have three possible types of Mori fibre space (or, briefly, Mori fibration) $\mu : X \rightarrow S$:

1) $\mathbb{Q}$-Fano: $\dim S = 0$, i.e. S is a point;
2) del Pezzo fibration of degree d: $\dim S = 1$, and over the generic point of S, the fibre is a del Pezzo surface of degree d.
3) conic bundle: $\dim S = 2$ and over the generic point of S, the fibre is a plane conic.

In what follows, we often denote a Mori fibration $\rho : V \rightarrow S$ by V/S, or even V if the base and structure morphisms are clear.

Factorization of birational maps between Mori fibrations is given by the Sarkisov program (which has been established in dimension 3, see [6]). The essential claim of this program is the following. Suppose we have two Mori fibrations V/S and U/T and a birational map

$$
\begin{array}{ccc}
V & \overset{\chi}{\dashrightarrow} & U \\
\downarrow & & \downarrow \\
S & & T
\end{array}
$$

Then there exists a finite chain of birational maps

$$
X_0 \overset{\chi_1}{\dashrightarrow} X_1 \overset{\chi_2}{\dashrightarrow} X_2 \overset{\chi_3}{\dashrightarrow} \cdots \overset{\chi_{N-1}}{\dashrightarrow} X_{N-1} \overset{\chi_N}{\dashrightarrow} X_N
$$
$$
\begin{array}{cccccc}
\downarrow & \downarrow & \downarrow & & \downarrow & \downarrow \\
S_0 & S_1 & S_2 & & S_{N-1} & S_N
\end{array}
$$

where $X_0/S_0, X_1/S_1, \ldots, X_N/S_N$ are Mori fibrations, $X_0/S_0 = V/S$, $X_N/S_N = U/T$, such that $\chi = \chi_N \circ \chi_{N-1} \circ \ldots \circ \chi_2 \circ \chi_1$ and any of χ_i belongs to one of the four types of elementary links listed below in Figure 1, (where, to avoid complicating the notation, χ_2 is taken as typical). For all types, X_1/S_1 and X_2/S_2 are Mori fibrations and ψ is an isomorphism in codimension 1 (actually, it is a sequence of log-flips). For type I, μ denotes a morphism with connected fibres and γ is an extremal divisorial contraction. Remark that $\rho(S_1/S_0) = 1$. For type II, γ_1 and γ_2 are extremal divisorial contractions and μ is a birational map. Type III is inverse to type I. Finally, for type IV, δ_1 and δ_2 are morphisms with connected fibres, T is a normal variety, and $\rho(S_1/T) = \rho(S_2/T) = 1$.

In dimension 2 the only Mori fibrations are the projective plane itself and ruled surfaces, isomorphisms in codimension 1 are biregular, and extremal divisorial contractions are blow-downs of (-1)-curves. Thus all elementary links are very simple. In dimension 3 the description just given is all we know about links; in consequence, up to the present the Sarkisov program mostly plays a theoretical rôle.

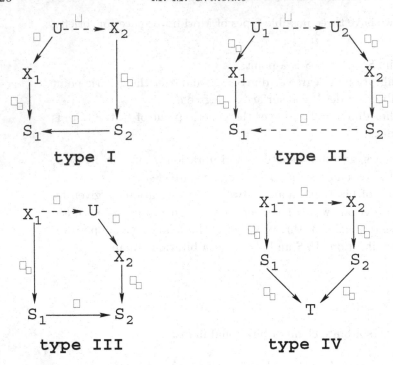

type I   type II

type III   type IV

Fig. 1.

The rationality problem for three-dimensional varieties also becomes extremely hard. For a long time, many mathematicians believed that it should be possible to find a simple rationality criterion, similar to those for lower dimensions. But in the early 70s, nearly simultaneously, three outstanding works of different authors were published. They gave examples of unirational but non-rational 3-dimensional varieties. These were works by Iskovskikh and Manin [18], Clemens and Griffiths [4], and Artin and Mumford [2]. Recall that an algebraic variety is said to be *unirational* if there exists a rational map from projective space which is finite at the generic point. All essential differential-geometric invariants vanish on unirational varieties as on projective spaces. Thus there is no hope of finding a rationality criterion in dimension 3 similar to the criteria for curves and surfaces which use these invariants. The reader can find an excellent survey of the rationality problem in higher dimensions in [17].

Nevertheless, during the last 10 years, considerable progress in birational classification problems in all dimensions has been achieved, mostly due to the concept of birationally rigid varieties. This was first formulated by A. V. Pukhlikov. In the original version, still actively used, birational rigidity is closely related to some technical features of the maximal singularities method (such as comparison with the canonical adjunction thresholds). Another version of birational rigidity arises from the theory of Mori fibrations. The two versions are not identical but coincide in many cases. In this paper, we use the second one since it has a clear geometric sense.

First we introduce the following useful notion:

Definition 0.1. We say that Mori fibrations V/S and U/T ($\dim S = \dim T > 0$) have the same *Mori structure* if there exist birational maps $\chi : V \dashrightarrow U$ and $\psi : S \dashrightarrow T$ that make the following diagram commute:

$$
\begin{array}{ccc}
V & \xdashrightarrow{\chi} & U \\
\downarrow & & \downarrow \\
S & \xdashrightarrow{\psi} & T
\end{array}
.
$$

In $\mathbb{Q}$-Fano cases ($\dim S = \dim T = 0$) we say that V and U have the same Mori structure if they are biregular to each other; in other words, if for any birational map $\chi : V \dashrightarrow U$ there exists a birational self-map $\mu \in \mathrm{Bir}(V)$ such that the composition $\chi \circ \mu$ is an isomorphism.

In the sequel we will often use the words 'to be birational over the base' instead of 'to have the same Mori structure'.

Since it is known that any threefold X can be birationally mapped onto a Mori fibration, one can formulate the classification task as follows: describe all Mori fibrations that are birational to X and not birational over the base to each other (that is, have different Mori structures). Clearly, varieties X and Y are birationally isomorphic if and only if they have the same set of Mori structures. It is worth noting that only links of type II join Mori fibrations with the same structure.

In the following cases the classification becomes especially easy:

Definition 0.2. A Mori fibration V/S is said to be *birationally rigid* if it has a unique Mori structure (that is, if any U/T that is birational to V is birational to V/S over the base).

Here are some examples of (birationally) rigid and non-rigid varieties. First let me note that no rigid variety is rational. Indeed, $\mathbb{P}^3$ is birational to $\mathbb{P}^1 \times \mathbb{P}^2$, so we have at least three different Mori structures: $\mathbb{Q}$-Fano ($\mathbb{P}^3$

itself), del Pezzo fibration $\mathbb{P}^1 \times \mathbb{P}^2 \to \mathbb{P}^1$, and conic bundle $\mathbb{P}^1 \times \mathbb{P}^2 \to \mathbb{P}^2$. The simplest example of a birationally rigid $\mathbb{Q}$-Fano variety is a smooth quartic 3-fold in $\mathbb{P}^4$; this is the result of Iskovskikh and Manin [18]. Note that they proved even more: any birational automorphism of a smooth quartic is actually biregular (and in the general case, there are no non-trivial automorphisms at all). For conic bundles, there exists the following result by Sarkisov [30, 31]: any standard conic bundle $V \to S$ over a smooth rational surface S with discriminant curve C is rigid if the linear system $|4K_S + C|$ is not empty (i.e. if V/S has enough degenerations). For example, if S is a plane, this holds if C has degree at least 12. A smooth cubic 3-fold in $\mathbb{P}^4$ is a Fano variety with many structures of conic bundle. Indeed, the projection from a line lying on the cubic realizes the cubic as a conic bundle over a plane with a quintic as the discriminant curve.

Non-singular rigid and non-rigid del Pezzo fibrations will be described in the main part of the paper. Almost all results were obtained using the maximal singularities method (see [18, 27]). We shall not give any proofs; the reader can find them in [9, 10].

We conclude this section with the remark that minimal models or Mori fibre spaces are not always the most convenient models. In many cases, there are preferable classes of varieties. For example, sometimes it is useful to consider Gorenstein terminal varieties with numerically effective anticanonical divisors [1, 7]. In other words, appropriate models depend on the context.

1 The rigidity problem for del Pezzo fibrations

Let $\rho : V \to S$ be a Mori fibration where S is a smooth curve, η the generic point of S, and V_η the fibre over the generic point. Thus V_η is a non-singular del Pezzo surface of degree $d = (-K_{V_\eta})^2$ over the field of functions of S. Moreover, it is a minimal surface since $\rho(V/S) = 1$, hence

$$Pic(V_\eta) \otimes \mathbb{Q} = \mathbb{Q}[-K_{V_\eta}].$$

Suppose we have another Mori fibration U/T and a birational map

$$
\begin{array}{ccc}
V & \xrightarrow{\;\chi\;} & U \\
\downarrow & & \downarrow \\
S & & T
\end{array}
$$

We would like to know as much as possible about this situation; in particular, whether χ is birational over the base or not.

First, we can assume that S is rational. Indeed, suppose S is a non-rational curve. Then U cannot be a $\mathbb{Q}$-Fano threefold. The simplest reason is that the Picard group of V or of any resolution of singularities must contain a continuous part arising from S, which is impossible for any birational model of U (including U itself). Or, we can use the fact that U is rationally connected (see [22]), which is not true for V: there can be no rational curves lying across the fibres of V/S (i.e. covering S).

Thus U/T is either a del Pezzo fibration or a conic bundle. In either case the fibres of U/T are covered by rational curves. Again, images of these curves on V cannot cover S and hence lie in the fibres of V/S. So there exists a rational map from T to S, which is actually a morphism since the curve S is smooth, and we have the following commutative diagram:

$$
\begin{array}{ccc}
V & \xrightarrow{\chi} & U \\
\downarrow & & \downarrow \\
S & \xleftarrow{\psi} & T
\end{array}
$$

Denote by ζ the generic point of T and by U_ζ the fibre over the generic point. Assuming U/T to be a del Pezzo fibration and taking into account that χ is birational, we easily see that ψ is an isomorphism, so it is possible to identify η and ζ. Thus χ induces a birational map χ_η between two del Pezzo surfaces over η. Now suppose that U/T is a conic bundle. Note that we may assume χ to be a link of type I. The composition $U \to T \to S$ represents U as a fibration over S, so we can define the fibre U_η over the generic point η. It is clear that U_η is a non-singular surface fibred in conics with $Pic(U_\eta) \simeq \mathbb{Z} \oplus \mathbb{Z}$ (see the construction of the link). Moreover, we may assume U_η to be minimal. Indeed, $\rho(U_\eta/T_\eta) = 1$, so there are no (-1)-curves in the fibres of $U_\eta \to T_\eta$. Hence any (-1)-curve on U_η has to be a section (which means that, birationally, $U \to T$ is a $\mathbb{P}^1$-bundle over T). Then by contracting of any of the (-1)-curves, one obviously gets a minimal del Pezzo surface, i.e. the previous case. So in all cases we have a birational map

$$
\chi_\eta : V_\eta \dashrightarrow U_\eta
$$

between two non-singular minimal surfaces defined over the field of functions of S, and U_η is either a del Pezzo surface or a conic bundle. This situation was completely studied in V. A. Iskovskikh's paper [16, Theorem 2.6]. In particular, if V_η is a del Pezzo surface of small degree

($d \leq 3$), then U_η is always a del Pezzo surface that is isomorphic to V_η, and χ_η is either an isomorphism if $d = 1$, or a composition of Bertini and/or Geiser involutions if $d = 2, 3$.

Thus in what follows we assume $S = \mathbb{P}^1$. Note that the field of functions $\mathbb{C}(S)$ of S is a C_1 field, and from [5] it follows that V_η always has a point over $\mathbb{C}(S)$ (i.e. $V \to S$ has a section). We may note that in fact V_η has a Zariski dense subset of such points [22]. Yu. I. Manin proved [24] that del Pezzo surfaces of degree 5 or greater with a point over a perfect field are always rational over the field itself. This is exactly our case. So if $d \geq 5$, V_η is rational over $\mathbb{C}(S) = \mathbb{C}(\mathbb{P}^1)$, hence V is rational over $\mathbb{C}$. Thus, V is birational to $\mathbb{P}^3$; in particular, V/S is always non-rigid. Now suppose $d = 4$. Blowing up a section of $V \to S$ and making some log flips in the fibres, we get a structure of conic bundle (or just blow up a point on V_η, as in Theorem 2.6 of [16]). Thus the case $d = 4$ is non-rigid too.

So we see that, from the viewpoint of the rigidity problem, only the cases $S = \mathbb{P}^1$ and $d \leq 3$, are interesting. The cases of degree 1 and 2 will be described in detail in the rest of this paper. For the case $d = 3$, which is less studied, we only outline some results.

Theorem 1.1 ([26]). *Let $V/\mathbb{P}^1$ be a non-singular Mori fibration in cubic surfaces with only the simplest degenerations (all singular fibres have at most one ordinary double point). If 1-cycles $N(-K_V)^2 - f$ are non-effective for all $N > 0$, where f is the class of a line in a fibre, then $V/\mathbb{P}^1$ is birationally rigid.*

The condition on 1-cycles above (called the K^2-condition) is also sufficient for degrees 1 and 2 (without assumptions on degenerations) but it is not necessary (see examples in [9]). The requirement on degenerations arises from technical reasons and can probably be omitted. The simplest examples of non-rigid fibrations in cubic surfaces are the following.

Example 1.2 (*Quartics with a plane*). Let $X \subset \mathbb{P}^4$ be a quartic three-fold containing a plane. It is easy to see that in the general case such a quartic contains exactly 9 ordinary double points lying in the plane. Note that X is a Fano but not a $\mathbb{Q}$-Fano variety (i.e. it is not in the Mori category). Indeed, hyperplane sections through the plane cut out residual cubic surfaces which intersect in the 9 singular points. This is impossible for $\mathbb{Q}$-factorial varieties. The plane and any of these residual cubics generate the Weil divisor group of X. Blow up the plane as a subvariety in $\mathbb{P}^4$, and let V be the strict transform of X. Then V is a

Mori fibration in cubic surfaces over $\mathbb{P}^1$, and the birational morphism $V \to X$ is a small resolution with 9 lines lying over the singular points of X. It is easy to check that one can produce a flop at all these 9 lines simultaneously and then contract the strict transform of the plane, which is $\mathbb{P}^2$ with normal bundle isomorphic to $\mathcal{O}(-2)$. One gets a $\mathbb{Q}$-Fano variety U, which has Fano index 1 and a singular point of index 2. This U is the complete intersection of two weighted cubic hypersurfaces in $\mathbb{P}(1,1,1,1,1,2)$. The transformation $V/\mathbb{P}^1 \dashrightarrow U$ is a link of type 3 (ψ is the flop and γ is the contraction of $\mathbb{P}^2$, in the notation of Figure 1). Conjecturally, $V/\mathbb{P}^1$ and U represent all Mori structures of X.

Example 1.3 (*Cubic threefolds*). Let $V \subset \mathbb{P}^4$ be a smooth cubic hypersurface. It is well known that V is non-rational [4]. Nevertheless, V is very far from being birationally rigid. Indeed, the projection from any line in V gives us a conic bundle over $\mathbb{P}^2$ with discriminant curve of degree 5. Then V is birational to a non-singular Fano variety of index 1 and degree 14, which is a section of $G(1,5) \subset \mathbb{P}^{14}$ by a linear subspace of codimension 5. Finally, one can blow up any plane cubic curve on V, and one gets a structure of fibration in cubic surfaces. Thus V has all possible types of dimension 3 Mori structure (i.e. $\mathbb{Q}$-Fano, conic bundle, del Pezzo fibration).

Example 1.4 (*Hypersurfaces of bidegree $(m, 3)$ in $\mathbb{P}^1 \times \mathbb{P}^3$*). Let $V = V_m \subset \mathbb{P}^1 \times \mathbb{P}^3$ be a non-singular hypersurface of bidegree $(m, 3)$. From the Lefschetz theorem on hyperplane sections it follows that $\mathrm{Pic}(V) \simeq \mathbb{Z}[-K_V] \oplus \mathbb{Z}[F]$, where F is the class of a fibre. Thus, $V/\mathbb{P}^1$ is a Mori fibration in cubic surfaces given by the projection $\mathbb{P}^1 \times \mathbb{P}^3 \to \mathbb{P}^1$. Consider the second natural projection $\pi : \mathbb{P}^1 \times \mathbb{P}^3 \to \mathbb{P}^3$. If $m = 1$ then, clearly, $\pi|_V$ gives a birational map $V \dashrightarrow \mathbb{P}^3$, so V is rational (and then it is non-rigid) in this case. If $m \geq 3$, then for general V the conditions of Theorem 1.1 are satisfied, and $V/\mathbb{P}^1$ is rigid. We consider the most interesting case $m = 2$. First, by the result of Bardelli [3], general V_m's are non-rational if $m \geq 2$. We want to show that $V = V_2$ is non-rigid. Note that a general $V/\mathbb{P}^1$ has exactly 27 sections which are fibres of the projection π. On the other hand, let t be any other fibre of π, then t intersects V at 2 points (not necessary different), and we can transpose these points. Thus, outside of the 27 sections, there exists an involution τ, and we have a birational self-map $\tau \in \mathrm{Bir}(V)$. On the other hand, we can simultaneously produce a flop χ centred at the indicated sections and get another Mori fibration $U/\mathbb{P}^1$, which is also a hypersurface of bidegree $(2, 3)$ in $\mathbb{P}^1 \times \mathbb{P}^3$. This χ is a link of type IV. Note that

$\dim |3(-K_V) - F| = 1$; moreover this pencil has no base components and the strict transforms of its elements become the fibres of $U/\mathbb{P}^1$. The same is true for the corresponding pencil on U. It only remains to notice that τ maps the pencil $|F|$ to $|3(-K_V) - F|$, so we can consider χ as the Sarkisov resolution of τ. In [32] it was shown that for general V_2 all Mori structures arises from any of these two structures by means of τ and a large group of fibrewise birational automorphisms (Bertini and Geiser involutions), so the set of Mori structures of V_2 contains a lot of fibrations in cubic surfaces and does not contain structures of $\mathbb{Q}$-Fano varieties and conic bundles.

We conclude this section with some remarks. First, in all non-rigid cases, it is easy to check that the linear systems $|n(-K_V) - F|$ are non-empty and are free from base components. In [9] the following conjecture was formulated.

Conjecture 1.5. *Suppose $V/\mathbb{P}^1$ is a non-singular Mori fibration in del Pezzo surfaces of degree 1, 2 or 3. Then $V/\mathbb{P}^1$ is birationally rigid if and only if the linear systems $|n(-K_V) - F|$ are either empty or not free from base components for all $n > 0$.*

This conjecture was completely proved for degree 1 [9] and, under some conditions of generality, for degree 2 [9, 12]. Moreover, for all cases that are known to be rigid or non-rigid, the statement of the conjecture holds. Finally, it is valid in one direction also for a general Mori fibration in del Pezzo surfaces of degree 1.

Theorem 1.6 ([11, Theorem 3.3]). *If a Mori fibration $V/\mathbb{P}^1$ in del Pezzo surfaces of degree 1 is birationally rigid, then for all $n > 0$ the linear systems $|n(-K_V) - F|$ are either empty or not free from base components.*

2 Projective models of del Pezzo fibrations

In this section we construct some simple projective models for non-singular (actually for Gorenstein) fibrations in del Pezzo surfaces of degrees 1 and 2.

First let me recall the simplest projective (and weighted projective) models of del Pezzo surfaces of degrees 1 and 2.

2.1 Models of del Pezzo surfaces of degrees 1 and 2

Consider first degree 1. Let X be a non-singular del Pezzo surface of degree $d = K_X^2 = 1$. It is easy to see that a general member of the linear system $|-K_X|$ is non-singular, all its elements are irreducible and reduced curves of genus 1, and $|-K_X|$ has a unique base point P. Choose a non-singular curve $C \in |-K_X|$. Using the exact sequence

$$0 \longrightarrow \mathcal{O}_X\left(-(i-1)K_X\right) \longrightarrow \mathcal{O}_X\left(-iK_X\right) \longrightarrow \mathcal{O}_C(-iK_X) \longrightarrow 0$$

and the Kodaira vanishing theorem, we see that

$$H^0\left(X, -iK_X\right) \longrightarrow H^0\left(C, (-iK_X)|_C\right) \longrightarrow 0$$

are exact for all $i \geq 0$. So we have a surjective map of the graded algebras

$$\mathcal{A}_X = \bigoplus_{i \geq 0} H^0\left(X, -iK_X\right) \longrightarrow \mathcal{A}_C = \bigoplus_{i \geq 0} H^0\left(C, (-iK_X)|_C\right)$$

which preserves the grading. Clearly, $\mathcal{A}_C$ is generated by elements of degree $r \leq 3$ (note that $\deg(-3K_X)|_C = 3$, i.e. $(-3K_X)|_C$ is very ample), so $\mathcal{A}_X$ is generated by elements of degree not higher than 3, too. Taking into account that

$$h^0(X, -K_X) = 2, \quad h^0(X, -2K_X) = 4, \quad h^0(X, -3K_X) = 7,$$

we can write down the generators as follows:

$$H^0(X, -K_X) = \mathbb{C} < x, y >,$$
$$H^0(X, -2K_X) = \mathbb{C} < x^2, xy, y^2, z >,$$
$$H^0(X, -3K_X) = \mathbb{C} < x^3, x^2y, xy^2, y^3, zx, zy, w >,$$

where x and y have weight 1, z has weight 2, and w has weight 3. Thus the homogeneous components of the graded algebra $\mathcal{A}_X$ are generated by monomials $x^i y^j z^k w^l$ of the corresponding degree. Now it is clear that X can be embedded as a surface into the weighted projective space $\mathbb{P}(1, 1, 2, 3)$:

$$X = \mathbf{Proj}\ \mathcal{A}_X = \mathbf{Proj}\ \mathbb{C}[x, y, z, w]/I_X \subset \mathbb{P}(1, 1, 2, 3),$$

where I_X is a principal ideal generated by a homogeneous element of degree 6. Indeed, $h^0(X, -6K_X) = 22$, but the dimension of the homogeneous component of $\mathbb{C}[x, y, z, w]$ of degree 6 is equal to 23, and there exists exactly one linear relation in it. This relation gives the equation of X. It only remains to notice that X necessarily avoids singular points of $\mathbb{P}(1,1,2,3)$, whence we can write down the equation of X as follows (stretching the coordinates if needed):

$$X = \{w^2 + z^3 + zf_4(x,y) + f_6(x,y) = 0\} \subset \mathbb{P}(1,1,2,3),$$

where f_i are homogeneous polynomials of the corresponding degree.

This embeds X into a well-known threefold as a surface. However, there is another useful model of X. Let

$$\mathbb{P}(1,1,2,3) \dashrightarrow \mathbb{P}(1,1,2)$$

be a (weighted) projection from the point $(0:0:0:1)$. This point does not belong to X. It follows from the equation of X, that the restriction of the projection on X gives us a morphism of degree 2

$$\varphi : X \to \mathbb{P}(1,1,2).$$

We can naturally identify $\mathbb{P}(1,1,2)$ with a non-degenerate quadratic cone $Q \subset \mathbb{P}^3$. The base point of $|-K_X|$ lies exactly over the vertex of the cone. The ramification divisor of φ is defined by an equation of degree 6 in $Q = \mathbb{P}(1,1,2)$. We can also describe this picture as follows: $|-2K_X|$ has no base points and defines a morphism, which is our φ. Thus, we can obtain X as a double cover of a quadratic cone in $\mathbb{P}^3$ branched along a non-singular cubic section that does not contain the vertex of the cone.

Now let us turn back to a non-singular Mori fibration $V/\mathbb{P}^1$ with fibres del Pezzo surfaces of degree 1. Clearly, its general fibre is smooth. Later (this section, after Lemma 2.2) it will be shown via relative Picard number arguments that $V/\mathbb{P}^1$ must have degenerations. Let X be a singular fibre. We can see that as 2-dimensional scheme X is projective, irreducible and reduced (since V is non-singular), Gorenstein – in particular Cohen-Macaulay – (the dualizing sheaf is invertible by the adjunction formula), K_X is anti-ample and $K_X^2 = 1$.

First suppose that X is normal. This case was completely studied in [15], and I can simply refer the reader to it. But in order to deal with projective models of X, we only need the fact that $|-K_X|$ contains a non-singular elliptic curve [15, Proposition 4.2], and then we can just repeat the arguments of the non-singular case. So X can be defined either by a homogeneous equation of degree 6 in $\mathbb{P}(1,1,2,3)$ (as above, X avoids singular points of the weighted projective space since they are non-Gorenstein), or as a double cover of a quadric cone in $\mathbb{P}^3$, branched along a (now singular) cubic section not passing through the vertex of the cone. This cubic section may be reducible, but all its components must be reduced. It only remains to add that the base point of $|-K_X|$

is always non-singular, and X has either only Du Val singularities, or a unique minimally elliptic singular point (defined in local coordinates by the equation $r^2 + p^3 + q^6 = 0$).

The non-normal case is much harder. Nevertheless, in [29] it was shown that the projective properties of the anticanonical linear systems on X are the same as in the non-singular case. In particular, the base point of $|-K_X|$ is non-singular, and either X can be embedded into $\mathbb{P}(1,1,2,3)$ exactly as in the non-singular case, or X doubly covers a quadric cone. The only difference is that the cubic section has to contain a non-reduced component.

We shall summarize all needed results concerning the case $d = 1$ a bit later. Now let us consider del Pezzo surfaces of degree 2. This case is fairly similar to the previous one, and we only outline the key points. Consider a non-singular del Pezzo surface X, $K_X^2 = 2$. Then $|-K_X|$ is base point free and contains a non-singular element C. As before, we have a surjective map $\mathcal{A}_X \to \mathcal{A}_C$, but now these algebras are generated by elements of degree not higher than 2. We see that $h^0(X, -K_X) = 3$ and $h^0(X, -2K_X) = 7$, and we can suppose that

$$H^0(X, -K_X) = \mathbb{C} < x, y, z >,$$
$$H^0(X, -2K_X) = \mathbb{C} < x^2, y^2, z^2, xy, xz, yz, w >,$$

where x, y, and z have weight 1, w has weight 2. Then,

$$X = \mathbf{Proj}\, \mathcal{A}_X = \mathbf{Proj}\, \mathbb{C}[x, y, z, w]/I_X \subset \mathbb{P}(1,1,1,2)$$

The principal ideal I_X is generated by a homogeneous element of degree 4; indeed, we have $h^0(X, -4K_X) = 21$, but the dimension of the homogeneous component of $\mathbb{C}[x, y, z, w]$ of degree 4 is equal to 22. Since X avoids the singular point of $\mathbb{P}(1,1,1,2)$, we get the following equation of X:

$$X = \{w^2 + f_4(x, y, z) = 0\} \subset \mathbb{P}(1,1,1,2),$$

where f_4 is a homogeneous polynomial of degree 4. The projection $\mathbb{P}(1,1,1,2) \dashrightarrow \mathbb{P}(1,1,1)$ from the point $(0 : 0 : 0 : 1)$, restricted to X, represents X as a double cover of $\mathbb{P}^2$ branched over a non-singular quartic curve. This morphism can be also defined by the linear system $|-K_X|$.

Projective models of singular (normal and non-normal) Gorenstein del Pezzo surfaces of degree 2 follow the same construction as in the non-singular case: they are defined by a quartic equation in $\mathbb{P}(1,1,1,2)$ and do not contain the point $(0 : 0 : 0 : 1)$. Alternatively, we can construct them as double coverings of $\mathbb{P}^2$ branched along quartic plane curves, but now these quartics have singular points and/or non-reduced components [15, 29]. In other words, the situation is the same as in the case $d = 1$. In particular, normal surfaces of this type may have either only Du Val singularities, or a unique minimally elliptic singularity locally defined by the equation $r^2 + p^4 + q^4 = 0$.

Now we summarize the results about projective models of del Pezzo surfaces of degrees 1 and 2:

Proposition 2.1. *Let X be a projective Gorenstein irreducible reduced del Pezzo surface of degree 1 or 2.*

If the degree is 1, then $|-K_X|$ has a unique base point, which is non-singular for X, the linear system $|-2K_X|$ is base point free, and $|-3K_X|$ is very ample and embeds X into $\mathbb{P}^6$ as a surface of degree 9. Under a suitable choice of coordinates $[x, y, z, w]$ of weights $(1, 1, 2, 3)$ in $\mathbb{P}(1, 1, 2, 3)$, one can define X by an equation of degree 6:

$$w^2 + z^3 + z f_4(x, y) + f_6(x, y) = 0,$$

where the f_i are homogeneous polynomials of degree i. The linear system $|-2K_X|$ defines a finite morphism

$$\varphi : X \xrightarrow[R_Q]{2:1} Q \subset \mathbb{P}^3$$

of degree 2 where Q is a non-degenerate quadric cone, $R_Q = R|_Q$ is the ramification divisor, and $R \subset \mathbb{P}^3$ is a cubic which does not pass through the vertex of Q.

If the degree is 2, then $|-K_X|$ is base point free and $|-2K_X|$ is very ample and embeds X into $\mathbb{P}^6$ as a surface of degree 8. The surface X is defined by an equation of degree 4:

$$w^2 + f_4(x, y, z) = 0,$$

where f_4 is homogeneous of degree 4 and $[x, y, z, w]$ are coordinates of weights $(1, 1, 1, 2)$ in $\mathbb{P}(1, 1, 1, 2)$. Finally, $|-K_X|$ defines a finite morphism of degree 2

$$\varphi : X \xrightarrow[R]{2:1} \mathbb{P}^2,$$

ramified over a curve $R \subset \mathbb{P}^2$ with $\deg_{\mathbb{P}^2} R = 4$.

It is clear that models of fibrations in del Pezzo surfaces over $\mathbb{P}^1$ can be obtained as the relative versions of the constructions introduced above.

2.2 Models of fibrations in del Pezzo surfaces of degree 1

Let $\rho : V \to \mathbb{P}^1$ be a non-singular fibration in del Pezzo surfaces of degree 1. Consider a sheaf of graded $\mathcal{O}_{\mathbb{P}^1}$-algebras

$$\mathcal{A}_V = \bigoplus_{i \geq 0} \rho_* \mathcal{O}_V(-iK_V).$$

Since $(-K_V)$ is ρ-ample, then clearly

$$V = \mathbf{Proj}_{\mathbb{P}^1} \mathcal{A}_V.$$

Comparing this situation with Proposition 2.1, we see easily that there exists an algebra $\mathcal{O}_{\mathbb{P}^1}[x, y, z, w]$ of polynomials over $\mathcal{O}_{\mathbb{P}^1}$ graded by the weights $(1, 1, 2, 3)$, and a sheaf of principal ideals $\mathcal{I}_V$ in it such that

$$V = \mathbf{Proj}_{\mathbb{P}^1} \mathcal{A}_V = \mathbf{Proj}_{\mathbb{P}^1} \mathcal{O}_{\mathbb{P}^1}[x, y, z, w] / \mathcal{I}_V \subset \mathbb{P}_{\mathbb{P}^1}(1, 1, 2, 3),$$

where $\mathbb{P}_{\mathbb{P}^1}(1, 1, 2, 3) = \mathbf{Proj}_{\mathbb{P}^1} \mathcal{O}_{\mathbb{P}^1}[x, y, z, w]$. Thus, $V/\mathbb{P}^1$ is defined by a (weighted) homogeneous polynomial (with coefficients in $\mathcal{O}_{\mathbb{P}^1}$) in the corresponding weighted projective space over $\mathbb{P}^1$. Note that $\mathbb{P}_{\mathbb{P}^1}(1, 1, 2, 3)$ has two distinguished sections along which it is singular, and V does not intersect these sections.

Now let us construct the second model of $V/\mathbb{P}^1$. All details and proofs can be found in [9, Section 2].

In what follows, we assume that

$$Pic(V) = \mathbb{Z}[-K_V] \oplus \mathbb{Z}[F],$$

where F denotes the class of a fibre. Notice that $V/\mathbb{P}^1$ has a distinguished section s_b that intersects each fibre at the base point of the anticanonical linear system. We can also define it as

$$s_b = \mathrm{Bas}\, | - K_V + lF|$$

for all $l \gg 0$. Then, $\rho_*(-2K_V + mF)$ is a vector bundle of rank 4 over $\mathbb{P}^1$. We can choose m such that

$$\mathcal{E} = \rho_*(-2K_V + mF) \simeq \mathcal{O} \oplus \mathcal{O}(n_1) \oplus \mathcal{O}(n_2) \oplus \mathcal{O}(n_3)$$

for some $0 \leq n_1 \leq n_2 \leq n_3$. Set

$$b = n_1 + n_2 + n_3.$$

Let $X = \mathbf{Proj}\, \mathcal{E}$ be the corresponding $\mathbb{P}^3$-fibration over $\mathbb{P}^1$, and let $\pi : X \to \mathbb{P}^1$ be the natural projection. Let M be the class of the tautological bundle on X (i.e. $\pi_* \mathcal{O}(M) = \mathcal{E}$), L the class of a fibre of π, t_0 the class of a section that corresponds to the surjection $\mathcal{E} \to \mathcal{O} \to 0$, and l the class of a line in a fibre of π. Note that t_0 is an effective irreducible section of π minimally twisted over the base. Then t_0 can be obtained from the conditions $M \circ t_0 = 0$, $L \circ t_0 = 1$. Thus, we have described all generators of the groups of 1- and 3-dimensional cycles on X.

Again, looking at Proposition 2.1, we see that there exists a threefold $Q \subset X$ fibred into quadratic cones (without degenerations) with a section t_b as the line of the cone vertices, together with a divisor R fibred into cubic surfaces and such that the restriction $R_Q = R|_Q$ does not intersect t_b, and a finite morphism $\varphi : V \to Q$ of degree 2 ramified along R_Q, such that the diagram

$$
\begin{array}{ccc}
V & \xrightarrow[\;R_Q,\,2:1\;]{\varphi} & Q \subset X \\[4pt]
\rho \downarrow & & \downarrow \pi = \pi|_Q \\[4pt]
\mathbb{P}^1 & =\!\!=\!\!= & \mathbb{P}^1
\end{array}
$$

is commutative. To make this construction precise, assume that $t_b \sim t_0 + \varepsilon l$. Obviously, ε is a non-negative integer.

Lemma 2.2 ([9, Lemma 2.2]). *Only the following cases are possible:*

1) $\varepsilon = 0$, *i.e.* $t_b = t_0$, *and then* $2n_2 = n_1 + n_3$, n_1 *and* n_3 *are even,* $Q \sim 2M - 2n_2L$, *and* $R \sim 3M$;
2) $\varepsilon = n_1 > 0$, *and then* $n_3 = 2n_2$, n_1 *is even,* $n_2 \geq 3n_1$, $Q \sim 2M - 2n_2L$, *and* $R \sim 3M - 3n_1L$.

This lemma shows that the numbers n_1, n_2, and n_3 completely define $V/\mathbb{P}^1$ (of course, up to moduli). Note that these numbers are not free from relations.

By the way, we can always assume that $b > 0$, i.e. $n_3 > 0$. Otherwise $Q \sim 2M$, $\varepsilon = 0$, and V is the direct product of $\mathbb{P}^1$ and a del Pezzo surface. This V is rational but it is not a Mori fibration, because $\rho(V/\mathbb{P}^1)$ is the Picard number of the del Pezzo surface and in this case is greater than 1.

It only remains to consider some formulae and relations which permit us to identify $V/\mathbb{P}^1$ more or less easily. First, we remark that a surface $G \sim (M - n_2L) \circ (M - n_3L)$ must lie in Q and is a ruled surface, minimally twisted over the base. Denote by $G_V = \varphi^*(G)$ its pre-image on V. Geometrically, G_V is a minimally twisted fibration in curves of genus 1. Then t_0 lies always on Q, and the fibres of Q contain lines of the class l, so the classes $s_0 = \frac{1}{2}\varphi^*(t_0)$ and $f = \frac{1}{2}\varphi^*(l)$ are well defined.

Note that f is the class of the anticanonical curves in the fibres of V, and s_0 always has an effective representative in 1-cycles on V. Incidentally it is easy to see that the Mori cone is generated by s_0 and f:

$$\overline{\mathrm{NE}}(V) = \mathbb{R}_+[s_0] \oplus \mathbb{R}_+[f].$$

It is not very difficult to compute the normal bundle of s_b:

$$\mathcal{N}_{s_b|V} \simeq \mathcal{O}(-\tfrac{1}{2}n_1) \oplus \mathcal{O}(-\tfrac{1}{2}n_3), \text{ if } \varepsilon = 0,$$
$$\mathcal{N}_{s_b|V} \simeq \mathcal{O}(n_1 - \tfrac{1}{2}n_3) \oplus \mathcal{O}(n_1), \text{ if } \varepsilon = n_1 > 0.$$

Thus we see that $V/\mathbb{P}^1$ is completely defined (again, up to moduli) by $\mathcal{N}_{s_b|V}$.

The following table contains the essential information about divisors and intersection indices on V:

$\varepsilon = 0$	$\varepsilon = n_1 > 0$
$n_1 + n_3 = 2n_2$	$n_3 = 2n_2,\ n_2 \geq 3n_1$
$s_b \sim s_0$	$s_b \sim s_0 + \tfrac{1}{2}f$
$\mathcal{N}_{s_b|V} \simeq \mathcal{O}(-\tfrac{1}{2}n_1) \oplus \mathcal{O}(-\tfrac{1}{2}n_3)$	$\mathcal{N}_{s_b|V} \simeq \mathcal{O}(n_1 - \tfrac{1}{2}n_3) \oplus \mathcal{O}(n_1)$
$K_V = -G_V + (\tfrac{1}{2}n_1 - 2)F$	$K_V = -G_V - (\tfrac{1}{2}n_1 + 2)F$
$K_V^2 = s_0 + (4 - n_2)f$	$K_V^2 = s_0 + (4 + \tfrac{3}{2}n_1 - n_2)f$
$s_0 \circ G_V = -\tfrac{1}{2}n_3$	$s_0 \circ G_V = -\tfrac{1}{2}n_3$
$(-K_V)^3 = 0 - 2n_2$	$(-K_V)^3 = 6 + 2n_1 - 2n_2$

Remark 2.3. The arguments still work if we assume V to be only Gorenstein, not necessary non-singular, provided we assume all fibres to be reduced. Indeed, all that is needed is the statement of Proposition 2.1. In the Gorenstein case it is easy to see that all fibres of $V/\mathbb{P}^1$ are irreducible (and are reduced by hypothesis), and this is enough to give the conclusions of Proposition 2.1, even if V is singular. In particular, we have the same projective models as above, the same formulae, and so on.

2.3 Models of fibrations in del Pezzo surfaces of degree 2

Let $\rho : V \to \mathbb{P}^1$ be a non-singular fibration in del Pezzo surfaces of degree 2. As in the case $d = 1$, consider the direct image of the anticanonical algebra

$$\mathcal{A}_V = \bigoplus_{i \geq 0} \rho_* \mathcal{O}_V(-iK_V).$$

This is a sheaf of graded algebras over $\mathcal{O}_{\mathbb{P}^1}$, and $V = \mathbf{Proj}_{\mathbb{P}^1} \mathcal{A}_V$. As before, we see that there exist a polynomial algebra $\mathcal{O}_{\mathbb{P}^1}[x, y, z, w]$ graded with the weights $(1, 1, 1, 2)$, and a sheaf of principal ideals $\mathcal{I}_V$ such that

$$V = \mathbf{Proj}_{\mathbb{P}^1} \mathcal{A}_V = \mathbf{Proj}_{\mathbb{P}^1} \mathcal{O}_{\mathbb{P}^1}[x, y, z, w]/\mathcal{I}_V \subset \mathbb{P}_{\mathbb{P}^1}(1, 1, 1, 2).$$

Clearly, $\mathcal{I}_V$ is generated by a (weighted) homogeneous element of degree 4, and $V \subset \mathbb{P}_{\mathbb{P}^1}(1, 1, 1, 2)$ avoids a section along which $\mathbb{P}_{\mathbb{P}^1}(1, 1, 1, 2)$ is singular.

The second model is obtained by taking a double cover. We suppose that $Pic(V) = \mathbb{Z}[-K_V] \oplus \mathbb{Z}[F]$. Consider

$$\mathcal{E} = \rho_* \mathcal{O}_V(-K_V + mF),$$

this is a vector bundle of rank 3 over $\mathbb{P}^1$. We may choose m such that

$$\mathcal{E} \simeq \mathcal{O} \oplus \mathcal{O}(n_1) \oplus \mathcal{O}(n_2),$$

where $0 \leq n_1 \leq n_2$. Set $b = n_1 + n_2$.

We have the natural projection $\pi : X = \mathbf{Proj}_{\mathbb{P}^1} \mathcal{E} \to \mathbb{P}^1$. Let M denote the class of the tautological bundle, let $\pi_* \mathcal{O}(M) = \mathcal{E}$, and let L be the class of a fibre of π, l the class of a line in a fibre, and t_0 the section corresponding to the surjection $\mathcal{E} \to \mathcal{O} \to 0$. Note that $t_0 \circ M = 0$. The diagram

$$
\begin{array}{ccc}
V & \xrightarrow{\ \varphi\ } & X \\
& {\scriptstyle R, 2:1} & \\
{\scriptstyle \rho} \downarrow & & \downarrow {\scriptstyle \pi} \\
\mathbb{P}^1 & =\!=\!= & \mathbb{P}^1
\end{array}
$$

is commutative. Here φ is a finite morphism of degree 2 ramified along a divisor $R \subset X$. The divisor R is fibred into quartic curves over $\mathbb{P}^1$, and we may suppose that

$$R \sim 4M + 2aL.$$

The set (n_1, n_2, a) defines $V/\mathbb{P}^1$, up to moduli: different sets correspond to different varieties. While n_1 and n_2 can be arbitrary, a is bounded from below; in fact, for R to exist, a cannot be much less than n_1.

We set $H = \varphi^*(M)$, $s_0 = \frac{1}{2}\varphi^*(t_0)$, $f = \frac{1}{2}\varphi^*(l)$; clearly, $F = \varphi^*(L)$. As in the case $d = 1$, the divisor

$$G_V \sim H - n_2 F = (-K_V) + (a + n_1 - 2)F$$

plays an important role in the geometry of V. This is a minimally twisted fibration in curves of genus 1 in V. Then it is clear that

$$\overline{\mathbf{NE}}(V) = \mathbb{R}_+[s_0] \oplus \mathbb{R}_+[f],$$

but now s_0 may not correspond to an effective 1-cycle on V. The table below summarizes the essential formulae on V (see [9, Section 3.1]):

$K_V = -H + (a + b - 2)F$
$K_V^2 = 2s_0 + (8 - 4a - 2b)f$
$(-K_V)^3 = 12 - 6a - 4b$

Remark 2.4. Again we see that all arguments and formulae work in the Gorenstein, possibly singular, case if we keep the assumption that fibres are reduced. The models just described both remain valid. The only difference with the case $d = 1$ is that Gorenstein singular $V/\mathbb{P}^1$ may have reducible fibres. Such fibres arise from a double covering of a plane with a double conic as the ramification divisor.

3 Fibre-to-fibre transformations

This section is devoted to describing some special transformations of del Pezzo fibrations. The birational classification problem is equivalent to finding the set of all Mori structures. In other words, we study birational maps modulo birational transformations over the base. From the viewpoint of the Sarkisov program, we only need links of type I, III, and IV, since these change the Mori structure. At first sight, we don't need links of type II, which don't change the structure of a fibration. Nevertheless, this type of Sarkisov link is especially important in the majority of questions related to birational classification problems, even if we are only interested in studying different Mori structures. The study of birational automorphisms is a typical question that involves fibre-to-fibre transformations (i.e. links of type II). Thus it is worth knowing as much as possible about these transformations, and this section gives some information about them.

Let V/C be a Mori fibration in del Pezzo surfaces of degree 1 or 2 over a curve C, and $\chi : V \dashrightarrow U$ a birational map onto another Mori fibration U/C that is birational over the base, i.e. we have the commutative diagram

$$
\begin{array}{ccc}
V & \overset{\chi}{\dashrightarrow} & U \\
\downarrow & & \downarrow \\
C & \overset{\simeq}{\longrightarrow} & C
\end{array} .
$$

Taking the specialization at the generic point η of C, we see that χ induces a birational map

$$
\chi_\eta : V_\eta \dashrightarrow U_\eta.
$$

It follows from [16, Theorem 2.6] that χ_η is an isomorphism if $d = K_{V_\eta}^2 = 1$, and it can be decomposed into Bertini involutions if $d = 2$. Let us consider the latter case in detail.

Bertini involutions can be constructed as follows. Let $A \in V_\eta$ be a rational point defined over the field of functions of C. There exists a morphism φ of degree 2 onto a plane (Proposition 2.1) branched over a (non-singular) quartic curve. Clearly, φ defines an involution $\tau \in \mathrm{Aut}(V)$ of V_η that transposes the sheets of the cover. Suppose $\varphi(A)$ does not lie on the ramification divisor. Then there exists a point $A^* \in V_\eta$ which is conjugate to A by τ, i. e. $\tau(A) = A^*$. Clearly, $A \neq A^*$. In the plane, take a pencil of lines through $\varphi(A)$. The inverse image in V_η of this pencil is a pencil of elliptic curves (with degenerations) having exactly two base points, A and A^*. Let us blow up these points. Let e and e^* be the exceptional divisors lying over A and A^* respectively. We obtain an elliptic surface S with 2 distinguished sections e and e^*. So we have two (biregular) involutions μ_A and μ_{A^*} on S defined by these sections. Indeed, the specialization of S at the generic point of e (or e^*, which is the same) gives us an elliptic curve with 2 points O and O^* that correspond to e and e^*. Each of these points can be viewed as a zero element of the group law on the elliptic curve. Thus, we get two reflection μ_A' and μ_{A^*}' defined as follows: for any point B, we have

$$
B + \mu_A'(B) \sim 2O^*,
$$
$$
B + \mu_{A^*}'(B) \sim 2O.
$$

(μ_A' is related to O^* and μ_{A^*}' to O). These reflections give the biregular involutions μ_A and μ_{A^*} on S (remark that S is relatively minimal, thus any fibre-wise birational map is actually biregular). Finally, we blow down e and e^*, and then μ_A and μ_{A^*} become the desired Bertini involutions. Another way to describe μ_A is the following. Blow up A^*, and you get a del Pezzo surface of degree 1. Its natural involution defined by the double cover of a cone (see Proposition 2.1) becomes birational on V_η, and this is exactly our μ_A.

Applying Theorem 2.6 from [16] to the case $d = 2$, we get that V_η is biregularly isomorphic to U_η. So we can consider χ_η as a birational automorphism of V_η, and then there exists a finite set of points $I = \{A_1, A_2, \ldots, A_n\}$ on V_η such that

$$\chi_\eta = \mu_{A_1} \circ \mu_{A_2} \circ \ldots \circ \mu_{A_n} \circ \psi,$$

where $\psi \in \mathrm{Aut}(V_\eta)$. Now ψ defines a fibre-wise birational transformation of $V/\mathbb{P}^1$. Applying it, we get a situation similar to degree 1 when we have an isomorphism of the fibres over the generic point.

Thus the cases $d = 1$ and $d = 2$, can both be reduced to the following: we have the commutative diagram

$$
\begin{array}{ccc}
V & \xrightarrow{\chi} & U \\
\downarrow & & \downarrow \\
C & \xrightarrow{\simeq} & C
\end{array}
\tag{3.1}
$$

where χ_η is an isomorphism of V_η and U_η:

$$\chi_\eta : V_\eta \xrightarrow{\simeq} U_\eta.$$

Now it is clear that if we throw out a finite number of points of C, say, $P_1, P_2, \ldots, P_k$, then χ gives an isomorphism of V and U over $C \setminus \{P_1, P_2, \ldots, P_k\}$. It follows that χ can be decomposed as

$$
\begin{array}{ccccccccc}
V & \xrightarrow{\chi_1} & V_1 & \xrightarrow{\chi_2} & V_2 & \xrightarrow{\chi_3} & \cdots & \xrightarrow{\chi_{k-1}} & V_{k-1} & \xrightarrow{\chi_k} & U \\
\downarrow & & \downarrow & & \downarrow & & & & \downarrow & & \downarrow \\
C & \xrightarrow{\simeq} & C & \xrightarrow{\simeq} & C & \xrightarrow{\simeq} & \cdots & \xrightarrow{\simeq} & C & \xrightarrow{\simeq} & C
\end{array}
$$

where χ_i is an isomorphism over $C \setminus \{P_i\}$. In order to distinguish such birational transformations from transformations like Bertini involutions, we will call them *fibre transformations*. It is natural to ask whether they are possible. To clarify the question, let us consider the 2-dimensional situation. First, for a ruled surface, we can take any elementary transformation of a fibre as an example of a fibre transformation: blow up a point in a fibre and then contract the strict transform of the fibre, which is a (-1)-curve. Remark that the original fibre will be replaced by the exceptional divisor. This transformation is not an isomorphism, though it gives an isomorphism of the fibres over the generic point of the base. On the other hand, consider any two relatively minimal elliptic surfaces which are birationally isomorphic over the base. It is well-known that any such birational map is actually an isomorphism. This is true for any relatively minimal fibration (i.e. no (-1)-curves in fibres) into curves of positive genus.

What about del Pezzo fibrations? They are fibrations by rational surfaces, and we may expect that they behave like ruled surfaces. On the other hand, it will follow from results of the next two sections, that their essential birational properties are different from those of rational varieties. In some sense they behave like elliptic fibrations. The exact picture is complicated.

The last part of this section is contained in Section 4 of [11] (the case $d = 1$) and in [10] (the case $d = 2$).

Let us first consider the case $d = 1$. Thus we have the commutative diagram (3.1) where χ is an isomorphism of the fibres over the generic points. This situation can be easily reduced to the following case. We assume C to be a germ of a curve with central point O, and V and U to be Gorenstein relatively projective varieties over C fibred into del Pezzo surfaces of degree 1; assume that V_0 and U_0, their central fibres (over the point O), are reduced. Algebraically, this means the following. Let $\mathcal{O}$ be a DVR (discrete valuation ring) with maximal ideal $\mathfrak{m} = (t)\mathcal{O}$, where t is a local parameter. If necessary, we may take the completion of $\mathcal{O}$. If $C = \mathbf{Spec}\,\mathcal{O}$ then V and U are non-singular del Pezzo surfaces of degree 1 defined over $\mathcal{O}$. Let K be the field of quotients of $\mathcal{O}$, and let $\eta = \mathbf{Spec}\,K$ be the generic point of C. Using Proposition 2.1, we may suppose that V and U are embedded respectively into copies P and R of the weighted projective space $\mathbb{P}_{\mathcal{O}}(1, 1, 2, 3)$. Denote $[x, y, z, w]$ and $[p, q, r, s]$ the coordinates in P and R with the weights $(1,1,2,3)$.

By our condition, $\chi_\eta : V_\eta \to U_\eta$ is an isomorphism. The key point is that χ_η induces an isomorphism between P_η and R_η. This easily follows from the Kodaira vanishing theorem and the exact sequences of restriction for the ideals that define V_η and U_η as surfaces in P_η and R_η respectively (see Section 4.1 in [11]).

Then we may choose coordinates in P and R in such a way that

$$
\begin{aligned}
V &= \{w^2 + z^3 + zf_4(x, y) + f_6(x, y) = 0\} \subset P, \\
U &= \{s^2 + r^3 + rg_4(p, q) + g_6(p, q) = 0\} \subset R,
\end{aligned}
\tag{3.2}
$$

where f_i and g_i are homogeneous polynomials of degree i. Since χ_η is an isomorphism of P_η and R_η, it is easy to see that χ and χ^{-1} can be defined as follows:

$$\chi = \left\{ \begin{array}{ccl} p & = & t^a x \\ q & = & t^b y \\ r & = & t^c z \\ s & = & t^d w \end{array} \right\}, \quad \chi^{-1} = \left\{ \begin{array}{ccl} x & = & t^\alpha p \\ y & = & t^\beta q \\ z & = & t^\gamma r \\ w & = & t^\delta s \end{array} \right\},$$

where each of the sets (a, b, c, d) and $(\alpha, \beta, \gamma, \delta)$ contains at least one zero. All these numbers have to respect the grading of P and R, and, moreover, we know that V and U avoid singular points of P and R. It follows that, for some integer $m > 0$, the conditions

$$\left\{ \begin{array}{rcl} a + \alpha & = & m \\ b + \beta & = & m \\ c + \gamma & = & 2m \\ d + \delta & = & 3m \\ 2d & = & 3c \\ 2\delta & = & 3\gamma \end{array} \right.$$

are satisfied. Using the symmetries of this situation, we may assume that $c = 2k$, $d = 3k$, $\gamma = 2l$, $\delta = 3l$ with the conditions $k + l = m$ and $k \leq l$ (i.e. $k \leq \frac{1}{2}m$). Now substituting these relations for the numbers in (3.2), we obtain

$$\begin{array}{l} f_4(x, y) = t^{-4k} g_4(t^a x, t^b y), \\ f_6(x, y) = t^{-6k} g_6(t^a x, t^b y). \end{array} \tag{3.3}$$

Suppose $k = 0$. Since the set (a, b, c, d) does not consist entirely of zeros (otherwise χ is already an isomorphism), one of the numbers a, b must be positive. Let it be a. Then (3.3) shows that the equation

$$w^2 + z^3 + z f_4(x, y) + f_6(x, y) = 0$$

defines a singularity of V at the point $(t, x, y, z) = (0, 1, 0, 0, 0)$: take an affine piece $x \neq 0$ (i.e. divide the equation by x^6) and check that the differentials vanish at the indicated point.

Now let $k > 0$. Since the set (a, b, c, d) has to contain at least one zero, we may assume $a = 0$. Then $\alpha = m$. Since $l \geq k > 0$, $\alpha = m > 0$, and the set $(\alpha, \beta, \gamma, \delta)$ contains zero, we must suppose $\beta = 0$. So $b = m - \beta = m$. Thus

$$\begin{array}{l} f_4(x, y) = t^{-4k} g_4(x, t^m y), \\ f_6(x, y) = t^{-6k} g_6(x, t^m y). \end{array}$$

It only remains to take into account that $k \leq \frac{1}{2}m$ and $f_i \in \mathcal{O}[x, y, z, w]$, and we see that the equation for V defines a singular point at

$$(t, x, y, z, w) = (0, 0, 1, 0, 0).$$

So we get the following assertion: V must be singular if χ is not an isomorphism!

For the case $d = 2$ (see [10]), we can repeat the above arguments with the only difference that P and R have the type $\mathbb{P}_{\mathcal{O}}(1, 1, 1, 2)$, and V and U are defined by

$$V = \{w^2 + f_4(x, y, z) = 0\} \subset P,$$
$$U = \{s^2 + g_4(p, q, r) = 0\} \subset R.$$

Then χ and χ^{-1} have the form

$$\chi = \left\{ \begin{array}{ccc} p & = & t^a x \\ q & = & t^b y \\ r & = & t^c z \\ s & = & t^d w \end{array} \right\}, \quad \chi^{-1} = \left\{ \begin{array}{ccc} x & = & t^\alpha p \\ y & = & t^\beta q \\ z & = & t^\gamma r \\ w & = & t^\delta s \end{array} \right\},$$

with the relations

$$m = a + \alpha = b + \beta = c + \gamma = \frac{1}{2}(d + \delta)$$

for some $m > 0$. Again, (a, b, c, d) and $(\alpha, \beta, \gamma, \delta)$ do not contain only zeros. By symmetry, we may assume that $\gamma = 0$ (thus $c = m$) and $d \geq \frac{1}{2}m$. Finally, the relation

$$f_4(x, y, z) = t^{-2d} g_4(t^a x, t^b y, t^m z) \in \mathcal{O}[x, y, z]$$

shows that the equation

$$w^2 + f_4(x, y, z) = 0$$

defines a singular point at

$$(t, x, y, z, w) = (0, 0, 0, 1, 0),$$

and V is always singular if χ is not an isomorphism.

These results can be summarized as follows.

Theorem 3.1 (Uniqueness of a smooth model). *Let V/C and U/C be non-singular Mori fibrations in del Pezzo surfaces of degree 1 or 2 that are birational by χ over the base*

$$\begin{array}{ccc} V & \overset{\chi}{\dashrightarrow} & U \\ \downarrow & & \downarrow \\ C & \overset{\simeq}{\longrightarrow} & C \end{array}$$

Then V and U are isomorphic. If $d = 1$ then χ is an isomorphism. If $d = 2$ then χ is either an isomorphism or a composition of Bertini involutions. In other words, any fibre transformation is trivial.

Remark 3.2. A similar result was proved for $d \leq 4$ by Park in [25] using Shokurov's complement and connectedness principles, and under the additional assumption that V and U have non-degenerate central fibres.

Thus the reader can see that though V/C is fibred into rational surfaces it behaves like an elliptic fibration in dimension 2: there are no non-trivial fibre transformations without loss of smoothness. Now it is time to give some examples.

Example 3.3 ('smooth case'). In these examples U is non-singular.
First let $d = 1$. Suppose $(a, b, c, d) = (0, 6, 2, 3)$ and $(\alpha, \beta, \gamma, \delta) = (6, 0, 10, 15)$. Then V and U are defined by

$$V : w^2 + z^3 + x^5 y + t^{24} x y^5 = 0, \quad U : s^2 + r^3 + p^5 q + p q^5 = 0.$$

It is easy to check that U is non-singular, V has a singular point of type cE_8 (the so-called compound E_8-singularity) at $(t, x, y, z, w) = (0, 0, 1, 0, 0)$. Note that the central fibre V_0 has a unique singular point of type E_8, and U_0 is non-singular.
Now let $d = 2$: suppose $(a, b, c, d) = (1, 4, 0, 2)$, $(\alpha, \beta, \gamma, \delta) = (3, 0, 4, 6)$. Then V and U are given by

$$V : w^2 + y z^3 + t x^4 + t^{12} y^4 = 0, \quad U : s^2 + q r^3 + t p^4 + q^4 = 0$$

Again, U is non-singular, V has a cE_8-singularity at $(t, x, y, z, w) = (0, 0, 1, 0, 0)$. The central fibre V_0 is a non-normal del Pezzo surface (the double cover of a cone branched over a triple plane section), and U_0 has an elliptic singularity.

Example 3.4 ('birational automorphism'). For the case $d = 1$, consider $(a, b, c, d) = (1, 0, 2, 3)$, $(\alpha, \beta, \gamma, \delta) = (0, 2, 2, 3)$, and

$$V : w^2 + z^3 + t^4 x^5 y + x y^5 = 0, \quad U : s^2 + r^3 + p^5 q + t^4 p q^5 = 0.$$

V and U have cE_8 singularities in the central fibres. Note that V and U are biregularly isomorphic: put $w = s$, $z = r$, $x = q$, and $y = p$. So we can assume χ to be defined as follows:

$$x \to t^{-1} y, \quad y \to t x, \quad z \to z, \quad w \to w.$$

Thus, χ gives an example of a fibre transformation that is a birational automorphism.

For $d = 2$: suppose $(a, b, c, d) = (1, 2, 0, 2)$, $(\alpha, \beta, \gamma, \delta) = (1, 0, 2, 2)$. Then V and U are given by

$$V : w^2 + t^2 y^3 z + yz^3 + x^4 = 0, \quad U : s^2 + q^3 r + t^2 qr^3 + p^4 = 0.$$

V and U have cD_4-singularities. As before they become isomorphic if you put $p = x$, $q = y$, $r = z$, $s = w$. So again χ is a fibre transformation that is a birational automorphism.

4 Mori structures on del Pezzo fibrations: the case $d = 1$

In this section we formulate known results on the rigidity of fibrations in del Pezzo surfaces of degree 1 and describe Mori structures for non-rigid cases. We use the projective model via double coverings described in Section 2.2.

Theorem 4.1 ([9, Theorem 2.6]). *Let $V/\mathbb{P}^1$ be a non-singular Mori fibration in del Pezzo surfaces of degree 1. Then $V/\mathbb{P}^1$ is birationally rigid except in two cases:*

1) $\varepsilon = 0$, $n_1 = n_2 = n_3 = 2$;
2) $\varepsilon = 0$, $n_1 = 0$, $n_2 = 1$, $n_3 = 2$.

In other words, Conjecture 1.5 holds for $d = 1$. Moreover, if $V/\mathbb{P}^1$ is rigid, then this is the unique non-singular Mori fibration in its class of birational equivalence, by Theorem 3.1.

Remark 4.2. It is easy to check that it is only in cases 1) and 2) that the linear system $|(-K_V) - F|$ is non-empty and free from base components (and hence the linear systems $|n(-K_V) - F|$ have the same properties for all n). The situation when $V/\mathbb{P}^1$ satisfies the K^2-condition was first proved in [26].

There are only two non-rigid cases. Let us consider them in detail.

4.1 The case $(\varepsilon, n_1, n_2, n_3) = (0, 2, 2, 2)$

First let us remark that the distinguished section s_b has the class s_0, and this is the unique section with this class. We have $\mathcal{N}_{s_b|V} \sim \mathcal{O}(-1) \oplus \mathcal{O}(-1)$, and, at least locally, there exists a flop centred at s_b. This flop exists in the projective category: it is enough to check that the linear system $n(-K_V)$ gives a birational morphism that contracts exactly s_b. Let $\psi : V \dashrightarrow U$ be such a flop. Consider the linear system $\mathcal{D} = |-K_V - F|$ and its strict transform $\mathcal{D}_U = \psi_*^{-1}\mathcal{D}$ on U. It is easy to see that $\text{Bas}\,\mathcal{D} = s_b$ and a general member of $\mathcal{D}$ is a del Pezzo surface of degree 1 blown up at the base point of its the anticanonical linear system. Moreover, $\dim \mathcal{D} = 1$, $\text{Bas}\,\mathcal{D}_U = \emptyset$, and $\mathcal{D}_U$ is a pencil of del Pezzo surfaces of degree 1. Thus, U is fibred over $\mathbb{P}^1$ by del Pezzo surfaces of degree 1. It follows that

$$
\begin{array}{ccc}
V & \overset{\psi}{\dashrightarrow} & U \\
\downarrow & & \downarrow \\
\mathbb{P}^1 & & \mathbb{P}^1
\end{array}
$$

is a link of type IV. It only remains to compute that the projective model of $U/\mathbb{P}^1$ (as a double cover) has the same structure constants $(\varepsilon, n_1, n_2, n_3)$ as $V/\mathbb{P}^1$. We are ready to formulate the result.

Proposition 4.3 ([9, Proposition 2.12]). *Let $V/\mathbb{P}^1$ be a non-singular Mori fibration in del Pezzo surfaces of degree 1 with the set of structure constants $(\varepsilon, n_1, n_2, n_3) = (0, 2, 2, 2)$. Then any other Mori fibration which is birational to V is birational over the base either to $V/\mathbb{P}^1$ itself, or to $U/\mathbb{P}^1$. The fibrations $V/\mathbb{P}^1$ and $U/\mathbb{P}^1$ are the only non-singular Mori fibrations in their class of birational equivalence. In the general case, the group $\text{Bir}(V) = \text{Aut}(V) \simeq \mathbb{Z}_2$, and is generated by the involution corresponding to the double cover.*

Remark 4.4. It may happen that V and U are isomorphic to each other (note that they have the same set of structure constants). In this case, $\psi \in \text{Bir}(V)$ and $\text{Bir}(V) \simeq \mathbb{Z}_2 \oplus \mathbb{Z}_2$.

4.2 The case $(\varepsilon, n_1, n_2, n_3) = (0, 0, 1, 2)$, or the double cone over the Veronese surface

Let $T \subset \mathbb{P}^5$ be the Veronese surface (i.e. $\mathbb{P}^2$ embedded into $\mathbb{P}^5$ by the complete linear system of conics), $Q \subset \mathbb{P}^6$ a cone over T with the vertex P, and $R \subset \mathbb{P}^6$ a cubic hypersurface such that $P \notin R$ and $R_Q = R \cap Q$ is non-singular. Then there exists a degree 2 finite morphism $\mu : U \to Q$ branched over R_Q. The variety U (the so-called double cone over the Veronese surface) is a well-known Fano variety of index 2 with $\rho(U) = 1$ and $(-K_V)^3 = 8$ (e.g. see [19] for classification of Fano 3-folds). Then U contains a two-dimensional family $\mathcal{S}$ of elliptic curves parametrized by T. These curves lie over the rulings of Q. $\mathcal{S}$ has a one-dimensional sub-family of degenerations consisting of rational curves with either a node or a cusp.

Let $l \in S$ be non-singular and let $\psi_l : V_l \to U$ be the blow-up of l. Then it is easy to see that V_l is a non-singular Mori fibration in del Pezzo surfaces of degree 1 with $(\varepsilon, n_1, n_2, n_3) = (0, 0, 1, 2)$. Conversely, take a del Pezzo fibration $V/\mathbb{P}^1$ with this set of structure constants. The linear system $|-K_V - 2F|$ consists of one element G_V which is the direct product of $\mathbb{P}^1$ and an elliptic curve [9, Lemma 2.9]. Then the linear system $|3(-K_V) - 3F|$ defines a birational morphism $\psi : V \to U$ that contracts G_V along the rulings.

What happens if $l \in S$ is singular? First, suppose that l is a rational curve with an ordinary double point B. Let $\psi_1 : U_1 \to U$ be the blow-up of B and $E_1 \sim \mathbb{P}^2$ the exceptional divisor. Note that l^1 (the strict transform of l) intersects E_1 at two points, and denote by t_1 the line on E_1 that contains these points. Now blow up l^1: $\psi_2 : U_2 \to U_1$. It is easy to check that the strict transform t_1^2 of t_1 on U_2 has normal bundle isomorphic to $\mathcal{O}(-1) \oplus \mathcal{O}(-1)$, and we may produce a flop $\psi_3 : U_2 \dashrightarrow U_3$ centred at t_1^2, without loss of projectivity. The strict transform E_1^3 of E_1 on U_3 becomes isomorphic to a non-singular quadric surface with normal bundle $\mathcal{O}(-1)$, so E_1^3 can be contracted to an ordinary double point: $\psi_4 : U_3 \to U_4 = V_l$. We have the birational map

$$\psi_l = (\psi_4 \circ \psi_3 \circ \psi_2 \circ \psi_1)^{-1} : V_l \dashrightarrow U$$

and the reader can easily check that V_l is a Gorenstein Mori fibration (over $\mathbb{P}^1$) in del Pezzo surfaces of degree 1 with a unique ordinary double point, and it has the structure set $(\varepsilon, n_1, n_2, n_3) = (0, 0, 1, 2)$ (take into account that the constructions in Section 2 are valid in the Gorenstein case as well, see Remark 2.3).

The case when l has a cusp at a point $B \in U$ is similar to the previous one but it is a little bit more complicated. As before, let $\psi_1 : U_1 \to U$ be the blow-up of B. We see that l^1 becomes non-singular and tangent to E_1 at some point $B_1 \in E_1$. The tangent direction to l_1 at B_1 defines a line $t_1 \subset E_1$. Now take the blow-up $\psi_2 : U_2 \to U_1$ of the curve l^1. The strict transform E_1^2 of E_1 becomes isomorphic to a quadric cone that is blown up at a point outside of the vertex of the cone. Moreover, the strict transform t_1^2 is exactly the unique (-1)-curve on E_1^2, and it has normal bundle $\mathcal{O}(-1) \oplus \mathcal{O}(-1)$. All this can be checked as follows: blow up the point B_1 with exceptional divisor E_2, then blow up the strict transform of l_1. The strict transform of E_2 is isomorphic to an $\mathbb{F}_1$-surface and can be contracted along its ruling. After that you get U_2.

So, now we produce a flop $\psi_3 : U_2 \dashrightarrow U_3$ centred at l_1^2. The strict transform E_1^3 of E_1^2 is isomorphic to a quadratic cone with the normal bundle $\mathcal{O}(-1)$, hence it can be contracted $\psi_4 : U_3 \to U_4 = V_l$ to a double point locally defined by the equation $x^2 + y^2 + z^2 + w^3 = 0$. We see that V_l is a Gorenstein Mori fibration over $\mathbb{P}^1$ in del Pezzo surfaces of degree 1 with a unique singular point. Moreover, V_l has the same structure set $(\varepsilon, n_1, n_2, n_3) = (0, 0, 1, 2)$ as before. We remark that fibres of $V_l/\mathbb{P}^1$ appear from elements of the pencil $|\frac{1}{2}(-K_V) - l|$ on U.

About twenty years ago, S. Khashin tried to show the non-rationality of U by proving the uniqueness of the structure of a Fano variety on it (see [21]). Unfortunately his arguments contain some mistakes and, up to now, we have no reliable proof of the non-rationality of U. Nevertheless it seems the following conjecture is true:

Conjecture 4.5. *U has exactly the following Mori structures: U itself, and $V_l/\mathbb{P}^1$ for any $l \in S$ (thus, we have a 'two-dimensional family' of different del Pezzo fibrations). In particular, U is not rational (because it has no conic bundles).*

It only remains to prove this conjecture† and the birational identification problem for non-singular Mori fibrations in del Pezzo surfaces of degree 1 will be completed.

5 Mori structures on del Pezzo fibrations: the case $d = 2$

We first formulate a rigidity result and then describe the non-rigid cases. The structure constants (a, n_1, n_2) are taken from subsection 2.3. Recall that $b = n_1 + n_2$.

Theorem 5.1 ([9, 12]). *Let $V/\mathbb{P}^1$ be a non-singular Mori fibration in del Pezzo surfaces of degree 2. Then $b + 2a > 0$, and if $b + 2a > 2$, $V/\mathbb{P}^1$ is birationally rigid.*

Suppose that $b + 2a = 2$. Then only the following cases are possible:

1) $a = 0$, $n_1 = 0$, $n_2 = 2$;
2) $a = -2$, $n_1 = 2$, $n_2 = 4$;
3) $a = -3$, $n_1 = 2$, $n_2 = 6$;
4) $a = 1$, $n_1 = 0$, $n_2 = 0$;
5) $a = 0$, $n_1 = 1$, $n_2 = 1$;
6) $a = -1$, $n_1 = 2$, $n_2 = 2$.

† The conjecture has now been proved by the present author: see [13].

General varieties of cases 1)–3) *are rigid. The cases* 4)–6) *are all non-rigid.*

Suppose $b + 2a = 1$, *then only the following cases are possible:*

7) $a = 0$, $n_1 = 0$, $n_2 = 1$;
8) $a = -1$, $n_1 = 1$, $n_2 = 2$.

Both of them are non-rigid.

Thus under the assumption of generality for the cases 1)–3) *Conjecture* 1.5 *holds for* $d = 2$, *and if* $V/\mathbb{P}^1$ *is rigid, then this is the unique non-singular Mori fibration in its class of birational equivalence (Theorem* 3.1).

Remark 5.2. The K^2-condition corresponds to the case $b + 2a \geq 4$, and this result was first obtained by A. Pukhlikov [26]. The reader should not be confused by the generality assumptions in the cases 1)–3) (see [12]). They arise from some technical problems of the maximal singularities method. The author believes that they can be finally omitted.

5.1 The case $(a, n_1, n_2) = (1, 0, 0)$

In this case, $X \simeq \mathbb{P}^1 \times \mathbb{P}^2$, so V is a double cover of $\mathbb{P}^1 \times \mathbb{P}^2$ ramified along a divisor R of bi-degree $(2, 4)$. Thus V is a del Pezzo fibration with respect to the projection onto $\mathbb{P}^1$. On the other hand the projection onto $\mathbb{P}^2$ represents V as a conic bundle. Indeed, a fibre of $V \to \mathbb{P}^2$ is a double cover of a line branched over 2 points, hence it is either a conic, or a couple of lines, or a double line. It is easy to see that the discriminant curve of $V \to \mathbb{P}^2$ has degree 8. So V has at least two different Mori structures. Note that $\mathbb{P}^1 \leftarrow V \to \mathbb{P}^2$ is a trivial example of a type IV link.

5.2 The case $(a, n_1, n_2) = (0, 1, 1)$, or a double quadric cone

In this case the linear system $|-2K_V|$ gives a small contraction onto the canonical model of V, which can be realized as a double covering of a non-degenerate quadric cone $Q \subset \mathbb{P}^4$ branched along a quartic section R_Q. It is easy to see that V has at most two curves of the class s_0. If a curve of the class s_0 is unique on V (this means that R_Q contains the vertex of Q), then s_0 is the so-called -2-curve of width 2 (in the terminology of [28]). Otherwise (if R_Q does not pass through the vertex of Q), there are two curves of the class s_0, which are disjoint and each has normal bundle $\mathcal{O}(-1) \oplus \mathcal{O}(-1)$. In both cases we obtain another structure of non-singular Mori fibration in del Pezzo surfaces of degree 2 after making a flop centred at these curves. Moreover, the second structure has the same set (a, n_1, n_2). Note that these Mori structures arise from two families (actually, two pencils) of planes on Q.

If R_Q does not pass through the vertex of the cone, then a general variety of this type has exactly the two Mori structures just described [8]. Actually, this should be always true, not only for general cases, and even when R_Q passes through the vertex of the cone, but a proof of this fact seems inaccessible at present.

5.3 The case $(a, n_1, n_2) = (-1, 2, 3)$

First, let us note that V has a unique curve of the class s_0 (say, C), which can be contracted (by a small contraction) using the linear system $|nH|$ for $n > 2$. The normal bundle of C is isomorphic to $\mathcal{O}(-1) \oplus \mathcal{O}(-2)$, hence there exists an anti-flip $\psi : V \dashrightarrow U$ centred at C (this, known as the Francia anti-flip, is the simplest example of an anti-flip: blow up C, make a flop centred at the minimal section of the corresponding ruled surface, and then contract the strict transform of the exceptional divisor, which is $\mathbb{P}^2$ with normal bundle $\mathcal{O}(-2)$). U has a unique (non-Gorenstein) singular point of index 2 (the latter means that $2K_U$ is a Cartier divisor). Moreover, it turns out that U is a Mori fibration over $\mathbb{P}^1$ in del Pezzo surfaces of degree 1. It is not very difficult to check that general elements of the pencil $|-K_V - F|$ are del Pezzo surfaces of degree 1 that are blown up at the base point of the anticanonical linear system, and C is exactly their common (-1)-curve. Thus via ψ they are the fibres of $U/\mathbb{P}^1$. Conjecturally, $V/\mathbb{P}^1$ and $U/\mathbb{P}^1$ are the only Mori structures in this case.

5.4 The case $(a, n_1, n_2) = (0, 0, 1)$, or a double space of index 2

This case is similar to a double cone over the Veronese surface. Let U be a double cover of $\mathbb{P}^3$ branched over a smooth quartic. This is a well known Fano variety of index 2, the so-called double space of index 2. Denote by H_U the generator of the Picard group, $\operatorname{Pic}(U) = \mathbb{Z}[H_U]$. Clearly, $K_U \sim -2H_U$. Let $l \subset U$ be a curve of genus 1 and degree 2 (i.e. $H \circ l = 2$), so l is a double cover of a line in $\mathbb{P}^3$ branched at four points. Then l is either a (non-singular) elliptic curve, or a rational curve with a double point (node or cusp), or a couple of lines with two points of intersection (the points may coincide).

Suppose l is non-singular, and $\psi_l : V_l \to U$ is the blow-up of l. Then V_l is a non-singular Mori fibration in del Pezzo surfaces of degree 2. The fibres of $V_l/\mathbb{P}^1$ arise from a pencil of planes in $\mathbb{P}^3$ that contain the image of l. It is not very difficult to compute that $V_l/\mathbb{P}^1$ has the structure set $(a, n_1, n_2) = (0, 0, 1)$. Conversely, given $V/\mathbb{P}^1$ with such a structure set, we can contract a unique divisor in $|-K_V - 2F|$, which is isomorphic to the direct product of an elliptic curve and a line, and get a double space of index 2.

If l is singular but irreducible, i.e. a rational curve with a node or a cusp, we can also obtain a Gorenstein Mori fibration in del Pezzo surfaces of degree 2, but now with a singular point of type respectively $x^2 + y^2 + z^2 + w^2 = 0$ or $x^2 + y^2 + z^2 + w^3 = 0$, by 'blowing up' the curve l in the same way as we obtained fibred structures from a double cone over the Veronese surface (see the previous section). The $V_l/\mathbb{P}^1$ has the same set of structure constants: $(a, n_1, n_2) = (0, 0, 1)$.

However, U has another type of Mori fibration. Suppose l is a couple of lines (say, $l = l_1 \cup l_2$) with two different points of intersection. Blow up one of these lines and make a flop centred at the strict transform of the second line, and you obtain a structure of a (non-singular) Mori fibration in cubic surfaces $V_{l_1}/\mathbb{P}^1$. As before, its fibres arise from a pencil of planes that contain the image of any of these lines in $\mathbb{P}^3$. The same construction works if l consists of two lines that are tangent to each other. Note that we can change the order of the lines (first blowing up l_2) and get $V_{l_2}/\mathbb{P}^1$. The point is that $V_{l_1}/\mathbb{P}^1$ is (biregularly) isomorphic to $V_{l_2}/\mathbb{P}^1$ over the base, and this change of order corresponds to a birational automorphism (the so-called Geiser involution, see [16]). So we may denote these fibration in cubic surfaces as $V_l/\mathbb{P}^1$ without any confusion.

It is not clear whether there are Mori structures different from $V_l/\mathbb{P}^1$ and U itself†. We know that U has a large group of birational automorphisms [20].

It remains to mention in this subsection that it is known that U is not rational. This follows from [33].

---

† At present new structures are found, see [14].

5.5 The case $(a, n_1, n_2) = (-1, 1, 2)$, or a singular double cone over the Veronese surface

Let us first remark about this case, that the linear system $|H - 2F|$ has a unique representative, which we denote by G_V. There is a unique curve of the class s_0 on V, and this curve has normal bundle $\mathcal{O}(-1) \oplus \mathcal{O}(-1)$. So we produce a flop (in the projective category) $\psi_1 : V \dashrightarrow V^+$. The strict transform G_V^+ of G_V on V^+ is isomorphic to either $\mathbb{P}^1 \times \mathbb{P}^1$ or a quadric cone, and it has normal bundle $\mathcal{O}(-1)$ in both these cases. Thus, there exists a contraction $\psi_2 : V^+ \to U$, which gives a Fano variety with a double point of type $x^2 + y^2 + z^2 + w^2 = 0$ or $x^2 + y^2 + z^2 + w^3 = 0$ respectively. It is easy to check that this U can be obtained as a double cover of the cone over the Veronese surface branched over a cubic section that does not pass through the vertex of the cone and has a unique du Val singular point of type A_1 or A_2. Conversely, given U with a singularity of this type, we can always obtain a structure of del Pezzo fibration, as indicated. Now it is clear that U can be transformed to a (singular Gorenstein) Mori fibration in del Pezzo surfaces of degree 1 as in the previous section. Conjecturally, these are all the Mori structures on U (i.e. one structure of Fano variety, one structure of fibration in del Pezzo surfaces of degree 2, and a '2-dimensional family' of structures of fibrations in del Pezzo surfaces of degree 1), and hence V is a unique non-singular Mori fibration in its birational equivalence class.

Bibliography

[1] V. Alexeev. General elephants of Q-Fano 3-folds. *Compositio Math.*, 91(1):91–116, 1994.

[2] M. Artin and D. Mumford. Some elementary examples of unirational varieties which are not rational. *Proc. London Math. Soc. (3)*, 25:75–95, 1972.

[3] F. Bardelli. Polarized mixed Hodge structure: on irrationality of three-folds via degeneration. *Annali di Math. pura e appl.*, 137:287–369, 1984.

[4] H. Clemens and P. Griffiths. The intermediate Jacobian of the cubic threefold. *Ann. of Math. (2)*, 95:281–356, 1972.

[5] J.-L. Colliot-Thélène. Arithmétique des variétés rationnelles et problèmes birationnels. In *Proceedings of the International Congress of Mathematicians, Vol. 1, 2 (Berkeley, Calif., 1986)*, pages 641–653, Providence, RI, 1987. Amer. Math. Soc.

[6] A. Corti. Factoring birational maps of threefolds after Sarkisov. *J. Algebraic Geom.*, 4(2):223–254, 1995.

[7] A. Corti. Del Pezzo surfaces over Dedekind schemes. *Ann. of Math. (2)*, 144(3):641–683, 1996.

[8] M. M. Grinenko. Birational automorphisms of a three-dimensional double cone. *Sbornik: Mathematics*, 189(7):991–1007, 1998.

[9] M. M. Grinenko. Birational properties of pencils of del Pezzo surfaces of degrees 1 and 2. *Sbornik: Mathematics*, 191(5):633–653, 2000.

[10] M. M. Grinenko. On fiber-to-fiber transformations of fibrations into del Pezzo surfaces of degree 2. *Russian Math. Surveys*, 56(4):753–754, 2001.

[11] M. M. Grinenko. On fibrations into del Pezzo surfaces. *Math. Notes*, 69(4):499–512, 2001.

[12] M. M. Grinenko. Birational properties of pencils of del Pezzo surfaces of degrees 1 and 2. Part II. *Sbornik: Mathematics*, 194(5–6):669–696, 2003.

[13] M. M. Grinenko. Mori structures on a Fano threefold of index 2 and degree 1. *Proc. of the Steklov Inst. of Math.*, 246:103–128, 2004.

[14] M. M. Grinenko. New Mori structures on a double space of index 2. *Russian Math. Surveys*, 59(3):573–574, 2004.

[15] F. Hidaka and K. Watanabe. Normal Gorenstein surfaces with ample anti-canonical divisor. *Tokyo J. Math.*, 4(2):319–330, 1981.

[16] V. A. Iskovskikh. Factorization of birational mappings of rational surfaces from the point of view of Mori theory. *Russian Math. Surveys*, 51(4):585–652, 1996.

[17] V. A. Iskovskikh. On the rationality problem for algebraic threefolds. *Proc. Steklov Inst. Math.*, 218:186–227, 1997.

[18] V. A. Iskovskikh and Yu. I. Manin. Three-dimensional quartics and counterexamples to the Lüroth problem. *Math. USSR, Sbornik*, 15:141–166, 1971.

[19] V. A. Iskovskikh and Yu. G. Prokhorov. Fano varieties. In *Algebraic geometry, V*, volume 47 of *Encyclopaedia Math. Sci.*, pages 1–247. Springer, Berlin, 1999.

[20] S. I. Khashin. Birational automorphisms of a Fano variety of index 2 and degree 2. Moskov. Gos. Univ., manuscript deposited at VINITI, Moscow 1985. (Russian).

[21] S. I. Khashin. Birational automorphisms of a double Veronese cone of dimension 3. *Moscow Univ. Math. Bull.*, 39(1):15–20, 1984.

[22] J. Kollar, Y. Miyaoka, and S. Mori. Rational connectedness and boundedness of Fano manifolds. *J. Differential Geom.*, 36(3):765–77, 1992.

[23] J. Kollár and S. Mori. *Birational geometry of algebraic varieties*, volume 134 of *Cambridge Tracts in Mathematics*. Cambridge University Press, Cambridge, 1998. With the collaboration of C. H. Clemens and A. Corti.

[24] Ju. I. Manin. Rational surfaces over perfect fields. *IHES Publ. Math.*, 30:55–113, 1966.

[25] J. Park. Birational maps of del Pezzo fibrations. *J. Reine Angew. Math.*, 538:213–221, 2001.

[26] A. V. Pukhlikov. Birational automorphisms of three-dimensional algebraic varieties with a pencil of del Pezzo surfaces. *Izv. Math.*, 62(1):115–155, 1998.

[27] A. V. Pukhlikov. Essentials of the method of maximal singularities. In *Explicit birational geometry of 3-folds*, volume 281 of *London Math. Soc. Lecture Note Ser.*, pages 73–100. Cambridge Univ. Press, Cambridge, 2000.

[28] M. Reid. Minimal models of canonical 3-folds. In *Algebraic varieties and analytic varieties (Tokyo, 1981)*, volume 1 of *Adv. Stud. Pure Math.*, pages 131–180. North-Holland, Amsterdam, 1983.

[29] M. Reid. Nonnormal del Pezzo surfaces. *Publ. Res. Inst. Math. Sci.*, 30(5):695–727, 1994.

[30] V. G. Sarkisov. Birational automorphisms of conic bundles. *Math. USSR, Izv.*, 17:177–202, 1981.

[31] V. G. Sarkisov. On conic bundle structures. *Math. USSR, Izv.*, 20:355–390, 1983.

[32] I. V. Sobolev. Birational automorphisms of a class of varieties fibred into cubic surfaces. *Izvestiya: Mathematics*, 66(1):201–222, 2002.

[33] A. S. Tikhomirov. The Abel-Jacobi mapping of sextics of genus 3 onto double $\mathbb{P}^3$s of index 2. *Sov. Math., Dokl.*, 33:204–206, 1986.

Steklov Mathematical Institute
Russian Academy of Sciences
117333 Moscow
Russia
grin@mi.ras.ru

Best Diophantine approximations: the phenomenon of degenerate dimension

Nikolai G. Moshchevitin

Introduction

This brief survey deals with multi-dimensional Diophantine approximations in the sense of linear forms and with simultaneous Diophantine approximations. We discuss the phenomenon of the degenerate dimension of linear subspaces generated by the best Diophantine approximations. Originally most of these results were established by the author in [15, 14, 16, 17, 18]. Here they are collected together and some new formulations are given. In contrast to our previous survey [17], this paper contains a wider range of results, especially results dealing with the best Diophantine approximations. It also includes proofs, or sometimes sketches of proofs. Some applications of these results and methods to the theory of small denominators can be found in [15, 19] and [13].

1 Best Diophantine approximations in the sense of linear forms

1.1 Notation

Let $\alpha_1, \ldots, \alpha_r$ be real numbers with $1, \alpha_1, \ldots, \alpha_r$ linearly independent over the rationals. For an integer point

$$m = (m_0, m_1, \ldots, m_r) \in \mathbb{Z}^{r+1} \setminus \{(0, \ldots, 0)\}$$

we define

$$\zeta(m) = |m_0 + m_1\alpha_1 + \cdots + m_r\alpha_r| \quad \text{and} \quad M = \max_{j=0,1,\ldots,r} |m_j|.$$

A point $m \in \mathbb{Z}^{r+1} \setminus \{0\}$ is defined to be the *best approximation (in*

158

the sense of linear forms) if

$$\zeta(m) = \min_{n \in \mathbb{Z}^{r+1} \setminus \{0\}: N \leq M} |\zeta(n)|$$

(here $N = \max_j |n_j|$). For the set of all best approximations m the corresponding values of $\zeta(m)$ and M can be ordered in descending (ascending) order:

$$\zeta_1 > \zeta_2 > \cdots > \zeta_\nu > \zeta_{\nu+1} > \dots,$$

$$M_1 < M_2 < \cdots < M_\nu < M_{\nu+1} < \dots.$$

(Here $m_\nu = (m_{0,\nu}, \dots, m_{r,\nu})$ is the νth best approximation and $\zeta_\nu = \zeta(m_\nu)$, $M_\nu = \max_j |m_{j,\nu}|$.) By the Minkowski convex body theorem it follows that $\zeta_\nu M_{\nu+1}^r \leq 1$. Let Δ_ν^r denote the determinant of the $r+1$ consecutive best approximations:

$$\Delta_\nu^r = \begin{vmatrix} m_{0,\nu} & m_{1,\nu} & \cdots & m_{r,\nu} \\ \cdots & \cdots & \cdots & \cdots \\ m_{0,\nu+r} & m_{1,\nu+r} & \cdots & m_{r,\nu+r} \end{vmatrix}.$$

1.2 Results on dimension

Here we observe some properties of the values Δ_ν^r discovered in [16].

The following statement is well known from the theory of continued fractions (see [9]).

Theorem 1.1. *Let $r = 1$ and let α_1 be an irrational number. Then for any natural number ν the determinant Δ_ν^1 is equal to $(-1)^{\nu-1}$.*

The next result deals with dimension 2. It follows from the Minkowski convex body theorem.

Theorem 1.2. *Let $r = 2$ and let $1, \alpha_1, \alpha_2$ be linearly independent over the rationals. Then there exist infinitely many values of ν for which $\Delta_\nu^2 \neq 0$.*

As was mentioned, Theorem 1.2 is a simple corollary of the Minkowski theorem and we shall give a sketch of the proof of Theorem 1.2 in the next section.

Now we formulate our main result in this area; it deals with the case $r > 2$.

Theorem 1.3. *Given $r \geq 3$ there exists an uncountable set of r-tuples $(\alpha_1, \ldots, \alpha_r)$ such that, for all large ν, the corresponding sequence of best approximations m_ν lies in a three-dimensional sublattice $\Lambda(\alpha_1, \ldots, \alpha_r)$ of the lattice $\mathbb{Z}^{r+1}$. Moreover, for each of these r-tuples, $1, \alpha_1, \ldots, \alpha_r$ are linearly independent over the rationals.*

Corollary . *For any r-tuple $(\alpha_1, \ldots, \alpha_r)$ in Theorem 1.3 there exists $\nu_0(\alpha)$ such that for all $\nu > \nu_0(\alpha)$ we have $\Delta_\nu^r = 0$.*

We shall give the proof of Theorem 1.3 in Section 1.5. It is based on the so-called singular r-tuples of Hinchin (see [7, 8] and Cassels' book [1]). Before this proof we discuss the properties of Hinchin's singular systems and their generalizations.

To finish this section we would like to emphasize once again that, for $r \geq 2$, for any r-tuple $(\alpha_1, \ldots, \alpha_r)$ of $\mathbb{Q}$-independent reals all but a finite number of the best approximation vectors never lie in a two-dimensional subspace but can lie in a three-dimensional subspace. We also would like to mention that there are many results related to various definitions, algorithmic calculating of the best approximations in general and for the algebraic numbers (see for example [4, 3, 10, 11, 12, 24]).

1.3 Sketch of the proof of Theorem 1.2.

Assume the contrary: suppose that for some α_1, α_2 with $1, \alpha_1, \alpha_2$ linearly independent over $\mathbb{Z}$, we know that all best approximations m_ν, $\nu \geq \nu_0$ lie in some two-dimensional linear subspace π. Then from the theory of continued fractions (compare with Theorem 1.1) we have

$$\zeta_\nu M_{\nu+1} \asymp \zeta_\mu M_{\mu+1} \qquad \text{for all } \nu, \mu > \nu_0. \qquad (1.1)$$

Consider the cube $E_H^3 = \{x = (x_1, x_2, x_3) \in \mathbb{R}^3 : |x_j| \leq H\}$ and the domain $L_\sigma^2 = \{x \in \mathbb{R}^3 : \rho(x; L^2) \leq \sigma\}$. Now the intersection $\Omega(\sigma, H) = E_H^3 \cap L_\sigma^2$ is a convex 0-symmetric body in π. As m_ν is a best approximation we conclude that there are no integer points in the set $\Omega(\zeta_\nu, M_{\nu+1})$. However from (1.1) it follows that

$$\text{Vol}\,\Omega(\zeta_\nu, M_{\nu+1}) \asymp \zeta_\nu M_{\nu+1}^2 \asymp M_{\nu+1} \to \infty, \quad \nu \to \infty,$$

and we have arrived at a contradiction to the Minkowski convex body theorem. □

1.4 Hinchin's ψ-singular linear forms

From the theory of continued fractions [9] we know that, in the case $r = 1$, we have

$$\zeta_\nu M_{\nu+1} \asymp 1. \tag{1.2}$$

Next we show that for linear forms in two or more variables the situation may be different: the values of $M_{\nu+1}$, corresponding to the best approximations $m_{\nu+1}$, may not be estimated from below in terms of the previous approximation ζ_ν.

Theorem 1.4. *Let $r \geq 2$ and $\psi(y)$ be a real valued function decreasing to 0 as $y \to \infty$. Then there exists an uncountable set of vectors $(\alpha_1, \ldots, \alpha_r)$ (with $1, \alpha_1, \ldots, \alpha_r$ linearly independent over the rationals) such that for all corresponding best approximations we have*

$$\zeta_\nu \leq \psi(M_{\nu+r-1}) \qquad \text{for all } \nu. \tag{1.3}$$

We would like to remind the reader of the definition of a ψ-*singular* linear form (in Hinchin's sense) [7, 8]. Let $\psi(y) = o(y^{-r})$, so that $\psi(y)$ decreases to 0 as $y \to \infty$. An r-tuple $(\alpha_1, \ldots, \alpha_r)$ is ψ-singular (in the sense of linear forms) if, for any $T > 1$, there is an integer r-tuple m which satisfies

$$\|m_1\alpha_1 + \cdots + m_r\alpha_r\| < \psi(T), \quad 0 < \max_{1 \leq j \leq r} |m_j| \leq T.$$

(Here $\| \cdot \|$ denotes the distance from the nearest integer.)

It is easy to verify that an r-tuple $(\alpha_1, \ldots, \alpha_r)$ is ψ-singular if and only if, for all natural numbers ν,

$$\zeta_\nu \leq \psi(M_{\nu+1}). \tag{1.4}$$

From this point of view, in the case $s \geq 3$ Theorem 1.4 establishes the existence of r-tuples which are "more singular" than Hinchin's singular linear forms.

Proof of Theorem 1.4

This proof was sketched in [16]. It is based on the following lemma.

Let

$$\nu_* \equiv \nu \pmod{r}, \quad 1 \leq \nu_* \leq r$$

and let σ be a large positive number (σ depends on r, and all the constants in the symbols $O(\cdot)$, $\ll$, $\gg$ below may depend on σ). Let

$$\sigma_{j,\nu} = \sigma\nu_*^j, \quad W = \max_{j,\nu} \sigma_{j,\nu}.$$

Lemma 1.5. *There exists an uncountable set of vectors $(\alpha_1, \ldots, \alpha_r)$ such that*

1) $1, \alpha_1, \ldots, \alpha_r$ *are linearly independent over* $\mathbb{Z}$;
2) *there exists a sequence of natural numbers* p_ν *with*

   (i) $\sigma_{j,\nu}\psi(p_\nu) \le \|p_\nu\alpha_j\| = p_\nu\alpha_j - a_{j,\nu} \le (\sigma_{j,\nu}+1)\psi(p_\nu)$ *for* $j = 1, \ldots, r$,
   (ii) $p_{\nu+1} \asymp p_\nu(\psi(p_\nu))^{-1}$.

Proof of Lemma 1.5. We construct the numbers $\alpha_1, \ldots, \alpha_r$ with simultaneous approximations of a special type.

Let λ be a sequence of zeros and ones: $\lambda = \{\lambda_\nu^2, \ldots, \lambda_\nu^s\}_{\nu=1}^\infty$; $\lambda_\nu^j \in \{0,1\}$. Now define the natural numbers p_ν, $a_{j,\nu}$, $j = 1, \ldots, r$; and segments $\Delta_{1,\nu}, \ldots, \Delta_{r,\nu}$ with lengths $|\Delta_{j,\nu}| = 2\psi(p_\nu)/p_\nu$ by the following recursive procedure.

The numbers p_0, $a_{1,0}, \ldots, a_{r,0}$ may be taken to be arbitrary. Define

$$\Delta_{j,0} = \left[\frac{a_{j,0}}{p_0} + \sigma_{j,\nu}\frac{\psi(p_0)}{p_0}, \frac{a_{j,0}}{p_0} + (\sigma_{j,\nu}+1)\frac{\psi(p_0)}{p_0}\right], \qquad j = 1, \ldots, r.$$

Suppose $p_0, \ldots, p_\nu$; $a_{j,0}, \ldots, a_{j,\nu}$ and $\Delta_{j,0}, \ldots, \Delta_{j,\nu}$ are already defined. We then construct $p_{\nu+1}$, $a_{j,\nu+1}$ and $\Delta_{j,\nu+1}$.

Let

$$p_{\nu+1} = \left[6p_\nu(\psi(p_\nu))^{-1}\right] + 1.$$

Then in any interval of the length $\psi(p_\nu)/(6p_\nu)$ one can find a number $a/p_{\nu+1}$, $a \in \mathbb{Z}$. Let

$$\frac{a_{j,\nu+1}^0}{p_{\nu+1}} \in \left[\frac{a_{j,\nu}}{p_\nu} + \left(\sigma_{j,\nu}+\frac{1}{6}\right)\cdot\frac{\psi(p_\nu)}{p_\nu}, \frac{a_{j,\nu}}{p_\nu} + \left(\sigma_{j,\nu}+\frac{2}{6}\right)\cdot\frac{\psi(p_\nu)}{p_\nu}\right],$$

$$\frac{a_{j,\nu+1}^1}{p_{\nu+1}} \in \left[\frac{a_{j,\nu}}{p_\nu} + \left(\sigma_{j,\nu}+\frac{4}{6}\right)\cdot\frac{\psi(p_\nu)}{p_\nu}, \frac{a_{j,\nu}}{p_\nu} + \left(\sigma_{j,\nu}+\frac{5}{6}\right)\cdot\frac{\psi(p_\nu)}{p_\nu}\right].$$

Now define

$$\Delta_{j,\nu+1}^\tau = \left[\frac{a_{j,\nu+1}^\tau}{p_{\nu+1}} + \sigma_{j,\nu+1}\frac{\psi(p_{\nu+1})}{p_{\nu+1}}, \frac{a_{j,\nu+1}^\tau}{p_{\nu+1}} + (\sigma_{j,\nu+1}+1)\frac{\psi(p_{\nu+1})}{p_{\nu+1}}\right],$$

where $\tau = 0, 1$; $j = 2, \ldots, r$.

We note that

$$\Delta_{j,\nu+1}^0 \cap \Delta_{2,\nu+1}^1 = \emptyset$$

and

$$\frac{a_{j,\nu}}{p_\nu} \notin \Delta_{j,\nu+1}^\tau, \qquad \tau = 0, 1.$$

Moreover, $|\Delta_{j,\nu+1}^{j}| = \psi(p_{\nu+1})/p_{\nu+1}$ and since

$$(\sigma_{j,\nu+1} + 1)\psi(p_{\nu+1})/p_{\nu+1} \le \psi(p_\nu)/6p_\nu$$

(here we suppose ψ decreases fast enough: for instance, $W\psi(p_{\nu+1}) < \psi(p_\nu)$), one has $\Delta_{j,\nu+1}^{\tau} \subset \Delta_{j,\nu}$, $\tau = 0, 1$. Put $a_{j,\nu+1} = a_{j,\nu+1}^{\lambda_j}$ and $\Delta_{j,\nu+1} = \Delta_{j,\nu+1}^{\lambda_\nu}$. We define an integer $a_{1,\nu+1}$ from the condition

$$\frac{a_{1,\nu+1}}{p_{\nu+1}} \in \left[\frac{a_{j,\nu}}{p_\nu} + \left(\sigma_{j,\nu} + \frac{1}{6} \right) \cdot \frac{\psi(p_\nu)}{p_\nu}, \frac{a_{j,\nu}}{p_\nu} + \left(\sigma_{j,\nu} + \frac{2}{6} \right) \cdot \frac{\psi(p_\nu)}{p_\nu} \right],$$

Now let

$$\Delta_{1,\nu+1} = \left[\frac{a_{1,\nu+1}}{p_{\nu+1}} \sigma_{j,\nu+1} \frac{\psi(p_{\nu+1})}{p_{\nu+1}}, \frac{a_{1,\nu+1}}{p_{\nu+1}} + (\sigma_{j,\nu+1} + 1) \frac{\psi(p_{\nu+1})}{p_{\nu+1}} \right].$$

We have $\Delta_{1,\nu+1} \subset \Delta_{1,\nu}$, $|\Delta_{1,\nu+1}| = \psi(p_{\nu+1})/p_{\nu+1}$ and $\frac{a_{1,\nu}}{p_\nu} \notin \Delta_{1,\nu+1}$.
To summarize: we have constructed a sequence of nested segments

$$\{\Delta_{1,\nu}\}_{\nu=0}^{\infty}$$

and, for an arbitrary 0,1-sequence λ, we have a sequence of nested segments $\{\Delta_{j,\nu}\}_{\nu=0}^{\infty}$. Denote

$$\alpha_1 = \bigcap_\nu \Delta_{1,\nu} ; \qquad \alpha_j = \alpha_j(\lambda) = \bigcap_\nu \Delta_{j,\nu}.$$

In the sequences of fractions $a_{1,\nu}/p_\nu$ and $a_{j,\nu}/p_\nu$ all elements are different, so that α_1, $\alpha_j(\lambda) \notin \mathbb{Q}$. Moreover, we can choose λ in such a way that $1, \alpha_1, \ldots, \alpha_r$ are linearly independent over the rationals. A similar procedure was performed in [21].
    So we construct $\alpha_1, \ldots, \alpha_r \in \mathbb{R}$ satisfying the conditions

A. $1, \alpha_1, \ldots, \alpha_r$ are linearly independent over $\mathbb{Z}$;
B. for a sequence of natural numbers p_ν,

$$\|p_\nu \alpha_j\| = \|p_\nu \alpha_j - a_{j,\nu}\| < \psi(p_\nu), \quad j = 1, \ldots, r,$$
$$3p_\nu \left(\psi(p_\nu) \right)^{-1} \le p_{\nu+1} \le 4p_\nu \left(\psi(p_\nu) \right)^{-1}.$$

One can easily verify that for any decreasing $\psi(y)$ the set

$$M_\psi = \{ (\alpha_1, \ldots, \alpha_r) \in \mathbb{R}^r : \alpha_1, \ldots, \alpha_r \text{ satisfy A, B} \}$$

is uncountable and dense in $\mathbb{R}^r$.

For large $\sigma = \sigma(r)$ and for any $\eta_{j,\mu} \in [-1,1]$, $j = 1,\ldots,r$; $\mu = \nu,\ldots,\nu+r-1$ we note that the determinant

$$\begin{vmatrix} \sigma_{1,\nu} & \sigma_{2,\nu} & \cdots & \sigma_{r,\nu} \\ \cdots & \cdots & \cdots & \cdots \\ \sigma_{1,\nu+r-1} & \sigma_{2,\nu+r-1} & \cdots & \sigma_{r,\nu+r-1} \end{vmatrix} = \pm\sigma^r \prod_{1\le u<v\le r}(v-u)$$

has

$$\begin{vmatrix} \sigma_{1,\nu}+\eta_{1,\nu} & \cdots & \sigma_{r,\nu}+\eta_{r,\nu} \\ \cdots & \cdots & \cdots \\ \sigma_{1,\nu+r-1}+\eta_{1,\nu+r-1} & \cdots & \sigma_{r,\nu+r-1}+\eta_{r,\nu+r-1} \end{vmatrix} = \pm\sigma^r + o(\sigma^r) \neq 0.$$

In the sequel this large value of $\sigma = \sigma(r)$ is fixed.

Moreover, it is easy to modify the construction in Lemma 1.6 to establish that for any value p_ν there exists a *best* approximation for the linear form $\|m_1^*\alpha_1 + \ldots + m_r^*\alpha_r\|$ where the vectors $(m_1^*,\ldots,m_r^*)$, $(p_\nu, a_{1,\nu}, 0,\ldots,0)$, $\ldots$, $(p_\nu, 0,\ldots,0,a_{r,\nu})$ are linearly dependent and $M^* \ll p_\nu^2$.

The proof is complete. □

Lemma 1.6. *For the numbers $\alpha_1,\ldots,\alpha_r$ constructed in Lemma 1.5, there exists an infinite sequence of values of the linear form*

$$\zeta(n_\nu) = n_{0,\nu} + n_{1,\nu}\alpha_1 + \cdots + n_{r,\nu}\alpha_r = |n_{1,\nu}\alpha_1 + \cdots + n_{r,\nu}\alpha_r|$$

with

$$0 < \zeta(n_{\nu+1}) < \zeta(n_\nu) \ll \psi(\kappa N_{\nu+s-1}),$$

where $N_\nu = \max|n_{j,\nu}|$ and $\kappa > 0$ is a constant.

Proof of Lemma 1.6. Let

$$\zeta(n_\nu) = \pm \begin{vmatrix} 1 & \alpha_1 & \cdots & \alpha_r \\ p_\nu & a_{1,\nu} & \cdots & a_{r,\nu} \\ \cdots & \cdots & \cdots & \cdots \\ p_{\nu+r-1} & a_{1,\nu+r-1} & \cdots & a_{r,\nu+r-1} \end{vmatrix}$$

$$= \pm \begin{vmatrix} a_{1,\nu}-p_\nu\alpha_1 & \cdots & a_{r,\nu}-p_\nu\alpha_r \\ \cdots & \cdots & \cdots \\ a_{1,\nu+r-1}-p_{\nu+r-1}\alpha_1 & \cdots & a_{r,\nu+r-1}-p_{\nu+r-1}\alpha_r \end{vmatrix}$$

$$= \psi(p_\nu)\cdots\psi(p_{\nu+r-1}) \begin{vmatrix} (\sigma+\eta)_{1,\nu} & \cdots & (\sigma+\eta)_{r,\nu} \\ \cdots & \cdots & \cdots \\ (\sigma+\eta)_{1,\nu+r-1} & \cdots & (\sigma+\eta)_{r,\nu+r-1} \end{vmatrix}$$

$$\asymp \psi(p_\nu)\cdots\psi(p_{\nu+r-1}).$$

(Here

$$\eta_{j,\mu} = \frac{a_{j,\mu} - p_\mu \alpha_j}{\psi(p_\mu)} + \sigma_{j,\mu} \in [-1,1]$$

and the sign $+$ or $-$ is taken to satisfy $\zeta(n_\nu) > 0$.) Then

$$\zeta(n_\nu) \asymp \prod_{\mu=\nu}^{\nu+s-1} \max_{j=1,r} |a_{j,\mu} - p_\mu \alpha_j| \asymp \prod_{\mu=\nu}^{\nu+r-1} \psi(p_\mu) < \psi(p_{\nu+r-1}). \quad (1.5)$$

For the coefficients $n_{j,\nu}$ we have

$$n_{j,\nu} = \pm \begin{vmatrix} p_\nu & a_{1,\nu} & \cdots & a_{j-1,\nu} & a_{j+1,\nu} & \cdots & a_{r,\nu} \\ \cdots & \cdots & \cdots & \cdots & \cdots & \cdots & \cdots \\ p_\nu & a_{1,\nu+r-1} & \cdots & a_{j-1,\nu+r-1} & a_{j+1,\nu+r-1} & \cdots & a_{s,\nu+r-1} \end{vmatrix}$$

$$= \pm \begin{vmatrix} p_\nu & a_{1,\nu} - p_\nu \alpha_1 & \cdots & a_{r,\nu} - p_\nu \alpha_r \\ \cdots & \cdots & \cdots & \cdots \\ p_{\nu+r-1} & a_{1,\nu+r-1} - p_{\nu+r-1}\alpha_1 & \cdots & a_{r,\nu+r-1} - p_{\nu+r-1}\alpha_r \end{vmatrix}.$$

Using (i) and (ii) we deduce

$$|n_{j,\nu+r-1}| \ll p_{\nu+r-1}\psi(p_{\nu+r-2}) \ll p_\nu \quad \text{for all } j. \quad (1.6)$$

(We note that $N_{\nu+r-1} = \max_j |n_{j,\nu+r-1}| \asymp p_\nu$.)

From (1.5) and (1.6) we have

$$0 < \zeta(n_\nu) \ll \psi(\kappa N_{\nu+r-1}).$$

We may suppose $\zeta(n_{\nu+1}) < \zeta(n_\nu)$. The proof is complete. $\qquad \square$

Theorem 1.4 follows immediately from Lemmas 1.5 and 1.6 since for the numbers constructed in Lemma 1.5 we have, by Lemma 1.6, approximations satisfying (1.3) and in this case the inequality (1.3) is also valid for the best approximations.

Theorem 1.4 is proved.

1.5 *Proof of Theorem 1.3*

We need the following notation: let $r \geq 2$; $\mathbb{R}^{r+1}$ be Euclidean space with Cartesian coordinates $(x_0, \ldots, x_r)$; $\mathbb{Z}^{r+1} \subset \mathbb{R}^{r+1}$ be the lattice of integers; L^r be the r-dimensional subspace in $\mathbb{R}^{r+1}$ orthogonal to the vector $(1, \alpha_1, \ldots, \alpha_r)$ and let the r-tuple $(\alpha_1, \ldots, \alpha_r)$ satisfy (1.4) with

$$\psi(y) = e^{-\gamma y}, \quad \gamma \in (0,1). \quad (1.7)$$

(So for our proof we need only ordinary Hinchin singular linear forms rather than the generalization from Theorem 1.4.)

Let $\mathbb{R}^{r+2} = \mathbb{R}^{r+1}(x_0, \ldots, x_r) \times \mathbb{R}^1(z)$ be the product of $\mathbb{R}^{r+1}$ and $\mathbb{R}^1$,

$$\mathcal{L}^{r+1} = L^r \times \mathbb{R}^1,$$

$$\mathcal{L}^{r+1}_\delta = \left\{ X \in \mathbb{R}^{r+2} : \rho(X, \mathcal{L}^{r+1}) \leq \delta \right\},$$

and let $E^{r+2}_H = \{(x_0, \ldots, x_r, z) : \max\{|x_0|, \ldots, |x_r|, |z|\} \leq H\}$,

$$\Pi(\delta; H) = E^{r+2}_H \cap \mathcal{L}^{r+1}_\delta,$$

$$\mathcal{K} = \bigcup_{t \geq 1} \Pi(2e^{-(\gamma-\varepsilon)t}; t), \quad \varepsilon \in (0, \gamma).$$

The infinite domain $\mathcal{K}$ has finite volume since

$$\mathrm{Vol}\,\mathcal{K} \ll \int_1^\infty t^{r+1} e^{-(\gamma-\varepsilon)t}\, dt < +\infty.$$

Moreover, from our choice of ψ by (1.7) we have $m_\nu \in \mathcal{K}$ for all $\nu \geq \nu_0$.

Let B_ϵ be a $(r+2)$-dimensional ball with radius $\epsilon < 1/2$ centred at $(0, \ldots, 0, 1) \in \mathbb{R}^{r+2}$. For a point $\xi \in B_\epsilon$ we put in the corresponding $(r+1)$-dimensional lattice $\Lambda_\xi = \mathbb{Z}^{r+1} \oplus \xi\mathbb{Z}$ generated by $\mathbb{Z}^{r+1}$ and the point ξ. Let $T_\xi : \mathbb{R}^{r+2} \to \mathbb{R}^{r+2}$ be the linear transformation preserving the lattice $\mathbb{Z}^{r+1}$ and transforming the vector ξ into the unit vector $(0, \ldots, 0, 1)$. Consider the $(r+1)$-dimensional subspace $T_\xi\mathcal{L}^{r+1}$ and define α_{r+1} in such a way that the vector $(1, \alpha_1, \ldots, \alpha_r, \alpha_{r+1})$ is orthogonal to the subspace $T_\xi\mathcal{L}^{r+1}$.

The proof of the following lemma is in general similar to the original proof of the Minkowski-Hlawka theorem (see for example [6]). It is based on a well-known metric procedure.

Lemma 1.7. *For almost all points $\xi \in B_\epsilon$ (in the sense of Lebesgue measure) we have*

1) $\Lambda_\xi \cap \mathcal{L}^{r+1} = \{0\}$,
2) *the intersection $\Lambda_\xi \cap \mathcal{K}$ contains the points m_ν and at most a finite number of other integer points.*

Corollary . *All but a finite number of the best approximations for the $(r+1)$-tuple $(\alpha_1, \ldots, \alpha_r, \alpha_{r+1})$ from the lattice $\mathbb{Z}^{r+2}$ coincide with the best approximations for the r-tuple $(\alpha_1, \ldots, \alpha_r)$ from the lattice $\mathbb{Z}^{r+1}$.*

Proof of Lemma 1.7. Let k be a natural number, e_j be unit vectors in $\mathbb{R}^{r+1}$ and $\chi(X)$ be the characteristic function of the domain $\mathcal{K}$. We consider the value

$$S_\xi(T) = \sum_{k=1}^T \sum_m \chi(m_0 e_0 + \cdots + m_r e_r + m_{r+1}\xi) \geq 0,$$

where the inner sum is taken over all

$$(m_0, \ldots, m_r) \in \mathbb{Z}^{r+1}, \quad m_{r+1} \in \mathbb{Z} \setminus \{0\}, \quad \max\{|m_0|, \ldots, |m_{r+1}|\} = k.$$

This sum calculates the number of points of the lattice Λ_ξ with norm not greater than T lying in $\mathcal{K}$ and distinct from the points of $\mathbb{Z}^{r+1} \subset \mathbb{R}^{r+1}$:

$$S_\xi(T) = \# \left\{ \begin{array}{c} m = m_0 e_0 + \cdots + m_r e_r + m_{r+1}\xi : \ m_{r+1} \neq 0, \\ 0 < \max\{|m_0|, \ldots, |m_{r+1}|\} \leq T \end{array} \right\}.$$

Observe that

$$\int_{B_\epsilon} S_\xi(T)d\xi = \sum_{k=1}^{T} \sum_m \mathrm{Vol}(B_\epsilon(m) \cap \mathcal{K}),$$

where

$$B_\epsilon(m) = \{X = m_0 e_0 + \cdots + m_r e_r + m_{r+1}\xi : \ \xi \in B_\epsilon\}.$$

It is clear that for $\max|m_j| = k$ we have

$$\mathrm{Vol}(B_\epsilon(m) \cap \mathcal{K}) < \mathrm{Vol}(\mathcal{K} \cap \{z \in R^{r+2} : \ \max_j |z_j| \geq k/2\} \ll e^{-\gamma_1 k},$$

where $0 < \gamma_1 < \gamma$. Hence, for any T

$$\int_{B_\epsilon} S_\xi(T)d\xi \ll \sum_{k=1}^{T} k^{r+2} e^{-\gamma_1 k} \ll 1.$$

Now we use Levi's theorem to establish that for almost all $\xi \in B_\epsilon$ the finite limit $\lim_{T \to \infty} S_\xi(T)$ exists. This means that for almost all ξ the intersection $\Lambda_\xi \cap \mathcal{K}$ consists of at most a finite number of points distinct from m_ν, and the proof is complete. ⊔

Now Theorem 1.3 can be proved by induction. For $r = 2$ we have Hinchin's singular vector (α_1, α_2) satisfying the singularity condition with $\psi(y) = e^{-y}$. The induction step is performed in Lemma 1.7 and the proof is complete.

2 Best simultaneous Diophantine approximations

2.1 Definitions

For an s-tuple of real numbers $\alpha = (\alpha_1, \ldots, \alpha_s) \in \mathbb{R}^s$ we define *the best simultaneous approximation* (briefly, b.s.a) as an integer point $\zeta = (p, a_1, \ldots, a_s) \in \mathbb{Z}^{s+1}$ such that

$$D(\zeta) := \max_{j=1,\ldots,s} |p\alpha_j - a_j| < \min_{j=1,\ldots,s}^* \max_{j=1,\ldots,s} |q\alpha_j - b_j|,$$

where $\min^*$ is taken over all $q, b_1, \ldots, b_s$ under the conditions

$$1 \leq q \leq p; \quad (b_1, \ldots, b_s) \in \mathbb{Z}^s \setminus \{(a_1, \ldots, a_s)\}.$$

In the case $\alpha_j \notin \mathbb{Q}$ all b.s.a. to α form infinite sequences

$$\zeta^\nu = (p^\nu, a_1^\nu, \ldots, a_s^\nu) \qquad (\nu = 1, 2, \ldots)$$

where

$$p^1 < \ldots < p^\nu < p^{\nu+1} < \ldots$$

and

$$D(\zeta^1) > \ldots > D(\zeta^\nu) > D(\zeta^{\nu+1}) > \ldots$$

Let

$$M_\nu[\alpha] = \begin{pmatrix} p^\nu & a_1^\nu & \cdots & a_s^\nu \\ \cdots & \cdots & \cdots & \cdots \\ p^{\nu+s} & a_1^{\nu+s} & \cdots & a_s^{\nu+s} \end{pmatrix}$$

and $\operatorname{rk} M_\nu[\alpha]$ be the rank of the matrix $M_\nu[\alpha]$. The natural number $R(\alpha)$, $2 \leq R(\alpha) \leq s+1$, is defined as follows:

$$R(\alpha) = \min \left\{ \begin{array}{l} n : \text{there exists a lattice } \Lambda \subseteq \mathbb{Z}^{s+1}, \dim \Lambda = n, \\ \text{and } \nu_0 \in \mathbb{N} \text{ such that, for all } \nu > \nu_0, \zeta^\nu \in \Lambda \end{array} \right\}.$$

The value $\dim_\mathbb{Z} \alpha$ is defined as the maximum number of $\alpha_{i_1}, \ldots, \alpha_{i_m}$ chosen from $(\alpha_0 = 1, \alpha_1, \ldots, \alpha_s) \in \mathbb{R}^{s+1}$ which are linearly independent over $\mathbb{Z}$.

Proposition 2.1. *For $s = 1$ and any ν we have the equality $\det M_\nu[\alpha] = \pm 1$ (which implies that for any ν we have $\operatorname{rk} M_\nu[\alpha] = 2$).*

Proposition 2.2. *For any $s \geq 1$ the following equality is valid:*

$$R(\alpha) = \dim_\mathbb{Z} \alpha.$$

Proposition 2.3. *Let $s = 2$ and $1, \alpha_1, \alpha_2$ be linearly independent over $\mathbb{Z}$. Then there exist infinitely many natural numbers ν such that*

$$\operatorname{rk} M_\nu[\alpha] = 3 = \dim_\mathbb{Z} \alpha$$

(hence the inequality $\det M_\nu[\alpha] \neq 0$ holds for infinitely many values of ν).

Propositions 2.1–2.3 are well-known and can be easily verified (compare [11]).

2.2 Counterexample to Lagarias' Conjecture

We formulate our result from [14] which deals with the degeneracy of the dimension of the spaces generated by successive b.s.a. It gives a counterexample to Lagarias' conjecture [11]. We would like to point out that this result was obtained due to discussion with Nikolai Dolbilin. We shall give a sketched proof in the next two sections.

Theorem 2.4. *Let $s \geq 3$. Then there exists an uncountable set of s-tuples $\alpha = (\alpha_1, \ldots, \alpha_s)$, with $1, \alpha_1, \ldots, \alpha_s$ linearly independent over $\mathbb{Z}$, such that $\operatorname{rk} M_\nu[\alpha] \leq 3$ for all $\nu \in \mathbb{N}$. (Hence for all ν the equality $\det M_\nu[\alpha] = 0$ is valid.)*

2.3 Inductive lemma for Theorem 2.4

Consider Euclidean space $\mathbb{R}^{s+1}$ with Cartesian coordinates $(x, y_1, \ldots, y_s)$. The letter ℓ will denote a ray from the origin of coordinates located in the half-space $\{x > 0\}$. For such a ray ℓ and for small enough positive ϵ the open cone $K_\epsilon(\ell)$ consists of all rays ℓ' such that the angle between ℓ and ℓ' is less than ϵ.

For a point ξ from the half-space $\{x > 0\}$ we define $\ell(\xi)$ to be the ray $\{\kappa\xi : \kappa \geq 0\}$. The subspace $\pi \subseteq \mathbb{R}^{s+1}$ is defined to be *absolutely rational* if the lattice $\Lambda = \pi \cap \mathbb{Z}^{s+1}$ has dimension equal to the dimension of the whole of π, that is $\dim \Lambda = \dim \pi$. Let ℓ be a ray parallel to a vector $\beta = (1, \beta_1, \ldots, \beta_s)$. The best approximation to the ray ℓ is defined as the point $\zeta \in \mathbb{Z}^{s+1}$ which is the b.s.a. to β.

In the case when each β_j is not half of an integer the sequence of all best approximations

$$\zeta^\nu = (p^\nu, a_1^\nu, \ldots, a_s^\nu), \qquad p^1 < \ldots, p^\nu < p^{\nu+1} < \ldots$$

to the ray ℓ is defined correctly. It is finite in the case when there exists an integer point on the ray ℓ different from the origin and is infinite in the opposite case. This sequence of the best approximations we write as

$$\mathcal{B}(\ell) = \{\zeta^1, \zeta^2, \ldots, \zeta^\nu, \ldots\}.$$

Moreover, we use the following notation:

$$\mathcal{B}_k^t(\ell) = \{\zeta^k, \zeta^{k+1}, \ldots, \zeta^t\}.$$

Lemma 2.5. *Let $\Lambda = \mathbb{Z}^{s+1} \cap \pi$ be a lattice located in an absolutely rational subspace π, $\dim \pi \geq 2$. Let a point $\zeta \in \Lambda$ satisfy the condition*

$$\mathcal{B}(\ell(\zeta)) = \{\zeta^1, \zeta^2, \ldots, \zeta^\tau, \ldots, \zeta^t\}, \qquad \zeta^t = \zeta$$

and in addition

$$\mathcal{B}_\tau^t(\ell(\zeta)) = \{\zeta^\tau, \ldots, \zeta^t\} \subset \Lambda \subset \pi.$$

Let $D(\zeta_\tau) < D(\xi)$ for any integer point ξ which does not belong to π.

Then for some $\epsilon > 0$ any ray $\ell' \subset K_\epsilon(\ell(\zeta))$ satisfies the following conditions:

1) $\mathcal{B}(\ell') \supset \mathcal{B}(\ell(\zeta))$;
2) *the sequence of the best approximations to the ray ℓ' between the approximations ζ^τ and ζ^t lies completely in the subspace π.*

We must remember that all the points in $\mathcal{B}_\tau^t(\ell(\zeta))$ obviously belong to the considered sequence of the best approximations to the ray ℓ' between the approximations ζ^τ and ζ^t, but it may happen that a number of new points appear.

Lemma 2.5 follows from two easy observations:

1. for a small perturbation ℓ' of the ray ℓ the first best approximations to ℓ remain the best approximations to ℓ';
2. a small perturbation of ℓ does not enable integer points not belonging to π to become best approximations between the approximations ζ^τ and ζ^t.

2.4 Sketch of the proof of Theorem 2.4

The proof uses two inductive steps.

The first step: Applying Lemma 2.5 many times we construct absolutely rational subspaces

$$\pi_1, \rho_1, \pi_2, \rho_2, \ldots, \pi_s, \rho_s$$

with dimensions $\dim \pi_j = 2$, $\dim \rho_j = 3$ and a ray $\ell = \ell(\zeta)$, $\zeta \in \pi_s$, such that

A) $\pi_j, \pi_{j+1} \subset \rho_j$,
B) $\mathcal{B}(\ell) = \{\zeta^1, \ldots, \zeta^{\tau_1}, \zeta^{\tau_1+1}, \ldots, \zeta^{t_1}, \zeta^{t_1+1}, \ldots, \zeta^{\tau_2}, \ldots, \zeta^{\tau_s+1}, \ldots, \zeta^{t_s}\}$,
   where

$$\zeta^{t_s} = \zeta, \quad t_0 = 1, \quad t_j - \tau_j \geq s + 1, \quad \tau_j - t_{j-1} \geq s + 1 \qquad \text{for all } j$$

   and

$$\zeta^{\tau_j+1}, \ldots, \zeta^{t_j} \in \pi_j \quad \text{and} \quad \zeta^{t_j+1}, \ldots, \zeta^{\tau_{j+1}} \in \rho_j \quad \text{for all } j,$$

C) $\mathcal{B}(\ell)$ (as well as the union $\bigcup_{j=1}^{s} \rho_j$) does not belong to any s-dimensional subspace of $\mathbb{R}^{s+1}$.

We perform the construction of such a ray $\ell(\zeta)$ for which the best approximations admit A), B), C) by means of Lemma 2.5 by an inductive procedure.

The beginning of the inductive procedure is trivial. Let the subspaces

$$\pi_1, \rho_1, \pi_2, \rho_2, \ldots, \pi_k, \rho_k$$

and the ray $\ell(\zeta^{t_k}), \zeta^{t_k} \in \pi_k$ be already constructed. Then by Lemma 2.5 we take $\zeta^{\tau_{k+1}}$ with the required properties and choose an absolutely rational subspace π_{k+1} such that the ray $\ell(\zeta^{\tau_{k+1}})$ lies in this subspace and the dimension of the subspace generated by all subspaces $\pi_1, \rho_1, \ldots, \pi_{k+1}$ is maximal. Using Lemma 2.5 again, we find a point $\zeta^{t_{k+1}}$ in π_{k+1} with the requested properties.

The second step: We must apply the procedure of the first step many times and construct a sequence of rays $\ell^k = \ell(\xi^k), \xi^k \in \mathbb{Z}^{s+1}, k = 1, 2, \ldots$ in such a way that, for any ray ℓ^k, the set of the best approximations $\mathcal{B}(\ell^k)$ consists of k successive blocks. Each of these blocks must satisfy the conditions A), B), C) from the first step.

The limit ray for the sequence of rays ℓ^k will correspond to the numbers $\alpha_1, \ldots, \alpha_s$ with the properties requested in Theorem 2.4: all successive $(s+1)$ b.s.a. for $\alpha_1, \ldots, \alpha_s$ will lie in two- or three-dimensional subspaces and Proposition 2.2 and the property C) leads to the independence of the reals $1, \alpha_1, \ldots, \alpha_s$ over the rationals. □

2.5 Best simultaneous approximations in different norms

We consider a convex 0-symmetric star function $f : \mathbb{R}^n \to \mathbb{R}_+$ satisfying the conditions

1) f is continuous,
2) $f(x) \geq 0$ for all $x \in \mathbb{R}^n$, $f(x) = 0 \iff x = 0$,
3) $f(-x) = f(x)$ for all $x \in \mathbb{R}^n$,
4) $f(tx) = tf(x)$ for all $x \in \mathbb{R}^n$ and all $t \in \mathbb{R}_+$,
5) the set $B_f^1 = \{y \in \mathbb{R}^n : f(y) \leq 1\}$ is convex and $0 \in \text{int } B_f^1$.

It is well known (see [2]) that f determines a norm in $\mathbb{R}^n$. The function (or norm) f is *strictly convex* if the set B_f^1 is strictly convex; that is, the boundary ∂B_f^1 does not contain segments of straight lines. We write

$B_f^\lambda(a)$ for the set

$$B_f^\lambda(a) = \{y \in \mathbb{R}^n : f(y - a) \leq \lambda\},$$

so $B_f^1 = B_f^1(0)$.

For an n-tuple $\alpha = (\alpha_1, \ldots, \alpha_n) \in \mathbb{R}^n$ we define an f-*best simultaneous approximation* (f-b.s.a.) as an integer point $\tau = (p, a_1, \ldots, a_n) \in \mathbb{Z}^{n+1}$ such that $p \geq 1$ and

$$f(\alpha q - b) > f(\alpha p - a)$$

for all

$$(q, b_1, \ldots, b_n) \in \mathbb{Z}^{n+1}, \quad 1 \leq q \leq p - 1$$

and for all

$$(p, b_1, \ldots, b_n) \in \mathbb{Z}^{n+1}, \quad b \neq a.$$

In the case when f determines the cube

$$B_f^1 = \{y = (y_1, \ldots, y_n) \in \mathbb{R}^n : \max_j |y_j| \leq 1\}$$

our definition leads to the classical definition of the b.s.a. considered in previous sections.

All the f-b.s.a. for α form the sequences

$$\tau_\nu = (p_\nu, a_\nu) \in \mathbb{Z}^{n+1}, \quad p_\nu \in \mathbb{N}, \quad a_\nu = (a_{1,\nu}, \ldots, a_{n,\nu}) \in \mathbb{Z}^n,$$

$$p_1 < p_2 < \ldots < p_\nu < \ldots,$$

$$f(\alpha p_1 - a_1) > f(\alpha p_2 - a_2) > \ldots . f(\alpha p_\nu - a_\nu) > \ldots .$$

These sequences are finite in the case $\alpha \in \mathbb{Q}^n$ and are infinite in the opposite situation.

Let $\xi_\nu = (\xi_{1,\nu}, \ldots, \xi_{n,\nu})$ denote the remainder vector $\xi_{j,\nu} = \alpha_j p_\nu - a_{j,\nu}$. Let

$$\Xi_\nu = (\Xi_{1,\nu}, \ldots, \Xi_{n,\nu}); \quad \Xi_{j,\nu} = \xi_{j,\nu} / f(\xi_\nu).$$

Obviously, $\Xi_\nu \in B_f^1$. For a given vector $\xi \in \mathbb{R}^n$ we also use the notation $\Xi(\xi) = \xi / f(\xi) \in B_f^1$. Moreover, for the integer vector $\zeta = (p, a_1, \ldots, a_n) \in \mathbb{R}^{n+1}$ we use the notation

$$\xi^\alpha(\zeta) = (p\alpha_1 - a_1, \ldots, p\alpha_n - a_n) \in \mathbb{R}^n.$$

2.6 The order of the best approximations

From the Minkowski convex body theorem applied to the cylinder

$$\Omega_\nu = \{z = (x, y_1, \ldots, y_n) \in \mathbb{R}^{n+1} : f(\alpha x - y) < f(\xi_\nu)\} \qquad (2.1)$$

(which does not contain nontrivial integer points) it follows that for any ν one has

$$f(\xi_\nu) \le C_1(f)p_{\nu+1}^{-1/n} \qquad (2.2)$$

with constant $C_1(f) = 2/(\operatorname{Vol} B_f^1)^{1/n}$. On the other hand, we can show that the following result is valid.

Theorem 2.6. *Let* $\dim_{\mathbb{Z}}(1, \alpha_1, \ldots, \alpha_n) \ge 3$. *Then*

$$f(\xi_\nu)p_{\nu+1} \to +\infty \qquad as \ \nu \to +\infty. \qquad (2.3)$$

Proof. 1) Let $\Lambda^2 \in \mathbb{Z}^{s+1}$ be a two-dimensional sublattice and $\det_2 \Lambda^2$ be the area of its fundamental domain. The set of all sublattices

$$\{\Lambda^2 \subset \mathbb{Z}^{s+1} : \det_2 \Lambda^2 \le \gamma\}$$

is finite for any γ.

2) Consider a two-dimensional lattice $\Lambda_\nu^2 = \langle \tau_\nu, \tau_{\nu+1} \rangle_{\mathbb{Z}}$. From

$$\operatorname{conv}(0, \tau_\nu, \tau_{\nu+1}) \subset \Omega_\nu$$

it follows that

$$\frac{1}{2} \det_2 \Lambda_\nu^2 = \operatorname{Vol}_2(\operatorname{conv}(0, \tau_\nu, \tau_{\nu+1})) \ll f(\xi_\nu)p_{\nu+1}.$$

3) From $\dim_{\mathbb{Z}}(1, \alpha_1, \ldots, \alpha_s) \ge 3$ it is easy to deduce (see Proposition 2.2) that the sequence of all f-b.s.a. cannot asymptotically lie in a two-dimensional sublattice and hence for a fixed sequence of natural numbers ν_k the embedding $\bigcup_{k=1}^{\infty} \tau_{\nu_k} \subset \bigcup_{n=1}^{\nu_0} \Lambda_\nu^2$ never holds.

Theorem 2.6 immediately follows from 1), 2), 3). □

We would like to refer to Hinchin once again [7, 8] as he actually proved that it is not possible to establish any specific rate of growth of the value $f(\xi_\nu)p_{\nu+1}$ in (2.3).

Proposition 2.7. *For any function* $\psi(y)$ *increasing to infinity (as slowly as one wishes) as* $y \to \infty$, *there exists an* n-*tuple* $\alpha \in \mathbb{R}^n$ *with* $\dim_{\mathbb{Z}}(1, \alpha_1, \ldots, \alpha_n) = n + 1$ *such that*

$$f(\xi_\nu)p_{\nu+1} = O(\psi(p_{\nu+1})) \qquad as \ \nu \to +\infty. \qquad (2.4)$$

Formula (2.4) shows that in the situation $n \geq 2$ there exist vectors α for which the lower estimate from (2.2,2.3) is the exact one. Of course, in the case $n = 1$ for any ν we have

$$C_2(f)p_{\nu+1}^{-1} \leq f(\xi_\nu) \leq C_1(f)p_{\nu+1}^{-1}$$

(see [9]).

2.7 The directions of the successive best approximations

Theorem 2.8. *For any natural number ν one has $\Xi_{\nu+1} \notin \operatorname{int} B_f^1(\Xi_\nu)$.*

Theorem 2.8 was actually proved by Rogers in [21] for signatures (see Section 2.11). It follows from the fact that in the cylinder (2.1) there are no nontrivial integer points and

$$\tau_{\nu+1} - \tau_\nu = (p_{\nu+1} - p_\nu, a_{1,\nu+1} - a_{1,\nu}, \ldots, a_{n,\nu+1} - a_{n,\nu})$$

does not belong to the cylinder Ω_ν. Then one observes that $0 < p_{\nu+1} - p_\nu < p_{\nu+1}$. Hence $\tau_{\nu+1} - \tau_\nu \notin \Omega$ implies that

$$\xi_{\nu+1} \notin \operatorname{int} B_f^{f(\xi_\nu)}(\xi_\nu). \tag{2.5}$$

Now $0 \in \partial S_f^{f(\xi_\nu)}(\xi_\nu)$ and due to convexity we have

$$\xi_{\nu+1} \frac{f(\xi_\nu)}{f(\xi_{\nu+1})} \notin \operatorname{int} B_f^{f(\xi_\nu)}(\xi_\nu),$$

which is exactly what is stated in the theorem.

We notice that the statement (2.5) is a little bit more general than Theorem 2.8.

2.8 Strictly convex norms

Theorem 2.9. *Let the norm f be strictly convex. Then there exists $\delta = \delta(f) > 0$ such that for any vector $\alpha \notin \mathbb{Q}^n$ there exist infinitely many values of ν for which*

$$\Xi_{\nu+1} \notin B_f^{1+\delta}(\Xi_\nu).$$

We remind the reader that the n-tuple $\alpha = (\alpha_1, \ldots, \alpha_n)$ is defined to be *badly approximable* if for some positive $D(\alpha) > 0$ the inequality

$$\max_{1 \leq j \leq n} \min_{a_j \in \mathbb{Z}} |p\alpha_j - a_j| \geq D(\alpha)p^{-1/n}$$

is valid for all $p \in \mathbb{N}$ (concerning the existence of the badly approximable vectors see [22]). It is easy to see that the vector α is badly approximable if and only if for any norm f there is a constant $D_1(f, \alpha)$ such that for any natural number p

$$\min_{a \in \mathbb{Z}^n} f(p\alpha - a) \geq D_1(f, \alpha)p^{-1/n} \qquad (2.6)$$

holds.

Theorem 2.10. *Let α be badly approximable and $D = D(\alpha)$ be the corresponding constant. Let the norm f be strictly convex. Then there exist $w = w(D, f) \in \mathbb{N}$ and $\delta = \delta(D, f) > 0$ with the following property:*

for any $\nu \geq 1$ there exists a natural number j from the interval $\nu \leq j \leq \nu + w$ such that

$$\Xi_{j+1} \notin B_f^{1+\delta}(\Xi_j). \qquad (2.7)$$

Theorem 2.9 shows that for a strictly convex norm the condition $\theta_{\nu+1} \notin \operatorname{int} B_f^1(\theta)$ for the sequence $\theta_\nu \in B_f^1$ is not sufficient for the existence of α such that $\lim_{\nu \to \infty} (\theta_\nu - \Xi_\nu) = 0$. Theorem 2.10 shows that for the badly approximable numbers the values of j for which we have (2.7) appear regularly. Probably, the result of Theorem 2.10 does not depend on the fact that α is badly approximable but we cannot prove it. The proofs of the theorems we give in next two sections. From the results of Section 2.11 it is clear that Theorem 2.9 *is not valid for non-strictly convex norms.*

2.9 Two lemmas

Lagarias [10] proved the following statement.

Lemma 2.11. *Define $h = 2^{n+1}$. Then for any norm f and for any natural number ν one has $p_{\nu+h} > 2p_\nu$.*

Corollary 2.12. *For any vector $\alpha \notin \mathbb{Q}^n$ and for all $\nu, j > 1$ one has*

$$f(\xi_{\nu+jh}) \leq C_1 p_\nu^{-1/n} \left(\frac{1}{2}\right)^{j/n}.$$

Corollary 2.13. *Let α be badly approximable. Then there exists $h^* = h^*(f, \alpha) \in \mathbb{N}$ such that*

$$f(\xi_{\nu+h^*}) < \frac{1}{2} f(\xi_\nu) \qquad \text{for all } \nu \geq 1. \qquad (2.8)$$

Proof of Corollary 2.13. From (2.2) and the condition (2.6) it follows that

$$D_1 p_\nu^{-1/n} \leq f(\xi_\nu) \leq C_1 p_\nu^{-1/n}$$

and by Lemma 2.2 p_ν grows exponentially. Now (2.8) follows. $\square$

Lemma 2.14. *Let f be strictly convex. Then for any $\varepsilon > 0$ there exists $\delta > 0$ such that for any $\theta \in \partial B_f^1(0)$ and any $\xi \in B_f^1(0) \setminus B_f^1(\theta)$ under the condition*

$$\Xi(\xi) \in \partial B_f^1(0) \bigcap \left(B_f^{1+\delta}(\theta) \setminus B_f^1(\theta) \right)$$

we have $f(\xi) > 1 - \varepsilon$.

Proof. Let $\eta \in \partial B_f^1(\theta) \bigcap \partial B_f^1(0)$. As f is strictly convex we have $(0; \eta) \subset$ int $B_f^1(\theta)$. Now if $\Xi \in \partial B_f^1(0) \setminus B_f^1(\theta)$ belongs to a small δ-neighbourhood of the point η then the segment $[0; \Xi]$ must intersect with $\partial B_f^1(\theta)$ at some point $\zeta(\Xi) = [0; \Xi] \cap (\partial B_f^1(\theta) \setminus 0)$ and $\zeta(\Xi) \to \eta$ when $\Xi \to \eta$. If $\xi \in B_f^1(0) \setminus B_f^1(\theta)$ then ξ is between $\Xi(\xi)$ and $\zeta(\Xi(\xi))$.

The lemma is proved. $\square$

2.10 Proofs of Theorems 2.9 and 2.10

We prove Theorem 2.9.

Suppose that Theorem 2.9 is not valid. Then for any $\delta > 0$ we have

$$\Xi_{\nu+1} \in B_f^{1+\delta}(\Xi_\nu)$$

for $\nu \geq \nu_0(\delta)$. Now from Theorem 2.8 and Lemma 2.14 we deduce that for any $\varepsilon > 0$

$$f(\xi_{\nu+1}) \geq (1 - \varepsilon) f(\xi_\nu)$$

when $\nu \geq \nu_0(\varepsilon)$. This means that

$$f(\xi_{\nu_0+j}) \geq (1 - \varepsilon)^j f(\xi_{\nu_0}). \tag{2.9}$$

But from Corollary 2.12 and Lemma 2.11 we see that

$$f(\xi_{\nu_0+j}) \leq C_1 p_{\nu_0}^{-1/n} (1/2)^{j/n}. \tag{2.10}$$

For small values of ε the inequalities (2.9) and (2.10) lead to a contradiction when $j \to \infty$. Thus Theorem 2.9 is proved.

Now we prove Theorem 2.10.

Suppose that Theorem 2.10 is not valid. Then for arbitrary large $w \in \mathbb{N}$ and for arbitrary small $\delta > 0$ there exists ν satisfying the condition

$$\Xi_{j+1} \in B_f^{1+\delta}(\Xi_j), \quad j = \nu, \nu+1, \ldots, \nu+w.$$

Applying Lemma 2.14 we see that

$$f(\xi_{\nu+w}) \geq (1 - \varepsilon)^w f(\xi_\nu), \tag{2.11}$$

and $\varepsilon > 0$ may be taken arbitrarily small. But at the same time from Corollary 2.13 and Lemma 2.11 we deduce that

$$f(\xi_{\nu+w}) \leq \left(\frac{1}{2}\right)^{[w/h^*]} f(\xi_\nu). \tag{2.12}$$

Again we take ε small enough and the inequalities (2.11) and (2.12) lead to a contradiction when $w \to \infty$.

The proofs are complete.

2.11 A result on signatures and illuminated points

For a vector $\eta = (\eta_1, \ldots, \eta_n)$ its *signature* is defined as

$$\operatorname{sign} \eta = (\operatorname{sign} \eta_1, \ldots, \operatorname{sign} \eta_n).$$

Rogers [21] showed that for ordinary b.s.a. (in the case $B_f^1 = \{y : \max_j |y_j| \leq 1\}$) the successive best approximations satisfy the condition

$$\operatorname{sign} \xi_\nu \neq \operatorname{sign} \xi_{\nu+1} \qquad \text{for all } \nu.$$

(This simple result was generalized in Theorem 2.8.) On the other hand, Sos and Szekeres [23] proved that for any sequence of signatures $\{\sigma_\nu\}$ with $\sigma_\nu \neq \sigma_{\nu+1}$ there exists a vector $\alpha = (\alpha_1, \ldots, \alpha_n) \in \mathbb{R}^n$ with $1, \alpha_1, \ldots, \alpha_n$ linearly independent over $\mathbb{Z}$ such that $\operatorname{sign} \xi_\nu = \sigma_\nu$. We give a generalization of this result.

Let $M \subset \mathbb{R}^n$ be a convex closed domain, $b \in \partial M$ and $a \notin M$. The point b (as a point of the boundary ∂M) is *illuminated from the point a* if there exists a positive λ such that $b + \lambda(b - a) \in \operatorname{int} M$.

Theorem 2.15. *Let the sequence of points $\{\theta_\nu\}_{\nu=1}^\infty \subset B_f^1$ satisfy the following condition: for each ν the point 0 as the point of the boundary $\partial B_f^1(\theta_\nu)$ is illuminated from the point $\theta_{\nu+1}$. Then there exists a vector $\alpha = (\alpha_1, \ldots, \alpha_n)$ with linearly independent components such that*

$$\lim_{\nu \to +\infty} |\Xi_\nu - \theta_\nu| = 0. \tag{2.13}$$

Remark 2.16. *The result of Sos and Szekeres on signatures immediately follows from our Theorem 2.15.*

Remark 2.17. *In (2.13) we can provide any rate of convergence to zero.*

We remark that the formulation of the conditions of Theorem 2.15 in terms of illuminated points is due to O. German, who has moreover proved some interesting results [5] on the distribution of directions for the best approximations in the sense of linear forms and on the rate of convergence to the asymptotic directions.

In the next section we shall give a sketch of the proof of this theorem and here we consider one example.

In the case $s = 2$ we consider the norm $f^*(x_1, x_2)$ with unit ball B_f^1 defined by the inequalities

$$|x_1 + x_2| \le 4, \quad |x_1 - x_2| \le 1.$$

Applying Theorem 2.15 and observing that the geometry of mutual configuration of the balls $B_f^1(0)$ and $B_f^1(\theta)$ we obtain the following statement.

Theorem 2.18. *For the norm $f^*(x)$ the set of all f-b.s.a. may have the constant sequence of signatures: $\sigma_\nu = (+, +)$ for all ν.*

Theorem 2.18 shows that the conclusion of Rogers' theorem from [21] is not true for the norm $f^*(x)$. One can easily construct the corresponding multi-dimensional example and an example with strictly convex norm.

We point out that we cannot construct an example of a norm f for which the sequence of all f-b.s.a. can have any given sequence of signatures. We may conjecture that the Euclidean norm $f(x_1, \ldots, x_n) = \sqrt{x_1^2 + \ldots + x_n^2}$ has this property.

2.12 Sketch of the proof of Theorem 2.15

The proof is performed in the same manner as the proof of Theorem 2.4. By means of an inductive procedure we construct a sequence of integer points

$$\tau_\nu = (p_\nu, a_{1,\nu}, \ldots, a_{n,\nu})$$

which form the sequence of all f-b.s.a. for the limit point

$$\lim_{\nu \to +\infty} (a_{1,\nu}/p_\nu, \ldots, a_{n,\nu}/p_\nu).$$

The base of the induction is trivial.

We sketch the induction step. Let the points $\tau_1, \ldots, \tau_\nu \in \mathbb{Z}^{n+1}$, where

$$\tau_j = (p_j; a_j) = (p_j, a_{1,j}, \ldots, a_{n,j}), \qquad 1 \le p_1 < p_2 < \ldots < p_\nu,$$

be constructed satisfying the following conditions:

1) $\tau_1, \ldots, \tau_\nu$ is the set of *all* f-b.s.a. to rational vector

$$\beta^\nu = (a_{1,\nu}/p_\nu, \ldots, a_{n,\nu}/p_\nu),$$

2) $\Xi(\xi^{\beta^\nu}(\tau_j)) - \theta_j$ is small for all $j = 1, \ldots, \nu - 1$,
3) θ_j illuminates the point 0 of the boundary of $\partial B_f^1(\Xi(\xi^{\beta^\nu}(\tau_{j-1})))$ for all $j = 1, \ldots, \nu$,
4) there are no integer points on the boundary of the cylinder

$$\{(x, y_1, \ldots, y_n) \in \mathbb{R}^{n+1} : |x| < p_\nu,\ f(\alpha x - y) \le f(\xi_{\nu-1})\}$$

but the best approximations.

We must show how one can determine an integer point

$$\tau_{\nu+1} = (p_{\nu+1}; a_{\nu+1}) = (p_{\nu+1}, a_{1,\nu+1}, \ldots, a_{n,\nu+1}), \quad p_\nu < p_{\nu+1}$$

such that

1*) $\tau_1, \ldots, \tau_{\nu+1}$ are *all* f-b.s.a. to the rational vector

$$\beta^{\nu+1} = (a_{1,\nu+1}/p_{\nu+1}, \ldots, a_{n,\nu+1}/p_{\nu+1}),$$

2*) $\Xi(\xi^{\beta^{\nu+1}}(\tau_j)) - \theta_j$ is small for all $j = 1, \ldots, \nu$,
3*) θ_j illuminates the point 0 of the boundary $\partial B_f^1(\Xi(\xi^{\beta^{\nu+1}}(\tau_{j-1}))$ for all $j = 1, \ldots, \nu + 1$,
4*) there are no integer points on the boundary of the cylinder

$$\{(x, y_1, \ldots, y_n) \in \mathbb{R}^{n+1} : |x| < p_{\nu+1},\ f(\alpha x - y) \le f(\xi_\nu + 1)\}$$

but the best approximations.

Consider a small neighbourhood of $B_f^\lambda(a_\nu)$. Let λ be small enough. Then for any $\beta \in B_f^\lambda(a_\nu)$ the integer points $\tau_1, \ldots, \tau_{\nu-1}$ form *all* the first successive $\nu - 1$ f-b.s.a to β. The main difficulty is that for any λ there are some $\beta \in B_f^\lambda(a_\nu)$ for which between $\tau_{\nu-1}$ and τ_ν there must arrive one new f-b.s.a. and it must be controlled.

We consider the ball B_f^1 and the point $\Xi(\xi^{\beta^\nu}(\tau_{\nu-1})) \in \partial B_f^1$. Let $B^* = B_f^1(\Xi(\xi^{\beta^\nu}(\tau_{\nu-1})))$. From the induction hypothesis 3) we know that θ_ν illuminates $0 \in B^*$.

Then near the point $\theta_\nu \cdot t$ for some positive t in the set $B_f^1 \cap \text{int } B^*$ there exists a point Ξ^* such that the two-dimensional subspace π^* generated by the points $\tau_\nu, \zeta^* = (p_\nu, \Xi^*)$ is absolutely rational. The point Ξ^* must be very close to $\theta_\nu \cdot t$. So due to continuity we can obtain $\theta_{\nu+1} \cdot t' \in$ int $B_f^1(\Xi^*)$ for some positive t'.

In the absolutely rational subspace π^* a point $\tau_{\nu+1} = (p_{\nu+1}; a_{\nu+1})$ must be taken in such a way that $\tau_{\nu+1}$ and ζ^* be on the same side of the line $0\tau_\nu$ (here we use the fact that π^* has dimension 2). As $\Xi^* \in$ int B^* we deduce from convexity that the whole segment $(0; \Xi^*)$ is in int B^*. Now we can choose $\tau_{\nu+1}$ very close to the line $0\tau_\nu$ and hence the sequence $\tau_1, \ldots, \tau_{\nu+1}$ really is the set of *all* f-b.s.a. to the rational vector $\beta^{\nu+1} = (a_{1,\nu+1}/p_{\nu+1}, \ldots, a_{n,\nu+1}/p_{\nu+1})$.

This concludes the inductive step.

The sequence of vectors β^ν converges due to the smallness of the difference $|\beta^{\nu+1} - \beta^\nu|$.

Linear independence over $\mathbb{Z}$ of the limit numbers $1, \alpha_1, \ldots, \alpha_n$ may be obtained by the application of Proposition 2.2.

2.13 Asymptotic directions

In this section we formulate, without proofs, some simple corollaries from our previous results in terms of the asymptotic directions for the best approximations.

The *asymptotic direction* for the f-b.s.a. sequence for a vector α is defined as a point $\theta \in \partial B_f^1(0)$ such that there exists a subsequence ν_j with the property $\lim_{j \to +\infty} \Xi_{\nu_j} = \theta$. The set of all asymptotic directions for α we denote by $\Gamma_f(\alpha)$. Obviously $\Gamma_f(\alpha) \subseteq B_f^1(0)$ is closed.

It seems to the author that C. Rogers was the first to define asymptotic direction for Diophantine approximations [20] but our definition differs from Rogers'.

A set $\mathcal{A} \subseteq B_f^1(0)$ is defined to be f-*asymptotically admissible* if there is an infinite sequence $\theta_0, \theta_1, \ldots, \theta_k, \ldots$ with $\theta_j \in \mathcal{A}$ such that

1) θ_j illuminates the point $0 \in \partial B_f^1(\theta_{j-1})$,
2) the set of all limiting points of the sequence $\{\theta_k\}$ is just $\mathcal{A}$.

Theorem 2.19. *Let $\mathcal{A} \subseteq B_f^1(0)$ be f-asymptotically admissible. Then there exists a vector $\alpha \in \mathbb{R}^n$ with components that are linearly independent over the rationals such that $\mathcal{A} = \Gamma_f(\alpha)$.*

Corollary . *If $\mathcal{A}$ is closed and there is $x \in \mathcal{A}$ such that $-x \in \mathcal{A}$ then there exists $\alpha \in \mathbb{R}^n$ with independent components such that $\mathcal{A} = \Gamma_f(\alpha)$.*

This result may be compared to Rogers' observation [20] that the set of all asymptotic directions is not necessarily 0-symmetric but there is some kind of symmetry.

The next theorem follows from Theorem 2.9.

Theorem 2.20. *Let the norm f be strictly convex. Then there exists a positive δ_1 depending on f such that in the case*

$$\mathcal{A} \subset \operatorname{int} B_f^{1+\delta_1}(\theta) \qquad \textit{for all } \theta \in \mathcal{A}$$

$\mathcal{A}$ *cannot be a set of the form* $\Gamma_f(\alpha)$.

Bibliography

[1] J. W. S. Cassels. *An introduction to Diophantine approximation*, volume 45 of *Cambridge Tracts in Mathematics and Mathematical Physics*. Cambridge University Press, New York, 1957.

[2] J. W. S. Cassels. *An introduction to the geometry of numbers*. Die Grundlehren der mathematischen Wissenschaften, Bd. 99. Springer-Verlag, Berlin-Göttingen-Heidelberg, 1959.

[3] T. W. Cusick. Diophantine approximation of linear forms over an algebraic number field. *Mathematika*, 20:16–23, 1973.

[4] T. W. Cusick. Best Diophantine approximation for ternary linear forms. *J. Reine Angew. Math.*, 315:40–52, 1980.

[5] O. N. German. Asymptotic directions for best approximations of an n-dimensional linear form. *Mat. Zametki*, 75(1):55–70, 2004. Translation in *Math. Notes*, 75(1–2) 51–65, 2004.

[6] P. M. Gruber and C. G. Lekkerkerker. *Geometry of numbers*, volume 37 of *North-Holland Mathematical Library*. North-Holland Publishing Co., Amsterdam, second edition, 1987.

[7] A. Y. Hinčin. Regulai systems of linear equations and a general problem of Čebyšev. *Izvestiya Akad. Nauk SSSR. Ser. Mat.*, 12:249–258, 1048

[8] A. Y. Khinchin. Über eine klasse linearer diophantische Approximationen. *Rendiconti Circ. Math. Palermo*, 50:170–195, 1926.

[9] A. Y. Khinchin. *Continued fractions*. The University of Chicago Press, 1964.

[10] J. C. Lagarias. Best simultaneous Diophantine approximations. I. Growth rates of best approximation denominators. *Trans. Amer. Math. Soc.*, 272(2):545–554, 1982.

[11] J. C. Lagarias. Best simultaneous Diophantine approximations. II. Behavior of consecutive best approximations. *Pacific J. Math.*, 102(1):61–88, 1982.

[12] J. C. Lagarias. Best Diophantine approximations to a set of linear forms. *J. Austral. Math. Soc. Ser. A*, 34(1):114–122, 1983.

[13] P. Loshak. Canonical perturbation theory: an approach based on joint approximations. *Uspekhi Mat. Nauk*, 47(6(288)):59–140, 1992. Translation in *Russian Math. Surveys*, 47(6):57–133, 1992.

[14] N. G. Moshchevitin. On best joint approximations. *Uspekhi Mat. Nauk*, 51(6(312)):213–214, 1996. Translation in *Russian Math. Surveys*, 51(6): 1214–1215, 1996.

[15] N. G. Moshchevitin. Multidimensional Diophantine approximations and dynamical systems. *Regul. Khaoticheskaya Din.*, 2(1):81–95, 1997.

[16] N. G. Moshchevitin. On the geometry of best approximations. *Dokl. Akad. Nauk*, 359(5):587–589, 1998.

[17] N. G. Moshchevitin. Continued fractions, multidimensional Diophantine approximations and applications. *J. Théor. Nombres Bordeaux*, 11(2):425–438, 1999.

[18] N. G. Moshchevitin. Best simultaneous approximations: norms, signatures, and asymptotic directions. *Mat. Zametki*, 67(5):730–737, 2000. Translation in *Math. Notes*, 67(5–6): 5-6, 618–624, 2000.

[19] N. G. Moshchevitin. Distribution of Kronecker sequences. In *Algebraic number theory and Diophantine analysis (Graz, 1998)*, pages 311–329. de Gruyter, Berlin, 2000.

[20] C. A. Rogers. The asymptotic directions of n linear forms in $n+1$ integral variables. *Proc. London Math. Soc. (2)*, 52:161–185, 1951.

[21] C. A. Rogers. The signatures of the errors of simultaneous Diophantine approximations. *Proc. London Math. Soc. (2)*, 52:186–190, 1951.

[22] W. M. Schmidt. *Diophantine approximation*, volume 785 of *Lecture Notes in Mathematics*. Springer, Berlin, 1980.

[23] G. Szekeres and V. T.-Sós. Rational approximation vectors. *Acta Arith.*, 49(3):255–261, 1988.

[24] G. F. Voronoi. On one generalization of continued fractions algorithm. In *Selected works, 1, AN USSR, Kiev*. 1952. (in Russian).

Department of Number Theory
Faculty of Mathematics and Mechanics
Moscow State University
Leninskie Gory
119992 Moscow
Russia
moshchevitin@rambler.ru

Projectively dual varieties
of homogeneous spaces

Evgueni A. Tevelev

Introduction

Various manifestations of projective duality have inspired research in algebraic and differential geometry, classical mechanics, invariant theory, combinatorics, etc. On the other hand, projective duality is simply a systematic way of recovering a projective variety from the set of its tangent hyperplanes. In this survey we have tried to collect together different aspects of projective duality and points of view on it. To save space we omit almost all proofs, but even this cannot save these notes from being incomplete. We hope that the interested reader will take a closer look at the many beautiful papers and books cited here.

An interesting feature of projective duality is given by the observation that the most important examples carry a natural action of a Lie group. This is especially true for projective varieties that have extremal properties from the point of view of projective geometry. We have tried to stress this phenomenon in this survey and to discuss many variants of it. However, one aspect is completely omitted – we do not discuss the dual varieties of toric varieties and the corresponding theory of A-discriminants. This theory is presented in the fundamental book [35] and we feel no need to reproduce it.

I would like to thank F. Zak for very inspiring discussions on projective geometry and S. Keel for many critical remarks that helped to improve the exposition. I am grateful to P. Aluffi, R. Muñoz, V. Popov, D. Saltman and A. J. Sommese for many helpful comments and encouragement. These lecture notes were written during my stay at the University of Glasgow and I would like to thank my hosts for their warm hospitality. I am grateful to the London Mathematical Society for financial support.

1 Projectively dual varieties

For any finite-dimensional complex vector space V we denote by $\mathbb{P}(V)$ its projectivization, that is, the set of 1-dimensional subspaces. If $U \subset V$ is a non-trivial linear subspace then $\mathbb{P}(U) \subset \mathbb{P}(V)$. We denote by V^* the dual vector space, the vector space of linear forms on V. Points of $\mathbb{P}(V)^* = \mathbb{P}(V^*)$ correspond to hyperplanes in $\mathbb{P}(V)$. Conversely, to any $p \in \mathbb{P}(V)$, we can associate a hyperplane in $\mathbb{P}(V)^*$, namely the set of all hyperplanes in $\mathbb{P}(V)$ passing through p. Therefore, $\mathbb{P}(V)^{**}$ is naturally identified with $\mathbb{P}(V)$. With any vector subspace $U \subset V$ we associate its annihilator $U^\perp = \{f \in V^* \mid f(U) = 0\}$. We have $(U^\perp)^\perp = U$. This corresponds to projective duality between projective subspaces in $\mathbb{P}(V)$ and $\mathbb{P}(V)^*$: for any projective subspace $L \subset \mathbb{P}(V)$ we denote by $L^* \subset \mathbb{P}(V)^*$ its dual projective subspace, parametrizing all hyperplanes that contain L.

Remarkably, projective duality between projective subspaces in $\mathbb{P}^N$ and $\mathbb{P}^{N^*}$ can be extended to an involutive correspondence between irreducible algebraic subvarieties in $\mathbb{P}^N$ and $\mathbb{P}^{N^*}$. Suppose that $X \subset \mathbb{P}^N$ is an irreducible algebraic subvariety, $\dim X = n$. For any $x \in X$, we denote by $\hat{T}_x X \subset \mathbb{P}^N$ the embedded projective tangent space. More precisely, we define $\mathrm{Cone}(X) \subset V$ as the conical variety formed by all lines l such that $\mathbb{P}(l) \in X$. If $x \in X$ is a smooth point then any non-zero point v of the corresponding line is a smooth point of $\mathrm{Cone}(X)$ and $\hat{T}_x(X)$ is defined as $\mathbb{P}(T_v \mathrm{Cone}(X))$, where $T_v \mathrm{Cone}(X)$ is the tangent space of $\mathrm{Cone}(X)$ at v considered as a linear subspace of V (it does not depend on the choice of v).

A hyperplane $H \subset \mathbb{P}^n$ is said to be *tangent* to X if it contains $\hat{T}_x X$ for some smooth point $x \in X_{sm}$. The closure of the set of all tangent hyperplanes is called the *dual variety* $X^* \subset \mathbb{P}^{N^*}$.

Consider the set $I_X^0 \subset \mathbb{P}^N \times \mathbb{P}^{N^*}$ of pairs (x, H) such that $x \in X_{sm}$ and H is a hyperplane tangent to X at x. The Zariski closure I_X of I_X^0 is called the conormal variety of X. The projection $\mathrm{pr}_1 : I_X^0 \to X_{sm}$ makes I_X^0 into a bundle over X_{sm} whose fibres are projective subspaces of dimension $N - n - 1$. Therefore, I_X^0 and I_X are irreducible varieties of dimension $N - 1$. By definition, X^* is the image of the projection $\mathrm{pr}_2 : I_X \to \mathbb{P}^{N^*}$. Therefore, X^* is an irreducible variety. Moreover, since $\dim I_X = N - 1$, we can expect that in 'typical' cases X^* will be a hypersurface.

The number $\mathrm{codim}_{\mathbb{P}^{N^*}} X^* - 1$ is called the defect of X, denoted by def X. If def $X = 0$ then X^* is defined by an irreducible homogeneous polynomial Δ_X, called the *discriminant* of X.

Example 1.1. The most familiar example of Δ_X is, of course, the discriminant of a binary form. In order to show that it actually coincides with some Δ_X we first give an equivalent definition of Δ_X. Suppose that $x_1, \ldots, x_{n+1}$ are some local coordinates on $\mathrm{Cone}(X) \subset V$. Any $f \in V^*$ (restricted to $\mathrm{Cone}(X)$) is an algebraic function in $x_1, \ldots, x_{n+1}$. Then Δ_X is just an irreducible polynomial, which vanishes at $f \in V^*$ whenever the function $f(x_1, \ldots, x_{n+1})$ has a multiple root, that is, vanishes at some $v \in \mathrm{Cone}(X)$, $v \neq 0$, together with all first derivatives $\partial f / \partial x_i$.

Consider now the d-dimensional projective space $\mathbb{P}^d = \mathbb{P}(V)$ with homogeneous coordinates $z_0, \ldots, z_d$, and let $X \subset \mathbb{P}^d$ be the Veronese curve

$$(x^d : x^{d-1}y : x^{d-2}y^2 : \ldots : xy^{d-1} : y^d), \ x, y \in \mathbb{C}, \ (x, y) \neq (0, 0)$$

(the image of the Veronese embedding $\mathbb{P}^1 \subset \mathbb{P}^d$). Any linear form $f(z) = \sum a_i z_i$ is uniquely determined by its restriction to $\mathrm{Cone}(X)$, which is a binary form $f(x, y) = \sum a_i x^{d-i} y^i$. Therefore, $f \in \mathrm{Cone}(X^*)$ if and only if $f(x, y)$ vanishes at some point $(x_0, y_0) \neq (0, 0)$ (so $(x_0 : y_0)$ is a root of $f(x, y)$) with its first derivatives (so $(x_0 : y_0)$ is a multiple root of $f(x, y)$). It follows that Δ_X is the classical discriminant of a binary form.

The following basic result is called the Reflexivity Theorem. Different proofs can be found in [35, 82, 62, 94].

Theorem 1.2. *For any irreducible projective variety $X \subset \mathbb{P}^n$, we have $X^{**} = X$. More precisely, If z is a smooth point of X and H is a smooth point of X^*, then H is tangent to X at z if and only if z, regarded as a hyperplane in $\mathbb{P}^{n*}$, is tangent to X^* at H.*

Up to linear projections, a projective embedding of an algebraic variety is determined by the corresponding invertible sheaf, see e.g. [39]. A pair $(X, \mathcal{L})$ of a projective variety and a very ample invertible sheaf on it is called a *polarized variety*. Any polarized variety admits a canonical embedding in a projective space with a linearly normal image, given by the complete linear system $|\mathcal{L}|$. Therefore we may speak about dual varieties, defect, discriminants, etc. of polarized varieties without any confusion.

2 Local calculations

The defect can be calculated locally. Let x_0 be a smooth point of X. Then one can choose linear functionals $T_0 \in V^* \setminus x_0^\perp$, $T_1, \ldots, T_n \in x_0^\perp$ such that the functions $t_1 = T_1/T_0$, $t_2 = T_2/T_0$, $\ldots$, $t_n = T_n/T_0$ are local coordinates on X in a neighbourhood of x_0. For every $U \in x_0^\perp$ the function $u = U/T^0$ on X near x_0 is an analytic function of $t_1, \ldots, t_n$ such that $u(0, \ldots, 0) = 0$. Consider the Hessian matrix

$$\mathrm{Hes}(u) = \mathrm{Hes}(U; T_0, T_1, \ldots, T_n; x_0) = \left(\frac{\partial^2 u}{\partial t_i \partial t_j}(0, \ldots, 0) \right)_{i,j=1,\ldots,n}.$$

Theorem 2.1 ([46]). $\mathrm{def}(X) = \min \mathrm{corank}\, \mathrm{Hes}(u)$, *the minimum over all possible choices of x_0 and U.*

If $X \subset \mathbb{P}^n$ is a hypersurface (or a complete intersection) then it is possible to rewrite Hessian matrices in homogeneous coordinates. In case of hypersurfaces the corresponding result was first formulated by B. Segre [82]. Namely, let $f(x_0, \ldots, x_N)$ be an irreducible homogeneous polynomial and let $X \subset \mathbb{P}^N$ be the hypersurface with the equation $f = 0$. Let m be the largest number with the following property: there exists a $(m \times m)$-minor of the Hessian matrix $(\partial^2 f/\partial x_i \partial x_j)$ that is not divisible by f. Then $\dim X^* = m - 2$.

The local machinery can be developed quite far, see [36, 54, 51, 96].

3 The contact locus and its normal bundle

We say that X is *ruled* in projective subspaces of dimension r if for any $x \in X$ there exists a projective subspace L of dimension r such that $x \in L \subset X$. By a standard closedness argument it is sufficient to check this property on some Zariski open dense subset $U \subset X$. The following result is an easy consequence of the Reflexivity Theorem.

Theorem 3.1. *Suppose that $\mathrm{def}\, X = r \geq 1$. Then X is ruled in projective subspaces of dimension r. If X is smooth then for any $H \in X^*_{sm}$ the contact locus $\mathrm{Sing}\, X \cap H$ is a projective subspace of dimension r and the union of these projective subspaces is dense in X.*

Suppose that X is smooth and non-linear. For any hyperplane $H \subset \mathbb{P}^N$ the contact locus $\mathrm{Sing}\, X \cap H$ is the subvariety of X consisting of all points $x \in X$ such that the embedded tangent space $\hat{T}_x X$ is contained in H. One can use the Jacobian ideal of $X \cap H$ to define a scheme structure on the contact locus; however, this scheme could be non-reduced. Clearly the contact locus is non-empty if and only if $H \in X^*$. If def $X = k$ then for any $H \in X^*{}_{sm}$ the contact locus $\mathrm{Sing}(H{\cap}X)$ is a projective subspace of dimension k and the union of these projective subspaces is dense in X. Suppose that q is a generic point of X and H is a generic tangent hyperplane of X at q, $L = \mathbb{P}^k$ is the contact locus of H with X.

Theorem 3.2 ([24]). *If $p \in L$, then the tangent cone of the hyperplane section $H \cap X$ at p is a quadric hypersurface of rank $n - k$ in $\hat{T}_p(X)$.*

Recall that any vector bundle on $\mathbb{P}^1$ has the form $\oplus_i \mathcal{O}(a_i)$ for some integers a_i (see e.g. [39]). A vector bundle E on a projective space $\mathbb{P}^N$ is called uniform if for any line $T \subset \mathbb{P}^N$ the restriction $E|_T$ is a fixed vector bundle $\oplus_i \mathcal{O}(a_i)$, cf. [69].

Theorem 3.3 ([24]). *The normal bundle $N_L X$ is uniform and $N_L X \cong (N_L X)^*(1)$. If $T = \mathbb{P}^1$ is a line in $L = \mathbb{P}^k$, then $N_L X|_T \cong \mathcal{O}_T^{\oplus(n-k)/2} \oplus \mathcal{O}_T(1)^{\oplus(n-k)/2}$.*

As a quite formal consequence of Theorem 3.3 we get the following parity theorem, which was first proved by A. Landman using the Picard-Lefschetz theory (unpublished):

Theorem 3.4 ([24]). *If def $X > 0$ then $\dim X \equiv \operatorname{def} X \mod 2$.*

It is possible to describe cases when the normal bundle splits as a sum of line bundles. The following result is well-known:

Theorem 3.5. *Let $X \subset \mathbb{P}^N$ be an n-dimensional scroll, i.e. a projective bundle $\mathbb{P}_Y(E)$ over a smooth variety Y such that all fibres are embedded linearly and let $\dim Y = m$. Suppose $n \geq 2m$. Then $\operatorname{def} X = n - 2m$ and $N_L^* X$ splits as a sum of line bundles.*

Moreover, the converse is almost true:

Theorem 3.6 ([23, 66]). *Let $X \subset \mathbb{P}^N$, $\dim X = n$, $\operatorname{def} X > 1$. If $N_L X$ splits as a sum of line bundles then X is a scroll.*

4 Fibrations and divisors

The following simple but very useful result first appeared in [55] (where it was attributed to the referee). It could be called the monotonicity theorem.

Theorem 4.1. *Let $X \subset \mathbb{P}^N$ be a smooth projective n-dimensional variety. Suppose that through its generic point there passes a smooth subvariety Y of dimension h and defect θ. Then $\operatorname{def} X \geq \theta - n + h$. In other words, $\dim X + \operatorname{def} X \geq \dim Y + \operatorname{def} Y$.*

Suppose that $X \subset \mathbb{P}^N$ is a smooth projective variety. The following theorem allows one to find the defect of smooth hyperplane or hypersurface sections of X.

Theorem 4.2 ([23, 40, 41]).

(a) *Assume that $Y = X \cap H$ is a smooth hyperplane section of X. Then we have $\operatorname{def} Y = \max\{0, \operatorname{def} X - 1\}$. Moreover, if X^* is not a hypersurface, then the dual variety Y^* is the linear projection of X^* with centre H.*

(b) *Assume that Y is a smooth divisor corresponding to a section of $\mathcal{O}_X(d)$ for $d \geq 2$. Then $\operatorname{def} Y = 0$.*

5 Dual varieties and jet bundles

Let $(X, \mathcal{L})$ be a smooth polarised projective variety. Consider the bundle $J(\mathcal{L})$ of first jets of sections of $\mathcal{L}$. If f is a section of $\mathcal{L}$, then $j = j(f)$ is the corresponding first jet. The relevance of jets to dual varieties is as follows. Any $f \in V^*$ is a linear function on V and hence can be regarded as a global section of $\mathcal{L}$. It is clear that $f \in V^*$ represents a point in the dual variety X^* if and only if $j(f)$ vanishes at some point $x \in X$. With the aid of simple properties of Koszul complexes, it is easy to see that this is equivalent to the non-exactness of any of the following complexes of sheaves:

$$\mathcal{K}_+(J(\mathcal{L}), j) = \left\{ 0 \to \mathcal{O}_X \xrightarrow{j} J(\mathcal{L}) \xrightarrow{\wedge j} \Lambda^2 J(\mathcal{L}) \xrightarrow{\wedge j} \ldots \xrightarrow{\wedge j} \Lambda^r J(\mathcal{L}) \to 0 \right\}$$

$$\mathcal{K}_-(J(\mathcal{L}), j) = \left\{ 0 \to \Lambda^r J(\mathcal{L})^* \xrightarrow{i_j} \ldots \xrightarrow{i_j} \Lambda^2 J(\mathcal{L})^* \xrightarrow{i_j} J(\mathcal{L})^* \xrightarrow{i_j} \mathcal{O}_X \to 0 \right\}.$$

Here the differential in $\mathcal{K}_+$ is given by exterior multiplication with j and the differential in $\mathcal{K}_-$ is given by contraction with j. After an appropriate twist, exactness of the complex of sheaves is equivalent to exactness of the complex of its global sections. However, it turns out that in our case much more is true. We use notations $C_+^i(X, \mathcal{M}) = H^0(X, \Lambda^i J(\mathcal{L}) \otimes \mathcal{M})$, $C_-^i(X, \mathcal{M}) = H^0(X, \Lambda^{-i} J(\mathcal{L})^* \otimes \mathcal{M})$ for an invertible sheaf $\mathcal{M}$.

Theorem 5.1 ([35]). *Suppose that all cohomology groups $H^i(X, \Lambda^\bullet J(\mathcal{L}) \otimes \mathcal{M})$ or $H^i(X, \Lambda^\bullet J(\mathcal{L})^* \otimes \mathcal{M})$ vanish for $i > 0$. If $f \in V^*$*

does not belong to the dual variety $X^* \subset \mathbb{P}(V^*)$, then $(C_+^*(X, \mathcal{M}), \partial_f)$ or $(C_-^*(X, \mathcal{M}), \partial_f)$ respectively is exact. Moreover, $\Delta_{X,\mathcal{M}}^- = \Delta_X$ or $(\Delta_{X,\mathcal{M}}^+)^{(-1)^{\dim X+1}} = \Delta_X$, where Δ_X is the discriminant of X and $\Delta_{X,\mathcal{M}}^-$ or $\Delta_{X,\mathcal{M}}^+$ respectively is the Cayley determinant of the exact complex.

The determinants of exact complexes (in the implicit form) were first introduced by Cayley [15]. A systematic early treatment of this subject was undertaken by Fisher [27] whose aim was to give a rigorous proof of Cayley's results. In topology determinants of complexes were introduced in 1935 by Reidermeister and Franz [28]. They used the word 'torsion' for the determinant-type invariants constructed. More details can be found in [35, 52, 21, 78, 79, 95]. We shall give a definition only, in a slightly non-standard way.

The base field k can be arbitrary. Suppose that V is a finite-dimensional vector space. Then the top-degree component of the exterior algebra $\Lambda^{\dim V} V$ is called the determinant of V, denoted by $\operatorname{Det} V$. If $V = 0$ then we set $\operatorname{Det} V = k$. It is easy to see that for any exact triple $0 \to U \to V \to W \to 0$ we have a natural isomorphism $\operatorname{Det} V \simeq \operatorname{Det} U \otimes \operatorname{Det} W$.

Suppose now that $V = V_0 \oplus V_1$ is a finite-dimensional supervector space. Then, by definition, $\operatorname{Det} V$ is set to be $\operatorname{Det} V_0 \otimes (\operatorname{Det} V_1)^*$. Once again, for any exact triple of supervector spaces $0 \to U \to V \to W \to 0$ we have $\operatorname{Det} V \simeq \operatorname{Det} U \otimes \operatorname{Det} W$. For any supervector space V we denote by $\tilde{V}$ the new supervector space given by $\tilde{V}_0 = V_1$, $\tilde{V}_1 = V_0$. Clearly, we have a natural isomorphism $\operatorname{Det} \tilde{V} = (\operatorname{Det} V)^*$.

Now let (V, ∂) be a finite-dimensional supervector space with a differential ∂ such that $\partial V_0 \subset V_1$, $\partial V_1 \subset V_0$, $\partial^2 = 0$. Then $\operatorname{Ker} \partial$, $\Im \partial$, and the cohomology space $H(V) = \operatorname{Ker} \partial / \Im \partial$ are again supervector spaces. There exists a natural isomorphism $\operatorname{Det} V \simeq \operatorname{Det} H(V)$. In particular, if ∂ is exact, $H(V) = 0$, then we have a natural isomorphism $\operatorname{Det} V \simeq k$.

Let us fix some bases $\{e_1, \ldots, e_{\dim V_0}\}$ in V_0 and $\{e_1', \ldots, e_{\dim V_1}'\}$ in V_1. Let $\{f_1', \ldots, f_{\dim V_1}'\}$ be a dual basis in V_1^*. Then we have the basis vector

$$e_1 \wedge \ldots \wedge e_{\dim V_0} \otimes f_1' \wedge \ldots \wedge f_{\dim V_1}' \in \operatorname{Det} V.$$

Therefore, if (V, ∂, e) is a based supervector space with an exact differential then using the natural isomorphism $\operatorname{Det} V \simeq k$ we get a number $\det(V, \partial, e)$ called the Cayley determinant of a based supervector space with an exact differential. If we fix other bases $\{\tilde{e}_1, \ldots, \tilde{e}_{\dim V_0}\}$ in V_0

and $\{\tilde{e}'_1, \dots, \tilde{e}'_{\dim V_1}\}$ in V_1 then, clearly,

$$\det(V, \partial, \tilde{e}) = \det A_0 (\det A_1)^{-1} \det(V, \partial, e)$$

where $(A_0, A_1) \in GL(V_0) \times GL(V_1)$ are transition matrices from bases e to bases $\tilde{e}$. The upshot of this is the fact that if bases e and $\tilde{e}$ are equivalent over some subfield $k_0 \subset k$ then the Cayley determinants with respect to these bases are equal up to a non-zero multiple from k_0.

Consider a finite complex $\dots \overset{\partial_{i-1}}{\to} V^i \overset{\partial_i}{\to} V^{i+1} \overset{\partial_{i+1}}{\to} \dots$ of finite-dimensional vector spaces. Then we can define a finite-dimensional supervector space $V = V_0 \oplus V_1$, $V_0 = \underset{i \equiv 0 \bmod 2}{\oplus} V^i$, $V_1 = \underset{i \equiv 1 \bmod 2}{\oplus} V^i$, with an induced differential ∂. In particular, all previous considerations are valid. Therefore, if the complex $(V^\bullet, \partial)$ is exact and there are some fixed bases $\{e^i_1, \dots, e^i_{\dim V^i}\}$ in each component V^i then we have the corresponding Cayley determinant $\det(V^\bullet, \partial, e) \in k^*$. For example, if L and M are based vector spaces and $A : L \to M$ is an invertible operator then the complex $0 \to L \overset{A}{\to} M \to 0$ is exact and the corresponding Cayley determinant is equal to $\det A$ (if L is located in even degree of the complex).

6 The Kac-Kleiman-Holme formula

For any vector bundle E on X we denote by $c_i(E)$ its ith Chern class (see e.g. [30]). For example, $c_1(\mathcal{L}) = H$, the hyperplane section divisor of X in $\mathbb{P}(V)$. For any zero-dimensional cycle Z on X we denote its degree ('the number of points' in Z) by $\int_X Z$. Using the classical representation of a Chern class of a spanned vector bundle, it is not very difficult to prove the following result.

Theorem 6.1 ([10]). $\deg \Delta_X = \int_X c_n(J(\mathcal{L}))$. *In particular, X^* is a hypersurface if and only if $c_n(J(\mathcal{L})) \neq 0$. Moreover, $\operatorname{def} X = k$ if and only if $c_r(J(\mathcal{L})) = 0$ for $r \geq n - k + 1$, and $c_{n-k}(J(\mathcal{L})) \neq 0$. In this case* $\deg X^* = \int_X c_{n-k}(J(\mathcal{L})) \cdot H^k$.

This formula can be rewritten in numerous ways. For instance, consider the Chern polynomial of X with respect to the given projective embedding $c_X(q) = \sum_{i=0}^n q^{i+1} \int_X c_{n-i}(\Omega_X^1) \cdot H^i$. It was shown, e.g. in [46, 49, 41, 42] that $\deg \Delta_X = c'_X(1) = \sum_{i=0}^n (i+1) \int_X c_{n-i}(\Omega_X^1) \cdot H^i$. And the codimension of X^* equals the order of the zero at $q = 1$ of the polynomial $c_X(q) - c_X(1)$. If this order is μ then $\deg X^* = c_X^{(\mu)}(1)/\mu!$.

7 Resultants

Classically, discriminants were studied in conjunction with resultants. Let X be a smooth irreducible projective variety and let E be a vector bundle on X of rank $k = \dim X + 1$. Set $V = H^0(X, E)$. We consider the variety $X_E = \mathbb{P}(E^*)$, the projectivization of the bundle E^*. There is a projection $p : X_E \to X$ whose fibres are projectivizations of fibres of E^*, and a natural projection $\pi : E^* \backslash X \to X_E$, where X is embedded into the total space of E^* as the zero section. We denote by $\xi(E)$ the tautological line bundle on X_E defined as follows. For open $U \subset X_E$, a section of $\xi(E)$ over U is a regular function on $\pi^{-1}(U)$ which is homogeneous of degree 1 with respect to dilations of E^*. The restriction of $\xi(E)$ to every fibre $p^{-1}(X) = \mathbb{P}(E_x^*)$ is the tautological line bundle $\mathcal{O}(1)$ of the projective space $\mathbb{P}(E_x^*)$. We shall assume that E is very ample, i.e. $\xi(E)$ is a very ample line bundle on X_E. In particular, $\xi(E)$ (and hence E) is generated by global sections. Notice that $H^0(X, E) = H^0(X_E, \xi(E))$, therefore $\xi(E)$ embeds X_E in $\mathbb{P}(V^*)$.

The *resultant variety* $\nabla \subset \mathbb{P}(V)$ is the set of all sections vanishing at some $x \in X$.

Example 7.1. Suppose that $X = \mathbb{P}^{k-1} = \mathbb{P}(\mathbb{C}^k)$ and $E = \mathcal{O}(d_1) \oplus \ldots \oplus \mathcal{O}(d_k)$. Then $V = S^{d_1}(\mathbb{C}^k)^* \oplus \ldots \oplus S^{d_k}(\mathbb{C}^k)^*$ and ∇ is the classical resultant variety parametrizing k-tuples of homogeneous forms on $\mathbb{C}^k$ of degrees $d_1, \ldots, d_k$ having a common non-zero root.

The following Theorem is sometimes called the Cayley trick, as Cayley first noticed that the resultant can be written as a discriminant. This theorem is very close to Theorem 3.5.

Theorem 7.2 ([35]). ∇ *is an irreducible hypersurface of degree* $\int_X c_{k-1}(E)$ *projectively dual to* X_E.

8 Dual varieties and Mori theory

Since projective varieties with positive defect are covered by projective lines, it is not very surprising that the machinery of the minimal model programme [18, 48, 97] can be used to study them.

A Cartier divisor D is said to be *nef* if $D \cdot Z \geq 0$ for any curve $Z \subset X$. Assume that K_X is not nef. Then $\tau = \min\{t \in \mathbb{R} \mid K_X + tL \text{ is nef}\}$ is called a *nef value* of (X, L); here we formally use fractional divisors. Kawamata's Rationality Theorem [47] asserts that τ is always a rational number.

Theorem 8.1 ([11, 10]). *Suppose that* $\text{def}(X, H) > 0$. *Then* K_X *is not nef and* $\tau = (\dim X + \text{def}(X, H))/2 + 1$.

Moreover, it is possible to describe quite explicitly the nef value morphism (morphism, associated with an appropriate tensor power of $K_X + \tau L$) and to calculate the defect of its generic fibres. Using this theory, one can get a list of smooth projective varieties X with positive defect such that $\dim X \leq 10$. The study of these varieties was initiated in [24, 23], continued in [55] and almost finished in [10], with contributions from many others.

9 The class formula

Consider a generic line $L \subset (\mathbb{P}^N)^*$. The intersection $L \cap X^*$ consists of d smooth on X^* points if $\text{def} X = 0$, where d is the degree of X^*. If $\text{def} X > 0$ then this intersection is empty and we set $d = 0$. L can be considered as a pencil (one-dimensional linear system) of divisors on X, and therefore it defines a rational map $F : X \to \mathbb{P}^1$. F is not defined along the subvariety $X \cap H_1 \cap H_2$ of codimension 2. We can blow it up and get the variety $\tilde{X}$ and the regular morphism $\tilde{F} : \tilde{X} \to \mathbb{P}^1$. Let $D \subset \tilde{X}$ be the preimage of $X \cap H_1 \cap H_2$. Now let us calculate the topological Euler characteristic $\chi(\tilde{X})$ in two ways. First,

$$
\begin{aligned}
\chi(\tilde{X}) &= \chi(\tilde{X} \setminus D) + \chi(D) \\
&= \chi(X \setminus X \cap H_1 \cap H_2) + \chi(D) \\
&= \chi(X \setminus X \cap H_1 \cap H_2) + 2\chi(X \cap H_1 \cap H_2) \\
&= \chi(X) + \chi(X \cap H_1 \cap H_2).
\end{aligned}
$$

Here we use the fact that D is a $\mathbb{P}^1$-bundle over $X \cap H_1 \cap H_2$, therefore $\chi(D) = 2\chi(X \cap H_1 \cap H_2)$. Notice that blowing up is an isomorphism on the complement of the exceptional divisor, hence

$$
\chi(\tilde{X} \setminus D) = \chi(X \setminus X \cap H_1 \cap H_2).
$$

On the other hand, we may use $\tilde{F}$ to calculate $\chi(\tilde{X})$. For $x \in \mathbb{P}^1$, we have $\chi(\tilde{F}^{-1}(x)) = \chi(X \cap H_x)$, where H_x is the hyperplane corresponding to x. Let $x_1, \ldots, x_d \in \mathbb{P}^1$ be points that correspond to the intersection of L with X^*. Then for any $x \in \mathbb{P}^1 \setminus \{x_1, \ldots, x_d\}$, $X \cap H_x$ is a smooth divisor. For any $x \in \{x_1, \ldots, x_d\}$, $X \cap H_x$ has a simple quadratic singularity. It is clear then that $\chi(X \cap H_x) = \chi(X \cap H_y) = \chi(X \cap H)$, where $x, y \in \mathbb{P}^1 \setminus \{x_1, \ldots, x_d\}$ and H is a generic hyperplane. For $x \in \{x_1, \ldots, x_d\}$, near a simple quadratic singularity of $X \cap H_x$, the family of divisors $X \cap H_y$, $y \to x$, looks like the family of smooth $(n-1)$-dimensional quadrics Q_ε with equations $T_1^2 + \ldots + T_{n-1}^2 = \varepsilon T_n^2$ near the unique singular point $(0 : \ldots : 0 : 1)$ of the quadric Q_0 given by $T_1^2 + \ldots + T_{n-1}^2 = 0$. Therefore,

$$\chi(\tilde{X}) = \chi(\mathbb{P}^1)\chi(X \cap H) + d(\chi(Q_0) - \chi(Q_1))$$
$$= 2\chi(X \cap H) + d(\chi(Q_0) - \chi(Q_1)).$$

After combining two formulas for $\chi(\tilde{X})$ and calculating $\chi(Q_0)$ and $\chi(Q_1)$, we finally get the class formula

$$d = (-1)^n \left[\chi(X) - 2\chi(X \cap H) + \chi(X \cap H \cap H') \right].$$

This simple topological approach has other applications. For instance, suppose that $\operatorname{def} X = 0$ and $\operatorname{Sing} X \cap H$ is finite. Then the multiplicity $m_H X^*$ of X^* at H can be found as follows.

Theorem 9.1 ([22, 68]). *If* $\operatorname{def} X = 0$, $H \in X^*$, *and* $\operatorname{Sing} X \cap H$ *is finite, then*

$$m_H X^* = \sum_{p \in \operatorname{Sing} X \cap H} \mu(X \cap H, p)$$

the sum over ordinary Milnor numbers [60]. In particular, if $\operatorname{Sing} X \cap H$ *is finite then* H *is smooth on* X^* *if and only if* $\operatorname{Sing} X \cap H = \{p_0\}$ *and* p_0 *is a simple quadratic singularity of* $X \cap H$.

Further developments can be found in [53, 1, 2, 22, 68, 71, 72].

10 Flag varieties

Let G be a connected simply-connected semisimple complex algebraic group with a Borel subgroup B and a maximal torus $T \subset B$. Let $\mathcal{P}$ be the character group of T (the weight lattice). Let $\Delta \subset \mathcal{P}$ be the set of roots of G relative to T. To every root $\alpha \in \Delta$ we can assign the 1-dimensional unipotent subgroup $U_\alpha \subset G$. We define the negative roots Δ^- as those roots α such that $U_\alpha \subset B$. The positive roots are $\Delta^+ = \Delta \setminus \Delta^- = -\Delta^-$. Let $\Pi \subset \Delta^+$ be simple roots, $\Pi = \{\alpha_1, \ldots, \alpha_n\}$, where $n = \operatorname{rank} G = \dim T$. Any root $\alpha \in \Delta$ is an integral combination $\sum_i n_i \alpha_i$ with nonnegative n_i (for $\alpha \in \Delta^+$) or nonpositive n_i (for $\alpha \in \Delta^-$). If G is simple then we use the Bourbaki numbering [13] of simple roots.

The weight lattice $\mathcal{P}$ is generated as a $\mathbb{Z}$-module by the fundamental weights $\omega_1, \ldots, \omega_n$ dual to the simple roots under the Killing form

$$\langle \omega_i | \alpha_j \rangle = \frac{2(\omega_i, \alpha_j)}{(\alpha_j, \alpha_j)} = \delta_{ij}.$$

A weight $\lambda = \sum_i n_i \omega_i$ is called dominant if all $n_i \geq 0$. We denote the dominant weights by $\mathcal{P}^+$. Dominant weights parametrize finite-dimensional irreducible G-modules: to any $\lambda \in \mathcal{P}^+$ we assign an irreducible G-module V_λ with highest weight λ. A weight $\lambda = \sum_i n_i \omega_i$ is called strictly dominant if all $n_i > 0$. We denote the strictly dominant weights by $\mathcal{P}^{++}$. There is a partial order on $\mathcal{P}$: $\lambda > \mu$ if $\lambda - \mu$ is an integral combination of simple roots with nonnegative coefficients. If $\lambda \in \mathcal{P}^+$ then we denote by $\lambda^* \in \mathcal{P}^+$ the highest weight of the dual G-module V_λ^*. Let W be the Weil group of G relative to T. If $w_0 \in W$ is the longest element then $\lambda^* = -w_0(\lambda)$.

The character group of B is identified with a character group of T. Therefore, for any $\lambda \in \mathcal{P}$ we can assign a 1-dimensional B-module $\mathbb{C}_\lambda$, where B acts on $\mathbb{C}_\lambda$ by a character λ. Now we can define the twisted product $G \times_B \mathbb{C}_\lambda$ to be the quotient of $G \times \mathbb{C}_\lambda$ by the diagonal action of B: $b \cdot (g, z) = (gb^{-1}, \lambda(b)z)$. Projection onto the first factor induces the map $G \times_B \mathbb{C}_\lambda \to G/B$, which realizes the twisted product as an equivariant line bundle $\mathcal{L}_\lambda$ on G/B with fibre $\mathbb{C}_\lambda$. It is well-known that the correspondence $\lambda \to \mathcal{L}_\lambda$ is an isomorphism of $\mathcal{P}$ and $\mathrm{Pic}(G/B)$. $\mathcal{L}_\lambda$ is ample if and only if $\mathcal{L}_\lambda$ is very ample if and only if $\lambda \in \mathcal{P}^{++}$. By the Borel-Weil-Bott theorem [12], for strictly dominant λ the vector space of global sections $H^0(G/B, \mathcal{L}_\lambda)$ is isomorphic as a G-module to V_λ, the irreducible G-module with highest weight λ. The embedding $G/B \subset \mathbb{P}(V_{\lambda^*})$ identifies G/B with the projectivization of the cone of highest weight vectors. The dual variety $(G/B)^*$ therefore lies in $\mathbb{P}(V_\lambda)$ and parametrises global sections $s \in H^0(G/B, \mathcal{L}_\lambda)$ such that the scheme of zeros $Z(s)$ is a singular divisor.

More generally, consider any flag variety of the form G/P, where $P \subset G$ is an arbitrary parabolic subgroup. The subgroup $P \subset G$ is called parabolic if G/P is a projective variety. P is parabolic if and only if it contains some Borel subgroup. Up to conjugacy, we may assume that P contains B. The combinatorial description is as follows. Let $\Pi_P \subset \Pi$ be some subset of simple roots. Let $\Delta_P^+ \subset \Delta^+$ denote the positive roots that are linear combinations of the roots in Π_P. Then P is generated by B and by the root groups U_α for $\alpha \in \Delta_P^+$. We denote $\Pi \setminus \Pi_P$ by $\Pi_{G/P}$

and $\Delta^+ \setminus \Delta_P^+$ by $\Delta_{G/P}^+$. A parabolic subgroup is maximal if and only if $\Pi_{G/P}$ is a single simple root.

The fundamental weights $\omega_{i_1}, \ldots \omega_{i_k}$ dual to the simple roots in $\Pi_{G/P}$ generate the sublattice $\mathcal{P}_{G/P}$ of $\mathcal{P}$. We denote $\mathcal{P}^+ \cap \mathcal{P}_{G/P}$ by $\mathcal{P}_{G/P}^+$. The subset $\mathcal{P}_{G/P}^{++} \subset \mathcal{P}_{G/P}^+$ consists of all weights $\lambda = \sum n_k \omega_{i_k}$ such that all $n_k > 0$. Any weight $\lambda \in \mathcal{P}_{G/P}$ defines a character of P, and therefore a line bundle $\mathcal{L}_\lambda$ on G/P. Then the following is well-known. The correspondence $\lambda \to \mathcal{L}_\lambda$ is an isomorphism of $\mathcal{P}_{G/P}$ and $\mathrm{Pic}(G/P)$. In particular, $\mathrm{Pic}(G/P) = \mathbb{Z}$ if and only if P is maximal. $\mathcal{L}_\lambda$ is ample if and only if $\mathcal{L}_\lambda$ is very ample if and only if $\lambda \in \mathcal{P}_{G/P}^{++}$. If $\lambda \in \mathcal{P}_{G/P}^+$ then the linear system corresponding to $\mathcal{L}_\lambda$ is base-point free. The corresponding map given by sections is a factorization $G/P \to G/Q$, where Q is a parabolic subgroup such that Π_Q is a union of Π_P and all simple roots in $\Pi_{G/P}$ orthogonal to λ. For strictly dominant $\lambda \in \mathcal{P}_{G/P}^{++}$ the vector space of global sections $H^0(G/P, \mathcal{L}_\lambda)$ is isomorphic as a G-module to V_λ, the irreducible G-module with highest weight λ.

The embedding $G/P \subset \mathbb{P}(V_{\lambda^*})$ identifies G/P with the projectivization of the cone of highest weight vectors. The dual variety $(G/P)^*$ therefore lies in $\mathbb{P}(V_\lambda)$ and parametrises global sections $s \in H^0(G/P, \mathcal{L}_\lambda)$ such that the scheme of zeros $Z(s)$ is a singular divisor.

11 Adjoint varieties

For the adjoint representation of $SL_n = SL(V)$ there is a natural notion of discriminant defined as follows. For any operator $A \in \mathfrak{sl}(V)$ let $P_A = \det(t\,\mathrm{Id} - A)$ be the characteristic polynomial. Then its discriminant, $D(A) = D(P_A)$ is a homogeneous form on $\mathfrak{sl}(V)$ of degree $n^2 - n$. Clearly $D(A) \neq 0$ if and only if all eigenvalues of A are distinct, that is, if A is a regular semisimple operator. This notion can be carried over to any simple Lie algebra. Before doing that, notice that $D(A)$ can be also defined as follows. Consider the characteristic polynomial

$$Q_A = \det(t\,\mathrm{Id} - \mathrm{ad}(A)) = \sum_{i=0}^{n^2-1} t^i D_i(A)$$

of the adjoint operator $\mathrm{ad}(A)$. If $\lambda_1, \ldots, \lambda_r$ are eigenvalues of A (counted with multiplicities) then the set of eigenvalues of $\mathrm{ad}(A)$ consists of $n-1$ zeros and differences $\lambda_i - \lambda_j$ for $i, j = 1, \ldots, n$, $i \neq j$. Therefore, $D_0(A) = \ldots = D_{n-2}(A) = 0$ and $D_{n-1}(A)$ coincides with $D(A)$ up to a non-zero scalar.

Suppose now that $\mathfrak{g}$ is a simple Lie algebra of rank r. Let $\dim \mathfrak{g} = n$. For any $x \in \mathfrak{g}$ let

$$Q_x = \det(t\,\mathrm{Id} - \mathrm{ad}(x)) = \sum_{i=0}^{n} t^i D_i(x)$$

be the characteristic polynomial of the adjoint operator. Then $D(x) = D_r(x)$ is called the discriminant of x. Clearly, D is a homogeneous Ad-invariant polynomial on $\mathfrak{g}$ of degree $n - r$. Since the dimension of the centralizer $\mathfrak{g}_x$ of any element $x \in \mathfrak{g}$ is greater than or equal to r, it follows that $D_0 = \ldots = D_{r-1} = 0$, and therefore $D(x) = 0$ if and only if $\mathrm{ad}(x)$ has the eigenvalue 0 with multiplicity $> r$. We claim that actually $D(x) \neq 0$ if and only if x is regular semisimple (recall that x is called regular if $\dim \mathfrak{g}_x = r$). Indeed, if x is semisimple then $\mathrm{ad}(x)$ is a semisimple operator, therefore $D(x) = 0$ if and only if the dimension of the centralizer $\mathfrak{g}_x$ is greater than r, i.e. x is not regular. If x is not semisimple then we take the Jordan decomposition $x = x_s + x_n$, where x_s is semisimple, x_n is nilpotent, and $[x_s, x_n] = 0$. Then x_s is automatically not regular, and therefore since $Q_x = Q_{x_s}$ we have $D(x) = D(x_s) = 0$.

To study $D(x)$ further, we can use the Chevalley restriction theorem (see [76]) $\mathbb{C}[\mathfrak{g}]^G = \mathbb{C}[\mathfrak{t}]^W$, where $\mathfrak{t} \subset \mathfrak{g}$ is any Cartan subalgebra and W is the Weil group. Let $\Delta \subset \mathfrak{t}^*$ be the root system, $\#\Delta = n - r$. Clearly, for any $x \in \mathfrak{t}$, $D(x) = 0$ if and only if x is not regular if and only if $\alpha(x) = 0$ for some $\alpha \in \Delta$. Since $\deg D = n - r$ and $D|_{\mathfrak{t}}$ is W-invariant, it easily follows that

$$D|_{\mathfrak{t}} = \prod_{\alpha \in \Delta} \alpha.$$

The Weyl group acts transitively on the set of roots of the same length. Therefore, $D|_{\mathfrak{t}}$, and hence D, is irreducible if and only if all roots in Δ have the same length, i.e. Δ is of type A, D, or E. If Δ is of type B, C, F, or G, we have $\Delta = \Delta_s \cup \Delta_l$, where Δ_s is the set of short roots and Δ_l is the set of long roots. Then we have $D = D_l D_s$, where D_l and D_s are irreducible polynomials and

$$D_l|_{\mathfrak{t}} = \prod_{\alpha \in \Delta_l} \alpha, \quad D_s|_{\mathfrak{t}} = \prod_{\alpha \in \Delta_s} \alpha.$$

In the $A - D - E$ case we also set $D_l = D$ to simplify notations.

We are going to show that D_l is also the discriminant in our regular sense. The adjoint representation $\mathrm{Ad} : G \to GL(\mathfrak{g})$ is irreducible. In $A - D - E$ case let $\mathcal{O} = \mathcal{O}_l$ be the Ad-orbit of any root vector, In

$B - C - F - G$ case let $\mathcal{O}_l \subset \mathfrak{g}$ or $\mathcal{O}_s \subset \mathfrak{g}$) be the Ad-orbit of any long or short root vector respectively. Then $\mathcal{O}_l$ is the orbit of the highest weight vector. Its projectivization is called the *adjoint variety*. Both orbits $\mathcal{O}_l$ and $\mathcal{O}_s$ are conical. Let $X_l = \mathbb{P}(\mathcal{O}_l)$, $X_s = \overline{\mathbb{P}(\mathcal{O}_s)}$.

Theorem 11.1. D_l *is the discriminant of* X_l *and* D_s *is the discriminant of* X_s.

Adjoint varieties have many interesting projective properties, see e.g. [54, 44].

12 The Pyasetskii pairing

In [77] Pyasetskii showed that if a connected algebraic group acts linearly on a vector space with a finite number of orbits then the dual action has the same property and the number of orbits is the same. This result easily follows from its projective version, which, in turn, follows from Reflexivity Theorem.

Theorem 12.1. *Suppose that a connected algebraic group* G *acts on a projective space* $\mathbb{P}^n$ *with a finite number of orbits. Then the dual action* $G : \mathbb{P}^{n*}$ *has the same number of orbits. Let* $\mathbb{P}^n = \overset{N}{\underset{i=1}{\sqcup}} \mathcal{O}_i$ *and* $\mathbb{P}^{n*} = \overset{N}{\underset{i=1}{\sqcup}} \mathcal{O}'_i$ *be the orbit decompositions. Let* $\mathcal{O}_0 = \mathcal{O}'_0 = \emptyset$. *Then the bijection is defined as follows:* $\mathcal{O}_i$ *corresponds to* $\mathcal{O}'_j$ *if and only if* $\overline{\mathcal{O}}_i$ *is projectively dual to* $\overline{\mathcal{O}'_j}$.

A large class of linear actions of reductive groups with finitely many orbits is provided by graded semisimple Lie algebras. Suppose that $\mathfrak{g} = \oplus_{k \in \mathbb{Z}} \mathfrak{g}_k$ is the graded semi-simple Lie algebra of the connected semi-simple group G. Then there exists a unique semisimple element $\xi \in \mathfrak{g}_0$ such that $\mathfrak{g}_k = \{x \in \mathfrak{g} \mid [\xi, x] = kx\}$. The connected component H of the centralizer $G_\xi \subset G$ is a reductive subgroup of G called the *Levi subgroup* (because it is the Levi part of the parabolic subgroup). $\mathfrak{g}_0$ is the Lie algebra of H. Further, H acts on each graded component $\mathfrak{g}_k$. Notice that the dual H-module $(\mathfrak{g}_k)^*$ is isomorphic to $\mathfrak{g}_{-k}$.

Theorem 12.2 ([80, 92]). H *acts on* $\mathfrak{g}_k$ *with finitely many orbits.*

If H acts on $\mathfrak{g}_k$ irreducibly, the corresponding linear group is called the θ-group of the first kind. It would be interesting to find the explicit description of Pyasetskii pairing for all those linear groups.

13 The multisegment duality

Apart from actions associated with $\mathbb{Z}$-graded semisimple Lie algebras, another class of actions with finitely many orbits is provided by the theory of representations of quivers (see [33]). These two classes overlap: the representations of quivers of type A give the same class of actions as the standard gradings of $SL(V)$. The Pyasetskii pairing in this case has been studied in a series of papers under the name of the *multisegment duality*, or the *Zelevinsky involution*.

We fix a positive integer r and consider the set $S = S_r$ of pairs of integers (i, j) such that $1 \leq i \leq j \leq r$. Let $\mathbb{Z}_+^S$ denote the semigroup of families $m = (m_{ij})_{(i,j) \in S}$ of non-negative integers indexed by S. We regard a pair $(i, j) \in S$ as a segment $[i, j] = \{i, i+1, \ldots, j\}$ in $\mathbb{Z}$. A family $m = m_{ij} \in \mathbb{Z}_+^S$ can be regarded as a collection of segments, containing m_{ij} copies of each $[i, j]$. Thus, elements of $\mathbb{Z}_+^S$ can be called multisegments. The weight $|m|$ of a multisegment m is defined as a sequence $\gamma = \{d_1, \ldots, d_r\} \in \mathbb{Z}_+^r$ given by $d_i = \sum\limits_{i \in [k,l]} m_{kl}$ for $i = 1, \ldots, r$. In other words, $|m|$ records how many segments of m contain any given number $i \in [1, r]$. For any $\gamma \in \mathbb{Z}_+^r$ we set $\mathbb{Z}_+^S(\gamma) = \{m \in \mathbb{Z}_+^S \mid |m| = \gamma\}$.

Another important interpretation of $\mathbb{Z}_S^+(\gamma)$ is that it parametrizes isomorphism classes of representations of quivers of type A. Let A_r be the quiver equal to the Dynkin diagram of type A_r, where all edges are oriented from the left to the right. Let A_r^* be a dual quiver with all orientations reversed. The representation of A_r with the dimension vector $\gamma = (d_1, \ldots, d_r) \in \mathbb{Z}_+^r$ is the collection of vector spaces $\mathbb{C}^{d_1}, \ldots, \mathbb{C}^{d_r}$ and linear maps $\varphi_1 : \mathbb{C}^{d_1} \to \mathbb{C}^{d_2}$, ..., $\varphi_{r-1} : \mathbb{C}^{d_{r-1}} \to \mathbb{C}^{d_r}$. The representation of A_r^* with the dimension vector $\gamma = (d_1, \ldots, d_r) \in \mathbb{Z}_+^r$ is the collection of vector spaces $\mathbb{C}^{d_1}, \ldots, \mathbb{C}^{d_r}$ and linear maps $\psi_1 : \mathbb{C}^{d_2} \to \mathbb{C}^{d_1}$, ..., $\psi_{r-1} : \mathbb{C}^{d_r} \to \mathbb{C}^{d_{r-1}}$. Therefore, representations of A_r with dimension vector γ are parametrized by points of a vector space $V(\gamma) = \bigoplus_{i=1}^{r-1} \mathrm{Hom}(\mathbb{C}^{d_i}, \mathbb{C}^{d_{i+1}})$. Representations of A_r^* with dimension vector γ are parametrized by points of a vector space $V(\gamma)^* = \bigoplus_{i=1}^{r-1} \mathrm{Hom}(\mathbb{C}^{d_{i+1}}, \mathbb{C}^{d_i})$. Notice that $V(\gamma)$ and $V(\gamma)^*$ are dual modules of the group $G(\gamma) = GL_{d_1} \times \ldots \times GL_{d_r}$ with respect to the natural action.

The orbits of $G(\gamma)$ on $V(\gamma)$, $V(\gamma)^*$ correspond to isoclasses of representations of A_r, A_r^* respectively with dimension vector γ. These orbits are parametrized by elements of $\mathbb{Z}_S^+(\gamma)$. Elements of S parametrize indecomposable A_r-modules (or A_r^*-modules). Namely, each $(i, j) \in S$ corresponds to an indecomposable module R_{ij} with dimension vector

$|(i,j)| = (0^{i-1}, 1^{j-i+1}, 0^{r-j})$ and all maps are isomorphisms, if it is possible, or zero maps otherwise. Then any family $(m_{ij}) \in \mathbb{Z}_S^+$ corresponds to an A_r (or A_r^*) module $\oplus_S R_{ij}^{m_{ij}}$.

By Pyasetskii's Theorem 12.1, there is a natural bijection of $G(\gamma)$-orbits in $V(\gamma)$ and $V(\gamma)^*$. Therefore, there exists a natural involution ζ of $\mathbb{Z}_+^S(\gamma)$, which can be extended to a weight-preserving involution of $\mathbb{Z}_+^S$ called the multisegment duality. The involution ζ can also be described in terms of irreducible finite-dimensional representations of affine Hecke algebras and in terms of canonical bases for quantum groups, etc. See [50, 61].

An explicit description of ζ was found in [50] using Poljak's theorem from the theory of networks [74]. To formulate it, we need the following definition. For any multisegment $m \in \mathbb{Z}_+^S$ the ranks $r_{ij}(m)$ are given by

$$r_{ij}(m) = \sum_{[i,j] \subset [k,l]} m_{kl}.$$

It is easy to see that the multisegment m can be recovered from its ranks by the formula $m_{ij} = r_{ij}(m) - r_{i-1,j}(m) + r_{i-1,j+1}(m)$. If the multisegment corresponds to the representation of A_r given by $(\varphi_1, \ldots, \varphi_{r-1}) \in V(\gamma)$ then r_{ij} is equal to the rank of the map $\varphi_{j-1} \circ \ldots \circ \varphi_{i+1} \circ \varphi_i$. In particular, $r_{ii} = d_i$.

For any $(i,j) \in S$ let T_{ij} denote the set of all maps $\nu : [1,i] \times [j,r] \to [i,j]$ such that $\nu(k,l) \le \nu(k',l')$ whenever $k \le k'$, $l \le l'$ (in other words, ν is a morphism of partially ordered sets, where $[1,i] \times [j,r]$ is supplied with the product order).

Theorem 13.1 ([50]). *For every $m = (m_{ij}) \in \mathbb{Z}_+^S$ we have*

$$r_{ij}(\zeta(m)) = \min_{\nu \in T_{ij}} \sum_{(k,l) \in [1,i] \times [j,r]} m_{\nu(k,l)+k-i, \nu(k,l)+l-j}.$$

14 Kashin's diagrams

In all previous examples we were considering a reductive group G acting on a vector space V with finitely many orbits. Though the Pyasetskii pairing can be very involved in these cases, the orbit decomposition itself for the actions $G : V$ and $G : V^*$ is the same. More precisely, there exists an involution θ of G such that the action of G on V^* is isomorphic to the action of G on V twisted by θ. For non-reductive groups this is, of course, no longer true.

A nice series of examples was considered by Kashin. Let G be a simple connected complex Lie group, $B \subset G$ a Borel subgroup. Let $\mathfrak{b} \subset \mathfrak{g}$ be the corresponding Lie algebras. Consider the action of B on $\mathfrak{g}/\mathfrak{b}$ and the dual action on $(\mathfrak{g}/\mathfrak{b})^*$. The latter B-module is isomorphic to $\mathfrak{b}_u$, the unipotent radical of $\mathfrak{b}$.

Theorem 14.1 ([45]). *The action of B on $\mathfrak{g}/\mathfrak{b}$ (or on $(\mathfrak{g}/\mathfrak{b})^*$) has finitely many orbits if and only if $\mathfrak{g}$ has type A_1, A_2, A_3, A_4, or C_2.*

Moreover, Kashin found all orbits in these cases, calculated the Pyasetskii pairing and described the natural order on the set of orbits given by $\mathcal{O}_1 \le \mathcal{O}_2$ if and only if $\mathcal{O}_1 \subset \overline{\mathcal{O}_2}$. Let us give the Hasse diagrams in case A_3. We enumerate orbits in such a way that an orbit $\mathcal{O}_i' \subset \mathfrak{g}/\mathfrak{b}$ corresponds to an orbit $\mathcal{O}_i'' \subset (\mathfrak{g}/b)^*$ via Pyasetskii pairing. The following diagram contains P-orbits on $(\mathfrak{g}/\mathfrak{p})^*$:

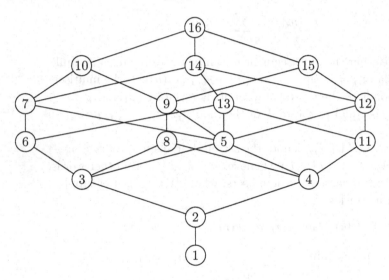

The following diagram contains P-orbits on $(\mathfrak{g}/\mathfrak{p})$:

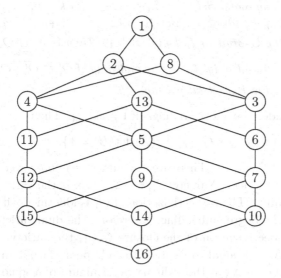

These diagrams have two interesting properties. First, we see that these graphs are networks with vertices located on several levels and edges going from level $i+1$ to level i. This is easy to explain: the level of the orbit is just its dimension. Since the group B is solvable, its orbits are affine varieties and therefore have divisorial boundaries.

The second property (noticed by Kashin) is much more mysterious. Namely, the Hasse diagrams corresponding to $\mathfrak{g}/\mathfrak{b}$ and $(\mathfrak{g}/\mathfrak{b})^*$ have the same number of edges! It is unknown, whether this is a mere coincidence.

15 Parabolic subgroups with Abelian unipotent radical

Let L be a simple algebraic group and $P \subset L$ a parabolic subgroup with abelian unipotent radical. In this case $\mathfrak{l} = \operatorname{Lie} L$ admits a $\mathbb{Z}$-grading with only three non-zero parts: $\mathfrak{l} = \mathfrak{l}_{-1} \oplus \mathfrak{l}_0 \oplus \mathfrak{l}_1$. Such a grading is said to be short. Here $\mathfrak{l}_0 \oplus \mathfrak{l}_1 = \operatorname{Lie} P$ and $\exp(\mathfrak{l}_1)$ is the abelian unipotent radical of P. There exists a unique semisimple element $\xi \in \mathfrak{l}_0$ such that $l_k = \{x \in \mathfrak{l} \mid [\xi, x] = kx\}$. We denote by G the connected component of the centralizer $L_\xi \subset L$. Then $\mathfrak{l}_0 = \mathfrak{g} = \operatorname{Lie} G$ and by Theorem 12.2 G acts on $\mathfrak{l}_{\pm 1}$ with a finite number of orbits. We denote by T the maximal torus in G (and hence in L), and write $\mathfrak{t}$ for $\operatorname{Lie} T$, Δ for the root system of $(\mathfrak{l}, \mathfrak{t})$ and $\Delta = \Delta_{-1} \cup \Delta_0 \cup \Delta_1$ for the partition corresponding to the short grading.

Theorem 15.1 ([70, 63]). *Let* $\alpha_1, \ldots, \alpha_r \in \Delta_1$ *be any maximal sequence of pairwise orthogonal long roots. Set* $e_k = e_{\alpha_1} + \ldots + e_{\alpha_k}$ *for* $k = 0, \ldots, r$. *Denote the G-orbit of* e_k *by* $\mathcal{O}_k \subset \mathfrak{l}_1$. *Set* $f_k = e_{-\alpha_1} + \ldots + e_{-\alpha_k}$ *for* $k = 0, \ldots, r$. *Denote the G-orbit of* f_k *by* $\mathcal{O}'_k \subset \mathfrak{l}_{-1}$. *Then* $\mathfrak{l}_1 = \overset{r}{\underset{i=0}{\bigsqcup}} \mathcal{O}_i$, $\mathfrak{l}_{-1} = \overset{r}{\underset{i=0}{\bigsqcup}} \mathcal{O}'_i$, $\mathcal{O}_i \subset \overline{\mathcal{O}_j}$ *if and only if* $i \leq j$ *if and only if* $\mathcal{O}'_i \subset \overline{\mathcal{O}'_j}$. $\mathcal{O}_k$ *corresponds to* $\mathcal{O}'_{r-k}$ *in the Pyasetskii pairing.*

Example 15.2. Consider the short gradings of $\mathfrak{l} = \mathfrak{sl}_{n+m}$. Then

$$G = \{(A, B) \in GL_n \times GL_m \mid \det(A)\det(B) = 1\}.$$

$\mathfrak{l}_1$ can be identified with $\mathbb{C}^n \times \mathbb{C}^m$. There are $r = \min(n, m)$ non-zero G-orbits (determinantal varieties). Namely, $\mathcal{O}_i$, $i = 1, \ldots, r$, is the variety of $m \times n$-matrices of rank i. The projectivization of $\mathcal{O}_1$ is identified with $X = \mathbb{P}^{n-1} \times \mathbb{P}^{m-1}$ in the Segre embedding. Therefore, the dual variety X^* is equal to the projectivization of the closure $\overline{\mathcal{O}_{r-1}}$, the variety of matrices of rank less than or equal to $r - 1$. X^* is a hypersurface if and only if $n = m$, in which case Δ_X is the ordinary determinant of a square matrix. Another interesting case is $n = 2$, $m \geq 2$: we see that the Segre embedding of $\mathbb{P}^1 \times \mathbb{P}^k$ is self-dual.

Example 15.3. Consider the short grading of $\mathfrak{l} = \mathfrak{so}_{n+2}$ such that $G = \mathbb{C}^* \times SO_n$ and $\mathfrak{l}_1 = C^n$ with a simplest action. There are two non-zero orbits: the dense one and the self-dual quadric hypersurface $Q \subset \mathbb{C}^n$ preserved by G.

Example 15.4. Consider the short grading of $\mathfrak{l} = \mathfrak{so}_{2n} = D_n$ such that $G = \mathbb{C}^* \times SL_n$ acts naturally on $\mathfrak{l}_1 = \Lambda^2 \mathbb{C}^n$. There are $r = [n/2]$ non-zero orbits, where $\mathcal{O}_i$, $i = 1, \ldots, r$, is the variety of skew-symmetric matrices of rank $2r$. The projectivization of $\mathcal{O}_1$ is identified with $X = \mathrm{Gr}(2, n)$ in the Plücker embedding. Therefore, the dual variety X^* is equal to the projectivization of the closure $\overline{\mathcal{O}_{r-1}}$, the variety of matrices of rank less than or equal to $2r - 2$. X^* is a hypersurface if and only if n is even, in which case Δ_X is the Pfaffian of a skew-symmetric matrix. If n is odd then $\mathrm{def}\, X = \mathrm{codim}\, \mathcal{O}_{r-1} = 2$.

Example 15.5. Consider the short grading of $\mathfrak{l} = \mathfrak{sp}_{2n} = C_n$ such that $G = \mathbb{C}^* \times SL_n$ acts naturally on $\mathfrak{l}_1 = S^2 \mathbb{C}^n$. There are n non-zero orbits, where $\mathcal{O}_i$, $i = 1, \ldots, n$, is the variety of symmetric matrices of rank n. The projectivization of $\mathcal{O}_1$ is identified with $\mathbb{P}^{n-1}$ in the second Veronese embedding. X^* is a hypersurface and Δ_X is the determinant of a symmetric matrix.

Example 15.6. The short grading of E_6 gives the following action. $G = \mathbb{C}^* \times SO_{10}$ and $\mathfrak{l}_1$ is the half-spinor representation. Except for the zero orbit and the dense orbit there is only one orbit $\mathcal{O}$. In particular, the projectivization of $\mathcal{O}$ is smooth and self-dual. It is a spinor variety $\mathbb{S}_5$ parametrizing one family of 5-dimensional isotropic subspaces of a non-degenerate quadratic form in $\mathbb{C}^{10}$. It could also be described via Cayley numbers. Let $\mathbb{C}a$ be the algebra of split Cayley numbers (therefore $\mathbb{C}a = \mathbb{O} \otimes \mathbb{C}$, where $\mathbb{O}$ is the real division algebra of octonions). Let $u \mapsto \bar{u}$ be the canonical involution in $\mathbb{C}a$. Let $\mathbb{C}^{16} = \mathbb{C}a \oplus \mathbb{C}a$ have octonionic coordinates u, v. Then the spinor variety $\mathbb{S}_5 \subset \mathbb{P}(\mathbb{C}^{16})$ is defined by homogeneous equations $u\bar{u} = 0$, $v\bar{v} = 0$, $u\bar{v} = 0$, where the last equation is equivalent to 8 complex equations.

Example 15.7. The final example appears from the short grading of E_7. Here $G = \mathbb{C}^* \times E_6$, and $\mathfrak{l}_1 = \mathbb{C}^{27}$ can be identified with the exceptional simple Jordan algebra (the Albert algebra), see [43]. Then E_6 is the group of norm similarities and $\mathbb{C}^*$ acts by homotheties. There are 3 non-zero orbits: the dense one, the cubic hypersurface (defined by the norm in the Jordan algebra), and the closed conical variety with smooth projectivization consisting of elements of rank one, i.e. the exceptional Severi variety (the model of the Cayley projective plane); see also [57].

16 Tangents and secants

The following theorem is called the Zak Theorem on tangents.

Theorem 16.1 ([98, 32])(a) *Let $X \subset \mathbb{P}^N$ be a smooth nondegenerate projective variety,* $\dim X = n$. *If L is a k-plane in $\mathbb{P}^N$ and $k \geq n$ then* $\dim \mathrm{Sing}(L \cap X) \leq k - n$.

b) *If $X \subset \mathbb{P}^N$ is a non-linear smooth projective variety, then $\dim X^* \geq X^*$. In particular, if X^* is smooth, then $\dim X = \dim X^*$.*

This theorem provides one of many links between the study of dual varieties and the study of secant and tangential varieties [100]. The *tangential variety* $\mathrm{Tan}(X)$ is the union of all (embedded) tangent spaces to X, and the *secant variety* $\mathrm{Sec}(X)$ is the closure of the union of all secant lines to X, i.e.

$$\mathrm{Tan}(X) = \bigcup_{x \in X} \hat{T}_x X \quad , \quad \mathrm{Sec}(X) = \overline{\bigcup_{x,y \in X} \mathbb{P}^1_{xy}},$$

where $\mathbb{P}^1_{xy}$ is a line connecting x and y. One proof of the following theorem is based on the Fulton-Hansen Connectedness Theorem [31].

Theorem 16.2 ([31, 98]). *Suppose that* $X \subset \mathbb{P}^N$ *is a smooth projective variety. Then either* $\dim \operatorname{Tan}(X) = 2n$, $\operatorname{Sec}(X) = 2n + 1$ *or* $\operatorname{Tan}(X) = \operatorname{Sec}(X)$.

It is possible to generalize this theorem in the following direction. Suppose that $Y \subset X \subset \mathbb{P}^N$ are arbitrary irreducible projective varieties. Then we can define relative secant and tangential varieties as follows. The relative secant variety $\operatorname{Sec}(Y, X)$ is the closure of the union of all secants $\mathbb{P}^1_{x,y}$, where $x \in X$, $y \in Y$. The relative tangential variety $\operatorname{Tan}(Y, X)$ is the union of tangent stars $T^*_y X$ for $y \in Y$, where the tangent star $T^*_x X$ for $x \in X$ is the union of limit positions of secants $\mathbb{P}^1_{x',x''}$, where $x', x'' \in X$ and $x', x'' \to x$.

Example 16.3. Let $V = \mathbb{C}^{k+1} \otimes \mathbb{C}^{l+1}$ denote the space of $(k + 1) \times (l+1)$ matrices. Let $X^r \subset \mathbb{P}(V)$ be the projectivization of the variety of matrices of rank at most r. Then $X^r \subset X^{r+1}$ and $\operatorname{Sec}(X^r, X^s)$ is equal to X^{r+s} if $r+s \leq \min(k+1, l+1)$ and to $\mathbb{P}(V)$ if $r+s > \min(k+1, l+1)$.

We have the following theorem:

Theorem 16.4 ([98]). *Suppose that* $Y \subset X \subset \mathbb{P}^N$ *are irreducible projective varieties. Then either* $\dim \operatorname{Tan}(Y, X) = \dim X + \dim Y$, $\dim \operatorname{Sec}(Y, X) = \dim X + \dim Y + 1$ *or* $\operatorname{Tan}(Y, X) = \operatorname{Sec}(Y, X)$.

Now the proof of Zak's theorem on tangents is quite simple. Indeed, in (a), let $Y = \operatorname{Sing} L \cap X$. Then $x \in Y$ if and only if $\hat{T}_x X \subset L$. Therefore, $\operatorname{Tan}(Y, X) \subset L$. In particular, $\dim \operatorname{Tan}(Y, X) \leq k$. On the other hand, since X is non-degenerate, $X \not\subset L$. Therefore $\operatorname{Sec}(Y, X) \not\subset L$. In particular, $\operatorname{Tan}(Y, X) \neq \operatorname{Sec}(Y, X)$. By Theorem 16.4 it follows that $\dim \operatorname{Tan}(Y, X) = \dim Y + n$. Finally, we have $\dim Y \leq k - n$. In (b), let $H \subset X^*_{sm}$. Then the contact locus $\operatorname{Sing} X \cap H$ is the projective subspace of dimension $\operatorname{def} X$. Therefore, by (a) we have $\operatorname{def} X \leq \dim H - \dim X = \operatorname{codim} X - 1$. It follows that $\dim X \leq \dim X^*$. If X^* is also smooth then using the reflexivity theorem and the same argument as above we get $\dim X^* \leq \dim X$, therefore $\dim X = \dim X^*$.

One can go even further and define the join $S(X, Y)$ of any two subvarieties $X, Y \subset \mathbb{P}(V)$, $S(X, Y) = \overline{\underset{x \in X, \ y \in Y}{\bigcup} \mathbb{P}^1_{xy}}$, where the closure is not necessary if the two varieties do not intersect. For example, if X and Y are projective subspaces then $S(X, Y) = \mathbb{P}(\operatorname{Cone}(X) + \operatorname{Cone}(Y))$. An important result about joins is the following lemma due to Terracini.

Theorem 16.5 ([87]). *Let $X, Y \subset \mathbb{P}(V)$ be irreducible varieties and let $x \in X$, $y \in Y$, $z \in \mathbb{P}^1_{xy}$. Then $\hat{T}_z S(X,Y) \supset S(\hat{T}_x X, \hat{T}_y Y)$. Moreover, if z is a generic point of $S(X,Y)$ then equality holds.*

If an algebraic group G acts on $\mathbb{P}^N$ with finitely many orbits, then for any two orbits $\mathcal{O}_1$ and $\mathcal{O}_2$ the join of their closures $S(\overline{\mathcal{O}_1}, \overline{\mathcal{O}_2})$ is G-invariant therefore coincides with some orbit closure $\overline{\mathcal{O}}$. This gives an interesting associative product on the set of G-orbits.

Example 16.6. Let L be a simple algebraic group and P a parabolic subgroup with abelian unipotent radical. In this case $\mathfrak{l} = \operatorname{Lie} L$ admits a short $\mathbb{Z}$-grading with only three non-zero parts: $\mathfrak{l} = \mathfrak{l}_{-1} \oplus \mathfrak{l}_0 \oplus \mathfrak{l}_1$. Let $G \subset L$ be a reductive subgroup with Lie algebra $\mathfrak{l}_0$. Recall that by Theorem 15.1 G has finitely many orbits in $\mathfrak{l}_1$ naturally labelled by integers from the segment $[0, r]$. Let $X_i = \overline{\mathbb{P}(\mathcal{O}_i)}$, $i = 0, \dots, r$. Then it is fairly easy to see that $S(X_i, X_j) = X_{\min(i+j,r)}$.

The following theorem is called *Zak's theorem on linear normality*.

Theorem 16.7 ([100]). *Suppose that $X \subset \mathbb{P}^N$ is a smooth non-degenerate projective variety, $\dim X = n$. If $\operatorname{codim} X < (N+4)/3$ then $\operatorname{Sec}(X) = \mathbb{P}^N$. If $\operatorname{codim} X < (N+2)/3$ then X is linearly normal.*

Zak [98] has also classified the varieties in the borderline case.

Theorem 16.8 ([100]). *Suppose that $X \subset \mathbb{P}^N$ is a smooth non-degenerate projective variety, $\dim X = n$. If $\operatorname{codim} X = (N+4)/3$ then $\operatorname{Sec}(X) = \mathbb{P}^N$ except the 4 following cases.*

- $n = 2$, $X = \mathbb{P}^2$, $X \subset \mathbb{P}^5$ is the Veronese embedding;
- $n = 4$, $X = \mathbb{P}^2 \times \mathbb{P}^2$, $X \subset \mathbb{P}^8$ is the Segre embedding;
- $n = 8$, $X = \operatorname{Gr}(2, \mathbb{C}^6)$, $X \subset \mathbb{P}^{14}$ is the Plücker embedding;
- $n = 16$, $X \subset \mathbb{P}^{26}$ is the projectivization of the highest weight vector orbit in the 27-dimensional irreducible representation of E_6.

The varieties listed in Theorem 16.8 are called *Severi varieties*, after Severi, who proved that the unique 2-dimensional Severi variety is the Veronese surface. Scorza and Fujita-Roberts [29] have shown that the unique 4-dimensional Severi variety is the Segre embedding of $\mathbb{P}^2 \times \mathbb{P}^2$. It is also shown in [29] that $n \equiv 0 \mod 16$ as $n > 8$. Finally, Tango [86] proved that if $n > 16$ then either $n = 2^a$ or $n = 3 \cdot 2^b$, where $a \geq 7$ and $b \geq 5$. The 16-dimensional Severi variety was discovered by Lazarsfeld [57]. In fact, all Severi varieties (1)–(4) arise from Example 16.6.

17 The degree of the dual variety to G/B

Let G be a simple algebraic group of rank r with Borel subgroup B. The minimal equivariant projective embedding of G/B corresponds to the line bundle $\mathcal{L}_\rho$, where $\rho = \omega_1 + \ldots + \omega_r$ is a half sum of positive roots.

Take all positive roots $\{\beta_1, \ldots, \beta_N\} = \Delta_+$ and coroots $\{\beta_1^\vee, \ldots, \beta_N^\vee\} = \Delta_+^\vee$. Consider the matrix

$$M = \left(\frac{(\beta_i^\vee, \beta_j)}{(\beta_i^\vee, \rho)} \right)_{i,j=1,\ldots,N}.$$

Let $P_s(M)$ be the sum of all permanents of $s \times s$ submatrices of M. Recall that the permanent of a $n \times n$ matrix (a_{ij}) is equal to $\sum_{\sigma \in S_n} \prod_{i=1}^{n} a_{i,\sigma(i)}$.

Theorem 17.1 ([20]). $\deg \Delta_{(G/B, \mathcal{L}_\rho)} = \sum_{s=0}^{N} (s+1)! P_{N-s}(M)$.

18 Degrees of hyperdeterminants

Consider the flag variety $\mathbb{P}^{k_1} \times \ldots \times \mathbb{P}^{k_r}$ of the group $SL_{k_1} \times \ldots \times SL_{k_r}$. The projectively dual variety of its 'minimal' equivariant projective embedding corresponds to a nice theory of hyperdeterminants that was initiated by Cayley and Schläffli [16, 14, 81].

Let $r \geq 2$ be an integer, and $A = (a_{i_1 \ldots i_r})$, $0 \leq i_j \leq k_j$ be an r-dimensional complex matrix of format $(k_1 + 1) \times \ldots \times (k_r + 1)$. The hyperdeterminant of A is defined as follows. Consider the product $X = \mathbb{P}^{k_1} \times \ldots \times \mathbb{P}^{k_r}$ of several projective spaces embedded into the projective space $\mathbb{P}^{(k_1+1)\times\ldots\times(k_r+1)-1}$ via the Segre embedding. Let X^* be the projectively dual variety. If X^* is a hypersurface then it is defined by a corresponding discriminant Δ_X, which in this case is called the *hyperdeterminant* (of format $(k_1 + 1) \times \ldots \times (k_r + 1)$) and denoted by Det. Clearly Det(A) is a polynomial function in the matrix entries of A invariant under the action of the group $SL_{k_1+1} \times \ldots \times SL_{k_r+1}$. If X^* is not a hypersurface then we set Det $= 1$. X^* is a hypersurface (and hence defines a hyperdeterminant) if and only if $2k_j \leq k_1 + \ldots + k_r$ for $j = 1, \ldots, r$. If for some j we have an equality $2k_j = k_1 + \ldots + k_r$ then the format is called a *boundary*.

Let $N(k_1, \ldots, k_r)$ be the degree of the hyperdeterminant of format $(k_1 + 1) \times \ldots \times (k_r + 1)$. The proof of the following theorem can be found in [35] or [34].

Theorem 18.1(a) *The generating function for the $N(k_1, \ldots, k_r)$ is given by*

$$\sum_{k_1, \ldots, k_r \geq 0} N(k_1, \ldots, k_r) z_1^{k_1} \ldots z_r^{k_r} = \frac{1}{\left(1 - \sum_{i=2}^{r} (i-1) e_i(z_1, \ldots, z_r)\right)^2},$$

where $e_i(z_1, \ldots, z_r)$ is the ith elementary symmetric polynomial.

(b) *The degree $N(k_1, \ldots, k_r)$ of the boundary format is (assuming that $k_1 = k_2 + \ldots + k_r$)*

$$N(k_2 + \ldots + k_r, k_2, \ldots, k_r) = \frac{(k_2 + \ldots + k_r + 1)!}{k_2! \ldots k_r!}.$$

(c) *The degree of the hyperdeterminant of the cubic format is given by*

$$N(k, k, k) = \sum_{0 \leq j \leq k/2} \frac{(j + k + 1)!}{(j!)^3 (k - 2j)!} \cdot 2^{k-2j}.$$

(d) *The exponential generating function for the degree N_r of the hyperdeterminant of format $2 \times 2 \times \ldots \times 2$ (r times) is given by*

$$\sum_{r \geq 0} N_r \frac{z^r}{r!} = \frac{e^{-2z}}{(1 - z)^2}.$$

19 Parametric formulas

In most circumstances, known formulas for the degree of the discriminant depend on a certain set of discrete parameters. There are two possibilities for these parameters. First, we may fix a projective variety and vary its polarizations. For instance, we have the following easy theorem:

Theorem 19.1. *Suppose that $\mathcal{L}$, $\mathcal{M}$ are very ample line bundles on X. Then $\det(X, \mathcal{L} \otimes \mathcal{M}) = 0$.*

Indeed, we have $\deg \Delta_{(X, \mathcal{L} \otimes \mathcal{M})} = \int_X c_n(J(\mathcal{L} \otimes \mathcal{M}))$ by Theorem 6.1. Therefore,

$$\deg \Delta_{(X, \mathcal{L} \otimes \mathcal{M})} = \int_X c_n(J(\mathcal{L}) \otimes \mathcal{M}) = \sum_{i=0}^{n} (i+1) \int_X c_1(\mathcal{M})^i \cdot c_{n-i}(J(\mathcal{L})).$$

Since $J(\mathcal{L})$ is spanned, all summands are non-negative; and since $\int_X c_1(\mathcal{M})^n > 0$ (being equal to the degree of X in the embedding determined by $\mathcal{M}$), the whole sum is positive.

For example, if X is embedded in $\mathbb{P}(V)$ and then re-embedded in $\mathbb{P}(S^d V)$ via the Veronese embedding then the dual variety X^* of this re-embedding is a hypersurface.

Suppose now that $\mathcal{L}_1, \ldots, \mathcal{L}_r$ are line bundles on X such that the corresponding linear systems $|\mathcal{L}_i|$ have no base points and such that for each i there is a representative of the linear system $|\mathcal{L}_i|$ which is a smooth divisor on X. Suppose further that any line bundle $\mathcal{L}$ of the form $\mathcal{L} = \mathcal{L}_1^{\otimes n_1} \otimes \ldots \otimes \mathcal{L}_r^{\otimes n_r}$ is very ample for positive n_i. Then the degree $\deg \Delta_{(X, \mathcal{L})}$ is a function of $n_1, \ldots, n_r$. Let us introduce new non-negative integers $m_i = n_i - 1$. Then $\deg \Delta_{(X, \mathcal{L})} = f(m_1, \ldots, m_r)$.

Theorem 19.2 ([20]).   *The function f is a non-trivial polynomial with non-negative coefficients. If each $n_i \geq 2$ then $(X, \mathcal{L})^*$ is a hypersurface.*

The second possibility is to change a variety and to fix a polarization (in a certain sense). For example, consider the irreducible representation of SL_{n_0} with the highest weight λ. Then this weight can be considered as a highest weight of SL_n for any $n \geq n_0$ with respect to the natural embedding $SL_n \subset SL_{n+1} \subset \ldots$. As a result, we shall obtain a tower of flag varieties with 'the same' polarization. The degree of the corresponding discriminants will be a function in n. This function can be very complicated. For example, in [56] a very involved combinatorial formula for this degree was found in the case λ is a fundamental weight (i.e. for the degree of a dual variety of the Grassmanian $Gr(k, n)$ in the Plücker embedding). However, sometimes this function has a closed expression. One example is given by the Boole formula $(n+1)(d-1)^n$ for the degree of the classical discriminant of homogeneous forms of degree d in $n+1$ variables (in other words, for the degree of the dual variety to the d-th Veronese embedding of $\mathbb{P}^n$). The following theorem is a slight shift of Boole's formula. We are not aware of any other similar formulas.

Theorem 19.3 ([91]).   *Let V be an irreducible SL_n-module with highest weight $(a-1)\varphi_1 + \varphi_2$, $a \geq 2$. Then the variety $X^* \subset \mathbb{P}(V^*)$ projectively dual to the projectivization X of the orbit of the highest vector is a hypersurface of degree*

$$\frac{(n^2 - n)a^{n+1} - (n^2 + n)a^{n-1} - 2n(-1)^n}{(a+1)^2}.$$

20 Polarized flag varieties with positive defect

Polarized flag varieties with positive defect were classified twice: the first time by Knop and Menzel [51] using local machinery, and later by Snow [83] using results from the minimal model programme.

Theorem 20.1. *Suppose that P_i is a maximal parabolic subgroup of a simple algebraic group corresponding to the simple root α_i. Let L be a very ample line bundle on G/P_i corresponding to the fundamental weight ω_i. Then $\operatorname{def}(G/P_i, L^{\otimes k}) = 0$ for $k > 1$ and $\operatorname{def}(G/P_i, L) > 0$ if and only if G/P_i is one of the following:*

- A_l/P_1, A_l/P_l. $\operatorname{def}(G/P) = \dim(G/P) = l$ *(projective space)*;
- A_l/P_2, A_l/P_{l-1}, $l \geq 4$ *even*; $\operatorname{def}(G/P) = 2$, $\dim(G/P) = 2l - 2$ *(Grassmanian of lines in the even-dimensional projective space)*;
- C_l/P_1. $\operatorname{def}(G/P) = \dim(G/P) = 2l - 1$ *(projective space)*;
- B_4/P_4, D_5/P_4, D_5/B_5. $\operatorname{def}(G/P) = 4$, $\dim(G/P) = 10$ *(spinor variety)*.

Notice that we do not follow the usual tradition and consider isomorphic polarized varieties with different group actions as distinct varieties. The next step is to consider flag varieties corresponding to arbitrary parabolic subgroups.

Theorem 20.2. *Let $(G/P, L)$ be a polarized flag variety of a complex semisimple algebraic group G. Then $\operatorname{def}(G/P, L) > 0$ if and only if either $(G/P, L)$ is one of polarized varieties described in Theorem 20.1 or the following conditions hold:*

- $G = G_1 \times G_2$, $P = P_1 \times P_2$, *where* $P_1 \subset G_1$, $P_2 \subset G_2$;
- $L = \operatorname{pr}_1^* L_1 \otimes \operatorname{pr}_2^* L_2$, *where* $\operatorname{pr}_i : G/P \to G_i/P_i$ *are projections and L_i are very ample line bundles on G_i/P_i*;
- $(G_1/P_1, L_1)$ *is one of the polarized varieties described in Theorem 20.1*;
- $\operatorname{def}(G_1/P_1, L_1) > \dim G_2/P_2$.

In this case $\operatorname{def}(G/P, L) = \operatorname{def}(G_1/P_1, L_1) - \dim G_2/P_2$.

Theorem 20.1 was used in [83] to recover the classification of homogeneous real hypersurface in a complex projective space due to [85], see also [17].

Theorem 20.3 ([85, 83]). *Let M be a homogeneous complete real hypersurface embedded equivariantly in $\mathbb{P}^N$. Then M is a tube over a linear projective space or one of the 4 self-dual homogeneous spaces $X \subset \mathbb{P}^N$ listed in Theorem 20.1.*

21 Varieties with small codegree

Let X be a projective variety. We define the *codegree* $\operatorname{codeg} X$ by the equality $\operatorname{codeg} X = \deg X^*$. If X^* is a hypersurface, then $\operatorname{codeg} X$ is just the class of X; this is a classical invariant playing an important role in enumerative geometry. If X^* is not a hypersurface and X' is a generic hyperplane section of X, then by Theorem 4.2 it is clear that $\operatorname{codeg} X' = \operatorname{codeg} X$.

With regard to codegree, the most simple nonsingular projective varieties are those whose codegree is small. The problem of classification of varieties of small codegree should be compared with that of classification of varieties of small degree. Much is known about this last problem. The case of varieties of degree two is classical (quadrics). The complete description of varieties of degree three was given by A. Weil. Swinnerton-Dyer [84] succeeded in classifying all varieties of degree four. After that, thanks particularly to Ionescu, there was considerable progress in classification of varieties of small degree, and now we have a complete list of nonsingular varieties up to degree eight. More generally, Hartshorne, Barth, Van de Ven and Ran proved that if the degree is sufficiently small with respect to dimension, then our variety is a complete intersection.

It seems worthwhile to consider the same question for codegree, but the flavour of the problem here is quite different. For example, in the case of varieties of small degree one can proceed by induction using the fact that degree is stable with respect to passing to hyperplane sections, whereas there is no such inductive procedure for codegree.

Theorem 21.1 ([99]). *There exist exactly ten non-degenerate nonsingular complex projective varieties of codegree three, namely*

- *the self-dual Segre threefold $\mathbb{P}^1 \times \mathbb{P}^2 \subset \mathbb{P}^5$,*
- *its hyperplane section $F_1 \subset \mathbb{P}^4$ obtained by blowing up a point in $\mathbb{P}^2$ by means of the map defined by the linear system of conics passing through this point,*
- *the four Severi varieties, i.e. the Veronese surface $v_2(\mathbb{P}^2) \subset \mathbb{P}^5$, the Segre variety $\mathbb{P}^2 \times \mathbb{P}^2 \subset \mathbb{P}^8$, the Grassmann variety $Gr_2(\mathbb{C}^6) \subset \mathbb{P}^{14}$ of lines in $\mathbb{P}^5$ and 16-dimensional variety $E \subset \mathbb{P}^{26}$ corresponding to the orbit of highest weight vector for the standard representation of the algebraic group of type E_6,*
- *the four varieties obtained by projecting the Severi varieties from generic points of their ambient linear spaces.*

The classification of smooth varieties of codegree 4 is still unknown.

All known examples arise from the Pyasetskii pairing (Theorem 12.1). The conjectural list consists of the Segre embedding of $\mathbb{P}^1 \times Q$, where Q is a quadric hypersurface; the twisted cubic curve $v_3(\mathbb{P}^1)$; the Plücker embedding of the Grassmanian $\mathrm{Gr}_3(\mathbb{C}^6)$; the isotropic Grassmanian $\mathrm{Gr}_3^0(\mathbb{C}^6)$ of isotropic 3 dimensional subspaces in the symplectic space $\mathbb{C}^3$ in its minimal equivariant embedding; the spinor variety $\mathbb{S}_6$; the 27-dimensional variety $E \subset \mathbb{P}^{55}$ corresponding to the orbit of highest weight vector for the minimal representation of the algebraic group of type E_7; the Segre embedding of $\mathbb{P}^1 \times \mathbb{P}^3$ (self-dual variety with defect 2), its non-singular hyperplane section and its non-singular section by two hyperplanes (which is either $\mathbb{P}^1 \times Q^1$ or the blow-up of a projective quadratic cone in its vertex).

22 The Matsumura-Monsky theorem

Let $D \subset \mathbb{P}^n$ be a smooth hypersurface of degree d. It was first proved in [59] that the group of projective automorphisms preserving D is finite if $d > 2$. In fact, it was also proved that the group of biregular automorphisms of D is finite if $d > 2$ (except the cases $d = 3$, $n = 2$ and $d = 4$, $n = 3$). Though this generalization looks much stronger, actually it is an easy consequence of the 'projective' version and the Bart Theorem [8].

More generally, let G/P be a flag variety of a simple Lie group and $D \subset G/P$ be a smooth ample divisor. Let $L_\lambda = \mathcal{O}(D)$ be the corresponding ample line bundle, where $\lambda \in \mathcal{P}^+$ is a dominant weight. Then one might expect that the normalizer $N_G(D)$ of D in G is finite if λ is big enough. Indeed, if $N_G(D)$ (or actually any linear algebraic group of transformations of D) contains a one-parameter subgroup of automorphisms of D then D is covered by rational curves. However, if λ is big enough then the canonical class K_D is nef by the adjunction formula, and therefore D cannot be covered by rational curves. Unfortunately, this transparent approach does not give strong estimates on λ. Much better estimates can be obtained using an original proof of the Matsumura-Monsky theorem.

The problem can be reformulated as follows. Suppose that V_λ is an irreducible G-module with highest weight λ. Let $\mathcal{D} \subset V_\lambda$ be the discriminant variety. We shall see that if λ is big enough then any point $x \in V_\lambda \setminus \mathcal{D}$ has a finite stabilizer G_x and the orbit of x is closed, $Gx = \overline{Gx}$ (therefore x is a stable point of V_λ in the sense of Geometric Invariant Theory, see [64]). This result can be compared with the results of [4], where all irreducible modules of simple algebraic groups with infinite stabilizers of generic points were found (this classification was extended later in [25] and [26] to handle irreducible representations of semisimple groups and any representations of simple groups as well).

If $\mathcal{D}$ is not a hypersurface then an easy inspection using Theorem 20.2 shows that the stabilizer of any point is infinite. So from now on we shall assume that $\mathcal{D}$ is a hypersurface defined by the vanishing of the discriminant Δ.

Recall that a dominant weight λ is called self-dual if V_λ is isomorphic to V_λ^* as a G-module. Let $\mathcal{P}_S^+ \subset \mathcal{P}^+$ be the subcone of self-dual dominant weights. Let γ be the highest root.

Theorem 22.1. *Let V_λ be an irreducible representation of a simple algebraic group G with highest weight λ such that $\mathcal{D}$ is a hypersurface. Suppose that $(\lambda - \gamma, \mu) > 0$ for any $\mu \in \mathcal{P}_S^+$. Let $x \in V_\lambda \setminus \mathcal{D}$. Then G_x is finite and $Gx = \overline{Gx}$. Moreover, $G_{[x]}$ is also finite, where $[x]$ is the line spanned by x.*

If $G = SL_n$ and $\lambda = \sum n_i \omega_i$, where $\omega_1, \ldots, \omega_n$ are the fundamental weights, then the assumptions of the Theorem are satisfied if and only if $\sum n_i > 2$, for example if $\lambda = n\omega_1$, $n > 2$. In particular, we recover the original Matsumura-Monsky Theorem.

23 Quasiderivations of commutative algebras

Let $V = \mathbb{C}^n$. Consider the vector space $\mathcal{A} = S^2 V^* \otimes V$ parametrizing bilinear commutative multiplications in V. In the sequel we identify points of $\mathcal{A}$ with the corresponding commutative algebras.

Let $A \in \mathcal{A}$. A non-zero element $v \in A$ is called a *quadratic nilpotent* if $v^2 = 0$. Let $\mathcal{D}_1 \subset \mathcal{A}$ be a subset of all algebras containing quadratic nilpotents. A one-dimensional subalgebra $U \subset A$ is called *singular* if there exists linear independent vectors $u \in U$ and $v \in A$ such that

$$u^2 = \alpha u, \quad uv = \frac{\alpha}{2}v, \quad \text{where } \alpha \in \mathbb{C}.$$

Let $\mathcal{D}_2 \subset \mathcal{A}$ be a subset of all algebras containing singular subalgebras.

Theorem 23.1 ([90])(1) *$\mathcal{D}_1$ and $\mathcal{D}_2$ are irreducible hypersurfaces.*
(2) *Let $A \in \mathcal{A}$. Then A contains a one-dimensional subalgebra.*
(3) *Let $A \in \mathcal{A} \setminus (\mathcal{D}_1 \cup \mathcal{D}_2)$. Then A contains exactly $2^n - 1$ one-dimensional subalgebras; all these subalgebras are spanned by idempotents.*

The algebras $A \in \mathcal{A}$ that do not belong to discriminant varieties $\mathcal{D}_1$

and $\mathcal{D}_2$ will be called *regular*. Of course, both hypersurfaces $\mathcal{D}_1$ and $\mathcal{D}_2$ can be interpreted as ordinary discriminants. First, we can enlarge the symmetry group and consider $S^2(\mathbb{C}^n)^* \otimes \mathbb{C}^n$ as an $SL_n \times SL_n$-module. Then this module is irreducible and its discriminant variety (the dual variety of the projectivization of the highest weight vector orbit) coincides with $\mathcal{D}_1$. Now consider $\mathcal{A} = S^2 V^* \otimes V$ as an $SL(V)$-module. Then this module is reducible, $\mathcal{A} = \mathcal{A}_0 + \tilde{\mathcal{A}}$, where $\mathcal{A}_0$ is a set of algebras with zero trace and $\tilde{\mathcal{A}}$ is isomorphic to V^* as an $SL(V)$-module. Consider the discriminant of $\mathcal{A}_0$ as a function on $\mathcal{A}$ (forgetting other coordinates). Then the corresponding hypersurface is exactly $\mathcal{D}_2$. If we consider the set of linear operators $\mathrm{Hom}(V, V) = V^* \otimes V$ instead of $\mathcal{A}$, then this construction will give us the determinant and the discriminant of the linear operator.

Let $\mathfrak{g}$ be a Lie algebra with representation $\rho : \mathfrak{g} \to \mathrm{End}(V)$. Consider any $v \in V$. Then subalgebra $\mathfrak{g}_v = \{g \in \mathfrak{g} \mid \rho(g)v = 0\}$ is called the annihilator of v. The subset $Q\mathfrak{g}_v = \{g \in \mathfrak{g} \mid \rho(g)^2 v = 0\}$ is called the quasi-annihilator of v. Clearly, $\mathfrak{g}_v \subset Q\mathfrak{g}_v$. Of course, the quasi-annihilator is not a linear subspace in general.

For example, suppose that $\mathfrak{g} = \mathfrak{gl}_n$ and ρ is the natural representation in the vector space $V^* \otimes V^* \otimes V$ that parametrizes bilinear multiplications in V. Let A be any algebra. Then the annihilator $\mathfrak{g}_A$ is identified with the Lie algebra of derivations $\mathrm{Der}(A)$. Operators $D \in Q\mathfrak{g}_A$ are called *quasiderivations*. Of course, it is possible to write down explicit equations that determine $Q\,\mathrm{Der}(A) = Q\mathfrak{g}_A$ in $\mathrm{End}(A)$, but this formula is quite useless (see [93]).

Quasiderivations can be used to define naive deformations. Namely, suppose that A is any algebra with the multiplication $u \cdot v$ and D is its quasiderivation. Consider the algebra A_D with the same underlying vector space and with the multiplication given by

$$u \star v = u \cdot D(v) + D(u) \cdot v - D(u \cdot v).$$

Then if A satisfies any polynomial identity then A_D satisfies this identity as well. More generally, let $\rho : \mathfrak{g} \to \mathrm{End}(V)$ be the differential of a representation of an algebraic group G, $v \in V$, $D \in Q\mathfrak{g}_v$. Suppose that $H \subset V$ is a closed conical G-equivariant hypersurface (the vanishing set of a homogeneous G-semiinvariant) and $v \in H$. Then $\rho(D)v$ also belongs to H. Indeed, since H is equivariant, $\exp(\lambda\rho(D))v$ belongs to H for any $\lambda \in \mathbb{C}$. Since D is a quasiderivation, $\exp(\lambda\rho(D))v = v + \lambda\rho(D)v$. Since H is conical, $v/\lambda + \rho(D)v$ belongs to H. Since H is closed, $\rho(D)v$ also belongs to H.

Now suppose that A is a regular commutative algebra.

Theorem 23.2 ([90]). *Let $A \in \mathcal{A}$, $A \notin \mathcal{D}_1 \cup \mathcal{D}_2$. Then $Q \operatorname{Der}(A) = 0$.*

This gives the following corollary, first proved in [3] by combinatorial methods.

Corollary 23.3 ([3]). *Let A be an n-dimensional semisimple commutative algebra, i.e. A is a direct sum of n copies of $\mathbb{C}$, that is A is the algebra of diagonal $n \times n$ matrices. Then A has no nonzero quasiderivations.*

Indeed, it is sufficient to check that $A \notin \mathcal{D}_1 \cup \mathcal{D}_2$. Clearly, A has no nilpotents. Suppose that $U \subset A$ is a one-dimensional subalgebra spanned by an idempotent $e \in U$. Since the spectrum of the operator of left multiplication by e is integer-valued (actually consists of 0 and 1), it follows that U is not singular.

24 Around the Hartshorne conjecture

In [38] R. Hartshorne made a number of conjectures related to the geometry of projective varieties of small codimension. Undoubtedly the most famous conjecture from this paper is the so-called Hartshorne conjecture on complete intersections.

Conjecture 24.1 ([38]). *If X is a smooth n-dimensional projective variety in $\mathbb{P}^N$ and $\operatorname{codim} X < N/3$, then X is a complete intersection.*

This conjecture is still very far from being solved and only partial results are known. If Hartshorne's conjecture is true then for any smooth projective variety X in $\mathbb{P}^N$ such that $\operatorname{codim} X < N/3$ the dual variety X^* should be a hypersurface. (It is not known whether or not this is true.) In particular, if X is a smooth self-dual variety then, up to Hartshorne's Conjecture, either X is a quadric hypersurface or $\operatorname{codim} X \geq N/3$, so that the following theorem should give a complete list of smooth self-dual varieties.

Theorem 24.2 ([24]). *Let X be a nonlinear smooth projective variety in $\mathbb{P}^N$ with $\operatorname{codim} X \geq N/3$ and $\dim X = \dim X^*$. Then X is either a hypersurface in $\mathbb{P}^2$ or $\mathbb{P}^3$, or is one of the following varieties:*

(i)   *the Segre embedding of $\mathbb{P}^1 \times \mathbb{P}^{n-1}$ in $\mathbb{P}^{2n-1}$;*
(ii) *the Plücker embedding of $\operatorname{Gr}(2,5)$ in $\mathbb{P}^9$;*
(iii) *the 10-dimensional spinor variety $\mathbb{S}_5$ in $\mathbb{P}^{15}$.*

The proof of the following finiteness theorem is based on the results of Nagura [67] about the distribution of primes.

Theorem 24.3 ([65]). *Let p be a fixed positive integer. Then there exists N_0 such that for any $N > N_0$ the following holds. If $X \subset \mathbb{P}^N$ is a smooth non-degenerate nonlinear n-fold and $\dim X^* = n + (p - 1)$ then either $n > 2N/3$ or $\operatorname{def} X = k \geq n/2$, in which case X is a projective bundle $X \simeq \mathbb{P}_Y(F)$, where F is a vector bundle of rank $(n + k + 2)/2$ on a smooth $(n - k)/2$-dimensional variety Y and the fibres are embedded linearly.*

25 Self-dual nilpotent orbits

If $X \subset \mathbb{P}^n$ is a smooth projective variety then X is almost never self-dual. Moreover, up to Hartshorne's conjecture Theorem 24.2 provides a complete list of them. However, there are a lot of non-smooth self-dual varieties. Many equivariant self-dual varieties are provided by Pyasetskii – see Theorem 12.1. Perhaps, the most interesting examples of self-dual varieties are the Kummer surface in $\mathbb{P}^3$ [37] and the Coble quartic in $\mathbb{P}^7$ [73].

Other interesting examples were found in [75]. Let G be semisimple and let $V = \mathfrak{g}$ be the adjoint representation. Then the Killing form $(\cdot, \cdot)$ is G-invariant and $\mathcal{R}(\mathfrak{g})$ is the cone of nilpotent elements. We identify $\mathfrak{g}$ and $\mathfrak{g}^*$.

Theorem 25.1 ([75]). *A nilpotent orbit $\mathcal{O} = \operatorname{Ad}(G)x$ has a self-dual projectivization if and only if the centralizer $\mathfrak{g}_x$ has no semi-simple elements.*

Nilpotent orbits with this property are said to be *distinguished* [19]. For example, the regular nilpotent orbit (the orbit dense in $\mathcal{R}(\mathfrak{g})$) is distinguished, hence the null-cone $\mathcal{R}(\mathfrak{g})$ itself has a self-dual projectivization.

The distinguished nilpotent orbits are, in a sense, the building blocks for the set of all nilpotent orbits. Namely, by virtue of the Bala-Carter correspondence [5, 6] the set of all nilpotent orbits in a semisimple Lie algebra $\mathfrak{g}$ is in natural bijection with the set of isoclasses of pairs $(\mathfrak{l}, \mathcal{O})$, where $\mathfrak{l} \subset \mathfrak{g}$ is a Levi subalgebra and $\mathcal{O} \subset \mathfrak{l}$ is a distinguished nilpotent orbit in $[\mathfrak{l}, \mathfrak{l}]$.

For instance, $\mathfrak{g} = E_8$ has precisely 11 distinguished orbits $\mathbb{R}(\mathfrak{g})_i$ of dimension 240, 238, 236, 234, 232, 230, 228, 226, 224, 220, 208. The first two of them are respectively the regular and the subregular orbits. This gives 11 self-dual algebraic varieties in $\mathbb{P}^{247}$.

26 Discriminants of anticommutative algebras

If a certain set of objects is parametrized by an algebraic variety X (for example, by a vector space) then it makes sense to speak about generic objects. Namely, we say that a generic object satisfies some property if there exists a dense Zariski-open subset $X_0 \subset X$ such that all objects parametrized by points from X_0 share this property. However, sometimes it is possible to find some discriminant-type closed subvariety $Y \subset X$ and then to study properties of 'regular' objects parametrized by points from $X \setminus Y$. For example, instead of studying generic hypersurfaces it is simetimes worthwhile to study smooth hypersurfaces. In this section we implement this programme for the study of some quite non-geometric objects, namely anticommutative algebras, with the multiplication depending on a number of arguments.

Let $V = \mathbb{C}^n$. We fix an integer k, $1 < k < n-1$. Let $\mathcal{A}_{n,k} = \Lambda^k V^* \otimes V$ be the vector space of k-linear anticommutative maps from V to V. We identify the points of $\mathcal{A}_{n,k}$ with the corresponding algebras, that is, we assume that $A \in \mathcal{A}_{n,k}$ is the space V equipped with the structure of a k-argument anticommutative algebra.

Theorem 26.1 ([91]). *Let $A \in \mathcal{A}_{n,k}$ be a generic algebra. Then A contains no m-dimensional subalgebras with $k+1 < m < n$.*

The set of k-dimensional subalgebras is a smooth irreducible $(k-1)(n-k)$-dimensional subvariety in the Grassmanian $\mathrm{Gr}(k, A)$.

There are finitely many $(k+1)$-dimensional subalgebras, and their number is

$$\sum_{\substack{n-k-1 \geq \mu_1 \geq \ldots \geq \mu_{k+1} \geq 0 \\ n-k-1 \geq \lambda_1 \geq \ldots \geq \lambda_{k+1} \geq 0 \\ \mu_1 \leq \lambda_1, \ldots, \mu_{k+1} \leq \lambda_{k+1}}} (-1)^{|\mu|} \frac{(\lambda_1 + k)!(\lambda_2 + k - 1)!\ldots\lambda_{k+1}!}{(\mu_1 + k)!(\mu_2 + k - 1)!\ldots\mu_{k+1}!}(|\lambda| - |\mu|)! \times$$

$$\times \left| \frac{1}{(i - j + \lambda_j - \mu_i)!} \right|^2_{i,j=1,\ldots,k+1},$$

where $|\lambda| = \lambda_1 + \ldots + \lambda_{k+1}$, $|\mu| = \mu_1 + \ldots + \mu_{k+1}$, $1/N! = 0$ if $N < 0$. In particular A contains a $(k+1)$-dimensional subalgebra. If $k = n-2$, then the number of $(k+1)$-dimensional subalgebras is equal to $(2^n - (-1)^n)/3$.

An essential drawback of Theorem 26.1 is the fact that it does not enable us to study the structure of subalgebras of any particular algebra. To correct this, we shall need to introduce an explicit class of 'regular' algebras instead of the implicit class of generic algebras. The natural way to remove degeneracies is to consider discriminants.

Let $\mathcal{A}_{n,k}^{0\,*}$ be the GL_n-module dual to $\mathcal{A}_{n,k}^0$, and let S_D be the closure of the orbit of the highest vector, $S_D \subset \mathcal{A}_{n,k}^{0\,*}$. Let $PS_D \subset P\mathcal{A}_{n,k}^{0\,*}$ be its projectivization, let $P\mathcal{D} \subset P\mathcal{A}_{n,k}^0$ be the subvariety projectively dual to the subvariety PS_D, and let $\mathcal{D} \subset \mathcal{A}_{n,k}^0$ be the cone over it. We shall call $\mathcal{D}$ the *D-discriminant subvariety*. The algebras $A \in \mathcal{D}$ are said to be *D-singular*. The algebras $A \notin \mathcal{D}$ are said to be *D-regular*.

Theorem 26.2 ([91]). *$\mathcal{D}$ is a hypersurface. Let A be a D-regular algebra. Then the set of k-dimensional subalgebras of A is a smooth irreducible $(k-1)(n-k)$-dimensional subvariety in $\mathrm{Gr}(k, A)$. Let $k = n - 2$. Then the degree of $\mathcal{D}$ is equal to $\left((3n^2 - 5n)2^n - 4n(-1)^n\right)/18$.*

Hence, the D-singularity of A is determined by the vanishing of the SL_n-invariant polynomial D that defines $\mathcal{D}$. This polynomial is called the *D-discriminant*.

We define the E-discriminant and E-regularity only for $(n-2)$-argument n-dimensional anticommutative algebras. Let $\mathcal{A} = \mathcal{A}_{n,n-2}^0$. Consider the projection $\pi : \mathrm{Gr}(n-1, V) \times P\mathcal{A} \to P\mathcal{A}$ on the second summand and the incidence subvariety $Z \subset \mathrm{Gr}(n-1, V) \times P\mathcal{A}$ that consists of pairs $S \subset P\mathcal{A}$, where S is a subalgebra in A. Let $\tilde{\pi} = \pi|_Z$. By Theorem 26.1, we have $\tilde{\pi}(Z) = \mathcal{A}$. Let $\tilde{\mathcal{E}} \subset Z$ be the set of critical points of $\tilde{\pi}$, let $P\mathcal{E} = \tilde{\pi}(\tilde{\mathcal{E}})$ be the set of critical values of $\tilde{\pi}$, and let $\mathcal{E} \subset \mathcal{A}$ be the cone over $P\mathcal{E}$. Then $\mathcal{E}$ is called the *E-discriminant subvariety*. The algebras $A \in \mathcal{E}$ are said to be *E-singular*. The algebras $A \notin \mathcal{E}$ are said to be *E-regular*.

Theorem 26.3 ([91]). *$\mathcal{E}$ is an irreducible hypersurface. Let A be an E-regular algebra. Then A has precisely $(2^n - (-1)^n)/3$ $(n-1)$-dimensional subalgebras. The map $\tilde{\pi} : \tilde{\mathcal{E}} \to P\mathcal{E}$ is birational.*

Hence, the E-singularity of A is determined by the vanishing of the SL_n-invariant polynomial that defines $\mathcal{E}$. This polynomial is called the *E-discriminant*. An $(n-2)$-argument n-dimensional anticommutative algebra is said to be *regular* if it is D-regular and E-regular. We consider 2-argument 4-dimensional algebras.

Theorem 26.4 ([91]). *Let A be a 4-dimensional regular anticommutative algebra. Then A has precisely five 3-dimensional subalgebras. The set of these subalgebras is a generic configuration of five hyperplanes. In particular, A has a pentahedral normal form, that is, it can be reduced by a transformation that belongs to GL_4 to an algebra such that the set of its five subalgebras is a Sylvester pentahedron $x_1 = 0$, $x_2 = 0$, $x_3 = 0$, $x_4 = 0$, $x_1 + x_2 + x_3 + x_4 = 0$. A has neither one- nor two-dimensional ideals. The set of two-dimensional subalgebras of A is a del Pezzo surface of degree 5 (a blowing up of $\mathbb{P}^2$ at four generic points). A has precisely 10 fans, that is, flags $V_1 \subset V_3$ of 1-dimensional and 3-dimensional subspaces such that every intermediate subspace U, $V_1 \subset U \subset V_3$, is a two-dimensional subalgebra.*

Let us give some definitions. Let X be an irreducible G-variety (a variety with an action of algebraic group G), $S \subset X$ be an irreducible subvariety. Then S is called a *section* of X if $\overline{G \cdot S} = X$. The section S is called a *relative section* if the following condition holds: there exists a dense Zariski-open subset $U \subset S$ such that if $x \in U$ and $gx \in S$ then $g \in H$, where $H = N_G(S) = \{g \in G \,|\, gS \subset S\}$ is the normalizer of S in G (see [76]). In this case for any invariant function $f \in \mathbb{C}(X)^G$ the restriction $f|_S$ is well-defined and the map $\mathbb{C}(X)^G \to \mathbb{C}(S)^H$, $f \mapsto f|_S$, is an isomorphism. Any relative section defines a G-equivariant rational map $\psi : X \to G/H$: if $g^{-1}x \in S$ then $x \mapsto gH$. Conversely, any G-equivariant rational map $\psi : X \to G/H$ with irreducible fibres defines the relative section $\overline{\psi^{-1}(eH)}$.

We are going to apply Theorem 26.4 and to construct a relative section in the SL_4-module $\mathcal{A}_0$ (the module of 4-dimensional anticommutative algebras with zero trace). The action of SL_4 on 'Sylvester pentahedrons' is transitive with finite stabilizer H (which is the central extension of the permutation group S_5). In the sequel the Sylvester pentahedron will always mean the standard configuration formed by the hyperplanes $x_1 = 0$, $x_2 = 0$, $x_3 = 0$, $x_4 = 0$, $x_1 + x_2 + x_3 + x_4 = 0$. Let $S \subset \mathcal{A}_0$ be a linear subspace formed by all algebras such that the hyperplanes of Sylvester pentahedron are their subalgebras. Then Theorem 26.4 implies that S is a 5-dimensional linear relative section of SL_4-module $\mathcal{A}_0$. It is easy to see that the multiplication in algebras from S is given by formulas $[e_i, e_j] = a_{ij}e_i + b_{ij}e_j$, $1 \leq i < j \leq 4$, where a_{ij} and b_{ij} satisfy a certain set of linear conditions. Consider six algebras $A_1, \ldots, A_6$ with the following structure constants

	A_1	A_2	A_3	A_4	A_5	A_6
a_{12}	0	1	-1	1	0	-1
b_{12}	1	0	1	-1	-1	0
a_{13}	1	1	-1	0	-1	0
b_{13}	0	-1	0	1	1	-1
a_{14}	1	0	0	1	-1	-1
b_{14}	-1	1	-1	0	0	1
a_{23}	-1	-1	0	1	0	1
b_{23}	1	0	-1	0	1	-1
a_{24}	0	-1	-1	0	1	1
b_{24}	-1	1	0	1	-1	0
a_{34}	-1	1	1	-1	0	0
b_{34}	0	0	-1	1	-1	1

Then it is easy to see that $A_i \in S$ for any i. Moreover, algebras A_i satisfy the unique linear relation $A_1 + \ldots + A_6 = 0$. It follows that any $A \in S$ can be written uniquely in the form $\alpha_1 A_1 + \ldots + \alpha_6 A_6$, where $\alpha_1 + \ldots + \alpha_6 = 0$. The coordinates α_i are called dodecahedral coordinates and S is called the dodecahedral section (this name will be clear later).

The stabilizer of the standard Sylvester pentahedron in PGL_4 is isomorphic to $\mathbb{S}_5$ represented by permutations of its hyperplanes. The group S_5 is generated by the transposition (12) and the cycle (12345). The preimages of these elements in GL_4 are given by matrices

$$\sigma = \begin{pmatrix} 0 & 1 & 0 & 0 \\ 1 & 0 & 0 & 0 \\ 0 & 0 & 1 & 0 \\ 0 & 0 & 0 & 1 \end{pmatrix}, \qquad \tau = \begin{pmatrix} -1 & -1 & -1 & -1 \\ 1 & 0 & 0 & 0 \\ 0 & 1 & 0 & 0 \\ 0 & 0 & 1 & 0 \end{pmatrix}.$$

The preimage of S_5 in SL_4 is the group H of 480 elements. The representation of H in S induces the projective representation of S_5 in $\mathbb{P}^4$. We have the following

Theorem 26.5. *This projective representation is the projectivization of either of the two possible 5-dimensional irreducible representations of S_5.*

Remark 26.6. The section S is called *dodecahedral* for the following reason. Though the surjection $H \to S_5$ does not split, the alternating group A_5 can be embedded in H. The induced representation of A_5 in S has the following description. $\mathbb{A}_5$ can be realised as a group of rotations of the dodecahedron. Let $\{\Gamma_1, \ldots, \Gamma_6\}$ be the set of pairs of opposite faces of the dodecahedron. Consider the vector space of functions $f : \{\Gamma_1, \ldots, \Gamma_6\} \to \mathbb{C}$, $\sum_{i=1}^{6} f(\Gamma_i) = 0$. Then this vector space is an A_5-module. It is easy to see that this module is isomorphic to S via the identification $A_i \mapsto f_i$, where $f_i(\Gamma_i) = 5$, $f_i(\Gamma_j) = -1$, $j \neq i$.

The following theorem follows from the discussion above.

Theorem 26.7. *The restriction of invariants induces an isomorphism of invariant fields* $\mathbb{C}(\mathcal{A}_0)^{GL_4} \simeq \mathbb{C}(\mathbb{C}^5)^{\mathbb{C}^* \times \mathbb{S}_5}$, *where* $\mathbb{C}^*$ *acts on* $\mathbb{C}^5$ *by homotheties and* $\mathbb{S}_5$ *acts via any of two 5-dimensional irreducible representations.*

Remark 26.8. The Sylvester pentahedron also naturally arises in the theory of cubic surfaces [58]. The SL_4-module of cubic forms $S^3(\mathbb{C}^4)^*$ admits the relative section (the so-called Sylvester section, or Sylvester normal form). Namely, a generic cubic form in a suitable system of homogeneous coordinates $x_1, \ldots, x_5$, $x_1 + \ldots + x_5 = 0$, can be written as a sum of 5 cubes $x_1^3 + \ldots + x_5^3$. The Sylvester pentahedron can be recovered from a generic cubic form f in a very interesting way: its 10 vertices coincide with 10 singular points of a quartic surface $\det \mathrm{Hes}(f)$. The Sylvester section has the same normalizer H as our dodecahedral section. It can be proved [9] that in this case the restriction of invariants induces an isomorphism $\mathbb{C}(S^3(\mathbb{C}^4)^*)^{GL_4} \simeq \mathbb{C}(\mathbb{C}^5)^{\mathbb{C}^* \times \mathbb{S}_5}$, where $\mathbb{C}^*$ acts via homotheties and $\mathbb{S}_5$ via permutations of coordinates (i.e. via the *reducible* 5-dimensional representation). Other applications of the Sylvester pentahedron to moduli varieties can be found in [7].

These results were used in [89] in order to prove that the field of invariant functions of the 5-dimensional irreducible representation of S_5 is rational (is isomorphic to the field of invariant functions of a vector space). From this result it is easy to deduce that in fact the field of invariant functions of any representation of S_5 is rational (see [88]).

Bibliography

[1] P. Aluffi. Singular schemes of hypersurfaces. *Duke Math. J.*, 80:325–351, 1995.

[2] P. Aluffi and F. Cukierman. Multiplicities of discriminants. *Manuscripta Math.*, 78:245–258, 1993.

[3] A. Z. Anan'in. Quasiderivations of the algebra of diagonal matrices. *Tret'ya Sibirskaya shkola po algebre i analizu, Irkutsk University*, pages 8–15, 1990. In Russian.

[4] E. M. Andreev, E. B. Vinberg, and A. G. Elashvili. Orbits of greatest dimension in semisimple linear Lie groups. *Functional Anal. Appl.*, 1:257–261, 1967.

[5] P. Bala and R. W. Carter. Classes of unipotent elements in simple algebras, I. *Math. Proc. Camb. Phil. Soc.*, 79:401–425, 1976.

[6] P. Bala and R. W. Carter. Classes of unipotent elements in simple algebras, II. *Math. Proc. Camb. Phil. Soc.*, 80:1–18, 1976.

[7] F. Bardelli. The Sylvester pentahedron and some applications to moduli spaces. *Preprint.*

[8] W. Barth. Transplanting cohomology classes in complex projective space. *Amer. J. Math.*, 92:951–961, 1970.

[9] N. D. Beklemishev. Classification of quaternary cubic forms in non-general position. In *Voprosy teorii grupp i gomologicheskoy algebry*, pages 3–17. Yaroslav. Gos. Univ., 1981. In Russian.

[10] M. C. Beltrametti, M. L. Fania, and A. J. Sommese. On the discriminant variety of a projective manifold. *Forum Math*, 4:529–547, 1992.

[11] M. C. Beltrametti, A. J. Sommese, and J. A. Wiśniewski. Results on varieties with many lines and their applications to adjunction theory. In *Complex algebraic varieties (Bayreuth, 1990)*, volume 1507 of *Lect. Notes in Math.*, pages 16–38. Springer, Berlin, 1992.

[12] R. Bott. Homogeneous vector bundles. *Annals of Math. (2)*, 66:203–248, 1957.

[13] N. Bourbaki. *Éléments de mathématique. Fasc. XXXIV. Groupes et algèbres de Lie.* Hermann, Paris, 1968.

[14] A. Cayley. On linear transformations. In *Collected papers Vol. 1*, pages 95–112. Cambridge Univ. Press, 1889.

[15] A. Cayley. On the theory of elimination. In *Collected papers Vol. 1*, pages 370–374. Cambridge Univ. Press, 1889.

[16] A. Cayley. On the theory of linear transformations. In *Collected papers Vol. 1*, pages 80–94. Cambridge Univ. Press, 1889.

[17] T. Cecil and P. Ryan. Focal sets and real hypersurfaces in complex projective space. *Trans. Amer. Math. Soc.*, 269:481–499, 1982.

[18] H. Clemens, J. Kollar, and S. Mori. Higher dimensional complex geometry. *Astérisque*, 166, 1988.

[19] D. Collingwood and W. McGovern. *Nilpotent orbits in semisimple Lie algebras.* Van Nostrand Reinhold Company, 1993.

[20] C. de Concini and J. Weiman. A formula with nonnegative terms for the degree of the dual variety of a homogeneous space. *Proc. Amer. Math. Soc.*, 125(1):1–8, 1997.

[21] P. Deligne. Le déterminant de la cohomologie. In *Current trends in arithmetic algebraic geometry*, volume 67 of *Contemporary Math.*, pages 93–177. Amer. Math. Soc., 1987.

[22] A. Dimca. Milnor numbers and multiplicities of dual varieties. *Rev. Roumaine Math. Pures*, pages 535–538, 1986.

[23] L. Ein. Varieties with small dual varieties, II. *Duke Math. J.*, 52:895–907, 1985.

[24] L. Ein. Varieties with small dual varieties, I. *Invent. Math.*, 86:63–74, 1986.

[25] A. G. Elashvili. Canonical form and stationary subalgebras of points of general position for simple linear Lie groups. *Functional Anal. Appl.*, 6:44–53, 1972.

[26] A. G. Elashvili. Stationary subalgebras of points of general position for irreducible linear Lie groups. *Functional Anal. Appl.*, 6:139–148, 1972.

[27] E. Fisher. Über die Cayleyshe Eliminationsmethode. *Math. Zeit.*, 26:497–550, 1927.

[28] W. Franz. Die torsion einer überdeckung. *J. Reine Angew. Math.*, 173:245–254, 1935.

[29] T. Fujita and J. Roberts. Varieties with small secant varieties: the extremal case. *Amer. J. Math.*, 103:953–976, 1981.

[30] W. Fulton. *Intersection Theory*, volume 2 of *Ergebnisse der Mathematik*. Springer, 1984.

[31] W. Fulton and J. Hansen. A connectedness theorem for projective varieties, with applications. *Ann. of Math. (2)*, 110(1):159–166, 1979.

[32] W. Fulton and R. Lazarsfeld. Connectivity and its applications in algebraic geometry. In *Proc. Midwest Algebraic Geometry Conf. Univ. Illinois, 1980*, volume 862 of *Lect. Notes in Math*, pages 26–92. Springer-Verlag, Berlin-Heidelberg-New York, 1981.

[33] P. Gabriel. Unzerlegbare Darstellungen, I. *Manuscripta Math.*, 6:71–103, 1972.

[34] I. M. Gelfand, M. M. Kapranov, and A. V. Zelevinski. Hyperdeterminants. *Adv. Math.*, 84:237–254, 1990.

[35] I. M. Gelfand, M. M. Kapranov, and A. V. Zelevinski. *Discriminants, resultants, and multidimensional determinants*. Birkhäuser, Boston, 1994.

[36] P. Griffiths and J. Harris. Algebraic geometry and local differential geometry. *Ann. Sci. Ec. Norm. Super.*, 12:355–432, 1979.

[37] P. Griffiths and J. Harris. *Principles of Algebraic Geometry*. Wiley-Interscience, London-New-York, 1979.

[38] R. Hartshorne. Varieties of low codimension in projective space. *Bull. Amer. Math. Soc.*, 80:1017–1032, 1974.

[39] R. Hartshorne. *Algebraic Geometry*, volume 52 of *Graduate Texts in Mathematics*. Springer-Verlag, New York-Heidelberg-Berlin, 1977.

[40] A. Hefez and S. Kleiman. Notes on duality for projective varieties.

[41] A. Holme. On the dual of a smooth variety. In *Algebraic geometry (Proc. Summer Meeting, Univ. Copenhagen, Copenhagen, 1978)*, volume 732 of *Lect. Notes in Math.*, pages 144–156. Springer–Verlag, 1979.

[42] A. Holme. The geometric and numerical properties of duality in projective algebraic geometry. *Manuscripta Math.*, 61:145–162, 1988.

[43] N. Jacobson. *Structure and representations of Jordan algebras*, volume 39 of *Publ. Amer. Math. Soc.* 1968.

[44] H. Kaji, M. Ohno, and O. Yasukura. Adjoint varieties and their secant varieties. *Indag. Math.*, 10:45–57, 1999.

[45] V. V. Kashin. Orbits of adjoint and coadjoint actions of Borel subgroups of semisimple algebraic groups. In *Problems in Group Theory and Homological Algebra, Yaroslavl' (Russian)*, pages 141–159. 1997.

[46] N. M. Katz. Pinceaux de Lefschetz; théorème d'existence. In *SGA 7 (II)*, volume 340 of *Lect. Notes in Math.*, pages 212–253. Springer-Verlag, 1973.

[47] Y. Kawamata. The cone of curves of algebraic varieties. *Ann. of Math. (2)*, 119:603–633, 1984.

[48] Y. Kawamata, K. Matsuda, and K. Matsuki. Introduction to the minimal model program. In *Algebraic Geometry, Sendai 1985*, volume 10 of *Adv. Stud. Pure Math.*, pages 283–360. North-Holland, 1987.

[49] S. Kleiman. Enumerative theory of singularities. In *Real and Complex Singularities (Proc. Ninth Nordic Summer School/NAVF Sympos. Math., Oslo, 1976)*, pages 297–396. 1977.

[50] K. Knight and A. Zelevinsky. Representations of quivers of type A and the multisegment duality. *Adv. Math*, 117:273–293, 1996.

[51] F. Knop and G. Menzel. Duale Varietäten von Fahnenvarietäten. *Comm. Math. Helv.*, 62:38–61, 1987.

[52] F. Knudsen and D. Mumford. Projectivity of moduli spaces of stable curves, I. Preliminaries on "det" and "div". *Math. Scand.*, 39:19–55, 1976.

[53] A. Landman. Examples of varieties with small dual varieties. Picard-Lefschetz theory and dual varieties, 1976. Two lectures at Aarhus University.

[54] J. M. Landsberg and L. Manivel. On the projective geometry of homogeneous varieties. *Preprint*, 2000.

[55] A. Lanteri and D. Struppa. Projective 7-folds with positive defect. *Compositio Math.*, 61:329–337, 1987.

[56] A. Lascoux. Degree of the dual Grassman variety. *Comm. Algebra*, 9(11):1215–1225, 1981.

[57] R. Lazarsfeld. An example of a 16-dimensional projective variety with a 25-dimensional secant variety. *Math. Letters*, 7:1–4, 1981.

[58] Yu. I. Manin. *Cubic forms. Algebra, geometry, arithmetic.* Izdat. "Nauka", Moscow, 1972. English translation: North-Holland, Amsterdam-London 1974.

[59] H. Matsumura and P. Monsky. On the automorphisms of hypersurfaces. *J. Math. Kyoto Univ.*, 3(3):347–361, 1964.

[60] J. Milnor. *Singular points of complex hypersurfaces*, volume 61 of *Ann. of Math. Studies*. Princeton Univ. Press, 1968.

[61] C. Moeglin and J. L. Waldspurger. Sur l'involution de Zelevinski. *J. Reine Angew. Math*, 372:136–177, 1986.

[62] B. G. Moishezon. Algebraic homology classes of algebraic varieties. *Izvestiya AN SSSR*, 31, 1967. In Russian.

[63] I. Muller, H. Rubenthaler, and G. Schiffmann. Structure des éspaces préhomogènes associés à certaines algèbres de Lie graduées. *Math. Ann.*, 274:95–123, 1986.

[64] D. Mumford, J. Fogarty, and F. Kirwan. *Geometric Invariant Theory*, volume 34 of *Ergebnisse der Mathematik und Grenzgebiete*. Springer-Verlag, New York-Heidelberg-Berlin, 1994. (3rd edition).

[65] R. Muñoz. Varieties with low dimensional dual variety. *Manuscripta Math.*, 94:427–435, 1997.

[66] R. Muñoz. Varieties with degenerate dual variety. *Forum Math.*, 13:757–779, 2001.

[67] J. Nagura. On the interval containing at least one prime number. *Proc. Japan Acad.*, 28:177–181, 1952.

[68] A. Némethi. Lefschetz theory for complex hypersurfaces. *Rev. Roumaine Math. Pures*, pages 233–250, 1988.

[69] C. Okonek, H. Spindler, and M. Schneider. *Vector bundles on complex projective spaces*, volume 3 of *Progress in Math*. Birkhäuser, 1980.

[70] D. I. Panyushev. Parabolic subgroups with abelian unipotent radical as a testing site for invariant theory. *Canad. J. Math.*, 51(3):616–635, 1999.

[71] A. Parusiński. A generalization of the Milnor number. *Math. Ann.*, 281:247–254, 1988.

[72] A. Parusiński. Multiplicity of the dual variety. *Bull. London Math. Soc.*, 23:428–436, 1991.

[73] C. Pauly. Self-duality of Coble's quartic hypersurface and applications. *Preprint.*

[74] S. Poljak. Maximum rank of powers of a matrix of a given pattern. *Proc. Amer. Math. Soc.*, 106(4):1137–1144, 1989.

[75] V. Popov. Self-dual algebraic varieties and nilpotent orbits. *Proceedings of the International Colloquium 'Algebra, Arithmetic, and Geometry', Bombay, 2001*, pages 496–520.

[76] V. L. Popov and E. B. Vinberg. Invariant theory. In *Algebraic Geometry (IV)*, volume 55 of *Encyclopaedia of Mathematical Sciences*, pages 123–278. Springer-Verlag, 1994.

[77] V. S. Pyasetskii. Linear Lie groups acting with finitely many orbits. *Funct. Anal. Appl.*, 9:351–353, 1975.

[78] D. Quillen. Determinants of Cauchy-Riemann operators on Riemann surfaces. *Funct. Anal. Appl.*, 19:31–34, 1985.

[79] N. Ray and I. Singer. *R*-torsion and the Laplacian for Riemannian manifolds. *Adv. Math.*, 7:145–210, 1971.

[80] R. W. Richardson. Finiteness theorems for orbits of algebraic groups. *Indag. Math.*, 88:337–344, 1985.

[81] L. Schläffli. In *Gesammelte Abhandlungen, Band. 2*, pages 9–112L. Birkhäuser-Verlag, Basel, 1953.

[82] B. Segre. Bertini forms and Hessian matrices. *J. London Math. Soc.*, 26:164–176, 1951.

[83] D. Snow. The nef value and defect of homogeneous line bundles. *Trans. Amer. Math. Soc.*, 340(1):227–241, 1993.

[84] H. P. F. Swinnerton-Dyer. An enumeration of all varieties of degree 4. *Amer. J. Math.*, 95:403–418, 1973.

[85] R. Takagi. On homogeneous real hypersurfaces in a complex projective space. *Osaka J. Math.*, 10:495–506, 1973.

[86] H. Tango. Remark on varieties with small secant varieties. *Bull. Kyoto Univ. Ed. Ser. B*, 60:1–10, 1982.

[87] A. Terracini. Alcune questioni sugli spazi tangenti e osculatori ad una varieta, I, II, III. *Societa dei Naturalisti e Matematici*, pages 214–247, 1913.

[88] E. Tevelev. *Generic algebras*. PhD thesis.

[89] E. Tevelev. Generic algebras. *Transformation Groups*, 1:127–151, 1996.

[90] E. Tevelev. Discriminants and quasi-derivations of commutative algebras. *Uspechi Matematicheskih Nauk.*, 54(2):189–190, 1999.

[91] E. Tevelev. Subalgebras and discriminants of anticommutative algebras. *Izvestiya RAN, Ser. Mat.*, 63(3):169–184, 1999.

[92] E. B. Vinberg. The Weyl group of a graded Lie algebra. *Math. USSR Izv.*, 10:463–495, 1976.

[93] E. B. Vinberg. Generalized derivations of algebras. *AMS Transl.*, 163:185–188, 1995.

[94] A. H. Wallace. Tangency and duality over arbitrary fields. *Proc. London Math. Soc.*, 6(3):321–342, 1956.

[95] J. Weyman. Calculating discriminants by higher direct images. *Trans. Amer. Math. Soc.*, 343:367–389, 1994.

[96] J. Weyman and A. Zelevinsky. Multiplicative properties of projectively dual varieties. *Manuscripta Math.*, 82(2):139–148, 1994.

[97] P. Wilson. Towards birational classification of algebraic varieties. *Bull. London Math. Soc.*, 19:1–48, 1987.

[98] F. L. Zak. Severi varieties. *Mat. Sbornik*, 126:115–132, 1985. In Russian.

[99] F. L. Zak. Some properties of dual varieties and their applications in projective geometry. In *Algebraic Geometry*, volume 1479 of *Lect. Notes in Math.*, pages 273–280. Springer-Verlag, New York, 1991.

[100] F. L. Zak. *Tangents and secants of algebraic varieties*, volume 127 of *Translations of Math. Monographs*. Amer. Math. Soc., 1993.

Department of Mathematics
University of Texas at Austin
1 University Station C1200
Austin, TX 78712-0257
USA
tevelev@math.utexas.edu

Equivariant embeddings of homogeneous spaces

Dmitri A. Timashev

Introduction

Homogeneous spaces of algebraic groups play an important rôle in various aspects of geometry and representation theory. We restrict our attention to linear, and even reductive, algebraic groups, because the most interesting interplay between geometric and representation-theoretic aspects occurs for this class of algebraic groups.

Classical examples of algebraic homogeneous spaces:

(i) The affine space $\mathbb{A}^n$ is homogeneous under GA_n, the general affine group;

(ii) The projective space $\mathbb{P}^n$ is homogeneous under GL_{n+1};

(iii) The sphere $S^{n-1} = SO_n/SO_{n-1}$;

(iv) Grassmannians $\mathrm{Gr}_k(\mathbb{P}^n)$ $(k \leq n)$ and flag varieties are homogeneous under GL_{n+1}.

(v) The space of non-degenerate quadrics $Q_n = PGL_{n+1}/PO_{n+1}$;

(vi) The space $\mathrm{Mat}_{m,n}^{(r)}$ of $(m \times n)$-matrices of rank r is homogeneous under $GL_m \times GL_n$.

The relations of algebraic homogeneous spaces to representation theory have their origin in the Borel-Weil theorem, realizing all simple modules of reductive groups as spaces of sections of line bundles on (generalized) flag varieties. This geometric approach to representation theory by the realization of representations in spaces of sections of line bundles on homogeneous spaces (or on their embeddings) is rather fruitful, and it also raises an interesting problem of the description of higher cohomology groups of line bundles (generalizing the Borel-Weil-Bott theorem).

As another motivation for studying embeddings of homogeneous spaces, consider enumerative geometry.

226

A classical enumerative problem: How many plane conic curves are tangent to five given conics in general position?

The natural approach is to compactify Q_2 by degenerate conics and to consider the compact embedding space $\mathbb{P}^5$, where each tangency condition determines a hypersurface of degree 6. However, the answer $6^5 = 7776$ suggested by the Bézout theorem is wrong! The reason is that these hypersurfaces do not intersect the boundary of Q_2 properly: each of them contains all double lines.

The approach of Halphen-De Concini-Procesi. More generally, consider a number of closed subvarieties $Z_1, \ldots, Z_s$ of a homogeneous space G/H (typically, varieties of geometric objects satisfying certain conditions) such that $\sum \operatorname{codim} Z_i = \dim G/H$. If Z_i are in sufficiently general position with respect to each other, then it is natural to expect that $Z_1 \cap \cdots \cap Z_s$ is finite. By Kleiman's transversality theorem [13, Thm. III.10.8], the translates $g_i Z_i$ are in general position for generic $g_1, \ldots, g_s \in G$, and the number $(Z_1, \ldots, Z_s) = |g_1 Z_1 \cap \cdots \cap g_s Z_s|$, called the *intersection number*, does not depend on the g_i.

To compute this number, one tries to embed G/H as an open orbit in a compact G-variety X with finitely many orbits, so that $\operatorname{codim}_Y(\overline{Z_i} \cap Y) = \operatorname{codim} Z_i$ for any G-orbit $Y \subseteq X$. If such an X exists, then $g_1 \overline{Z_1} \cap \cdots \cap g_s \overline{Z_s} \subset G/H$ for generic g_i, whence $(Z_1, \ldots, Z_s) = [\overline{Z_1}] \cdots [\overline{Z_s}]$, the product in $H^*(X)$. It is now clear that in order to solve enumerative problems on homogeneous spaces, one needs to have a good control on their compactifications or, more generally, equivariant embeddings.

The geometry of embeddings of a homogeneous space G/H under a reductive group G is governed by its complexity, which is the codimension of generic orbits of a Borel subgroup $B \subseteq G$. The complexity has also a representation-theoretic meaning: it characterizes the growth of multiplicities of simple G-modules in the spaces of sections of line bundles on G/H, see Subsection 1.5. Another important numerical invariant is the rank of a homogeneous space. Complexity and rank are discussed in Section 1.

A method for computing complexity and rank was developed by Knop and Panyushev. It involves equivariant symplectic geometry of the cotangent bundle $T^*(G/H)$ and gives formulæ for these numbers in terms of the coisotropy representation; see Subsection 1.3. Panyushev showed that the computations can be reduced to representations of reductive groups; see Subsection 1.4. Other contributions of Panyushev are formulæ for complexity and rank of double flag varieties, which are considered in Subsection 1.6. Double flag varieties arise in the problem of decomposing tensor products of simple G-modules, cf. Subsection 3.6.

There are two distinct approaches to embedding theory of homogeneous spaces. The first one is based on explicit constructions of embeddings in ambient spaces (determinantal varieties, complete quadrics, wonderful compactifications of de Concini-Procesi, projective compactifications of reductive groups; see Subsection 3.4, etc.) In Section 2 we discuss the second, intrinsic, approach to equivariant embeddings of arbitrary homogeneous spaces, due to Luna, Vust, and the author. An important rôle in the local description of embeddings is played by B-stable divisors and respective discrete valuations of $\mathbb{C}(G/H)$. However, the Luna-Vust theory provides a complete and transparent description of equivariant embeddings only for homogeneous spaces of complexity ≤ 1.

Homogeneous spaces of complexity 0 are called spherical. They are characterized by a number of particularly nice properties (Theorem 3.1). Many classical homogeneous varieties are in fact spherical: for instance, all above examples, except the first one, are spherical. Normal embeddings of spherical homogeneous spaces are called spherical varieties. For spherical varieties, the Luna-Vust theory provides an elegant description in terms of certain objects of combinatorial convex geometry (coloured cones and fans). The well-known theory of toric varieties is in fact a particular case. We study spherical varieties in Section 3.

The group G itself may be considered as a spherical homogeneous space $(G \times G)/\operatorname{diag} G$. We study its embeddings in Subsections 3.3–3.4. As an application, we obtain a classification of reductive algebraic semigroups due to Vinberg and Rittatore. We also study natural projective compactifications of G obtained by closing the image of G in the space of operators of a projective representation of G.

Divisors and line bundles on spherical varieties are discussed in Subsection 3.5. Following Brion, we describe the Picard group of a spherical variety and give criteria for a divisor to be Cartier, base point free, or ample. We also describe the G-module structure for the space of sections of a line bundle on a spherical variety in terms of lattice points of certain polytopes.

An interesting application of the divisor theory on spherical varieties is a geometric way to decompose certain tensor products of simple G-modules considered in Subsection 3.6. The idea is to view these simple modules as spaces of sections of line bundles on flag varieties. Then their tensor product is the space of sections of a line bundle on a double flag variety, cf. Subsection 1.6, and the above description of its *G*-module structure enters the game.

Another application is a formula for the degree of an ample divisor on a projective spherical variety, which leads to an intersection theory of divisors and to the 'Bézout theorem' on spherical homogeneous spaces; see Subsection 3.7.

Finally, we discuss the embedding theory for homogeneous spaces of complexity 1, due to the author. It is developed from the general Luna-Vust theory in a way parallel to the spherical case. However, the description of embeddings is more complicated. We try to emphasize the common features and the distinctions from the spherical case.

The aim of this survey is to introduce a reader to equivariant embeddings of homogeneous spaces under reductive groups, and to show how this subject links together algebra, geometry, and representation theory. There are several excellent monographs and surveys devoted to some of the topics discussed in these notes, see e.g. [19, 5] for spherical varieties, and [33] for complexity and rank. However, in this paper we hope to gather some useful results, which are scattered in the literature and never appeared in survey papers before, paying special attention to practical computation of important invariants of homogeneous spaces and to the general embedding theory.

For its introductory character, this survey does not cover all topics in this area, and some results are not considered in full generality, as well as the list of references is by no means complete. Also we have tried to avoid long and complicated proofs, so that *Proof* in the text often means rather *Sketch of a proof*, or even *Hints for a proof*.

Acknowledgements

These notes were written on the base of a mini-course which I gave in November 2002 at the Manchester University. Thanks are due to this institution for hospitality, to Prof. A. Premet for invitation and organization of this visit, and to the London Mathematical Society for financial support. The paper was finished during my stay at Institut Fourier in spring 2003, and I would like to express my gratitude to this institution and to Prof. M. Brion for the invitation, and for numerous remarks and suggestions, which helped to improve the original text. Thanks are also due to I. V. Arzhantsev for some helpful remarks.

Notation and terminology

All algebraic varieties and groups are considered over the base field $\mathbb{C}$ of complex numbers. Lowercase gothic letters always denote Lie algebras of respective 'uppercase' algebraic groups.

The unipotent radical of an algebraic group H is denoted by U_H. The centralizer in H or $\mathfrak{h}$ of an element or subset of H or $\mathfrak{h}$ is denoted by $Z(\cdot)$ or $\mathfrak{z}(\cdot)$, respectively. The *character group* $\Lambda(H)$ consists of homomorphisms $\chi : H \to \mathbb{C}^\times$ and is written additively. It is a finitely generated Abelian group, and even a lattice if H is connected. Any action of H on a set M is denoted by $H : M$, and M^H is the set of H-fixed points. If $H : M$ is a linear representation, then $M_\chi^{(H)}$ denotes the set of H-eigenvectors of eigenweight $\chi \in \Lambda(H)$.

Throughout the paper, G is a connected reductive group. We often fix a Borel subgroup $B \subseteq G$ and a maximal torus $T \subseteq B$. $U \subseteq B$ is the maximal unipotent subgroup, and B^- is the *opposite* Borel subgroup (i.e. such that $B^- \cap B = T$), with the maximal unipotent radical U^-. Denote by V_λ the simple G-module of B-dominant highest weight λ. If G is semisimple simply connected, then the character lattice $\Lambda(B) = \Lambda(T)$ is generated by the *fundamental weights* ω_i, $i = 1, \ldots, \mathrm{rk}\, G$, dual to the simple coroots w.r.t. B, and the dominant weights are the positive linear combinations of the ω_i.

$\mathbb{C}[X]$ is the coordinate algebra of a quasiaffine variety X, and $\mathbb{C}(X)$ is the field of rational functions on any variety X. The line bundle associated with a Cartier divisor δ on X is denoted by $\mathcal{O}(\delta)$. The divisor of a rational section s of $\mathcal{O}(\delta)$ is denoted by $\mathrm{div}_X s$, and s_δ is the canonical rational section with $\mathrm{div}_X s_\delta = \delta$.

An H-*line bundle* on an H-variety X is a line bundle equipped with a fibrewise linear H-action compatible with the projection onto the base. If X is normal and H is connected, then any line bundle on X can be $\widetilde{H}$-linearized for some finite cover $\widetilde{H} \to H$ [22]. Hence a sufficiently big power of any line bundle can be H-linearized. If $H \subset G$ is a closed subgroup, then $G \times^H X$ denotes the *homogeneous fibration* over G/H with fibre X, i.e. the quotient variety $(G \times X)/H$ modulo the action $h(g, x) = (gh^{-1}, hx)$ where $g \in G$, $h \in H$, $x \in X$. The image of (g, x) in $G \times^H X$ is denoted by $g * x$.

We shall frequently speak of *generic* points (or orbits) in X assuming thereby that we consider points (orbits) from a certain (sufficiently small for our purposes) dense open subset of X.

We use the notation $\mathrm{conv}\, \mathcal{C}$, $\mathrm{int}\, \mathcal{C}$ for the convex hull and the relative interior of a subset C in a vector space E over $\mathbb{Q}$ or $\mathbb{R}$. If $C \subseteq E$ is a convex polyhedral cone, then $C^\vee \subseteq E^*$ denotes the dual cone.

Our general references are: [13] for algebraic geometry, [14, 15] for linear algebraic groups, and [24, 35] for algebraic transformation groups and Invariant Theory.

1 Complexity and rank

There are two numerical invariants of a homogeneous space G/H which have proved their importance in its embedding theory as well as in other geometric, representation-theoretic and invariant-theoretic problems on G/H. Roughly speaking, the first one, the *complexity*, says whether the geometry and embedding theory of G/H can be well controlled. The second invariant, the *rank* (or more subtly, the weight lattice) of G/H provides an environment for certain combinatorial objects used in the description of equivariant embeddings and in the representation theory related to G/H.

Actually, these invariants can be defined for an arbitrary G-variety.

Definition 1.1. Let X be an (irreducible) algebraic variety equipped with a G-action. The *complexity* $c(X)$ is by definition the codimension of a generic B-orbit in X, or equivalently, tr.deg $\mathbb{C}(X)^B$.

If we denote by $d_H(X)$ the *generic modality* of X under an action of an algebraic group H, i.e. the codimension of generic H-orbits, then $c(X) = d_B(X)$.

The set of all weights of rational B-eigenfunctions on X forms the *weight lattice* $\Lambda(X) \subseteq \Lambda(B)$. The *rank* of X is $r(X) = \operatorname{rk} \Lambda(X)$.

If X is quasiaffine, then we have the isotypic decomposition of its coordinate algebra $\mathbb{C}[X] = \bigoplus_{\lambda \in \Lambda_+(X)} \mathbb{C}[X]_{(\lambda)}$, where $\mathbb{C}[X]_{(\lambda)}$ is the sum of all simple G-submodules of highest weight λ (w.r.t. B), and $\Lambda_+(X) = \{\lambda \mid \mathbb{C}[X]_{(\lambda)} \neq 0\}$ is the *weight semigroup* of X. Every rational B-eigenfunction can be represented as a ratio of two regular B-eigenfunctions, whence $\Lambda_+(X)$ generates $\Lambda(X)$.

1.1 Local structure

The complexity, rank, and weight lattice are visible in terms of the 'local structure' of the action $G : X$ described by Brion, Luna, and Vust [6].

We start with the following simple situation. Let $G : V$ be a rational finite-dimensional representation, $v \in V$ a lowest weight vector, and

$v^* \in V^*$ a highest weight vector such that $\langle v, v^* \rangle \neq 0$. Let $P \supseteq B$ be the projective stabilizer of v^* with a Levi decomposition $P = L \cdot U_P$, so that the opposite parabolic subgroup $P^- = L \cdot U_P^-$ is the projective stabilizer of v. Put $\mathring{V} = V \setminus \langle v^* \rangle^\perp$, $W = (\mathfrak{u}_P^- v^*)^\perp$, and $\mathring{W} = W \cap \mathring{V}$. (Here $\perp$ denotes the annihilator in the dual space.)

Theorem 1.2 ([6]). *There is a natural P-equivariant isomorphism*

$$\mathring{V} \simeq U_P \times \mathring{W} \simeq P \times^L \mathring{W}.$$

Proof. First note that $V = \mathfrak{u}_P v \oplus W$. Indeed, by dimension count it suffices to prove $\mathfrak{u}_P v \cap W = 0$. Otherwise there would exist a root vector $e_\alpha \in \mathfrak{u}_P$ such that $e_\alpha v \in W$, in particular, $\langle e_\alpha v, e_{-\alpha} v^* \rangle = \langle [e_\alpha, e_{-\alpha}] v, v^* \rangle = 0$, hence $[e_\alpha, e_{-\alpha}] v = 0$ and α is a root of L, a contradiction.

Also note that $W = \langle v \rangle \oplus W_0$, where $W_0 = (\mathfrak{g} v^*)^\perp$. The hyperplanes $V_c = \{x \in V \mid \langle x, v^* \rangle = c\} = \mathfrak{u}_P v + cv + W_0$ as well as W_0 are U_P-stable. Now it suffices to prove that U_P acts on V_c / W_0 transitively and freely for all $c \neq 0$.

Clearly, $cv \bmod W_0$ has a dense U_P-orbit in V_c / W_0 and trivial stabilizer. Being an affine space, this orbit cannot be embedded into another affine space as a proper open subset. (Otherwise the boundary is a hypersurface, and its equation yields an invertible regular function on the orbit, a contradiction.) This proves the required assertion. $\square$

This theorem applies to the description of the structure of an open subset of sufficiently general points in any G-variety X.

Theorem 1.3 ([6]). *There exist a parabolic subgroup $P = L \cdot U_P \supseteq B$, an intermediate subgroup $[L, L] \subseteq L_0 \subseteq L$, an open P-stable subset $\mathring{X} \subseteq X$, and a closed subset $C \subseteq \mathring{X}^{L_0}$ such that $\mathring{X} \simeq U_P \times A \times C \simeq P \times^{L_0} C$, where $A = L/L_0$ is the quotient torus.*

Proof. Replace X by a birationally isomorphic projective G-variety in $\mathbb{P}(V)$. In the notation of Theorem 1.2, put $\mathring{X} = \mathbb{P}(\mathring{V}) \cap X$, $Z = \mathbb{P}(\mathring{W}) \cap X$, then $\mathring{X} \simeq U_P \times Z \simeq P \times^L Z$. If the kernel L_0 of the action $L : Z$ contains $[L, L]$, then the effectively acting group is the torus $A = L/L_0$, and we may replace $\mathring{X}$ and Z by open subsets such that $Z \simeq A \times C$.

In order to arrive at this situation, take a B-stable hypersurface $D \subset X$ such that the parabolic subgroup $P(D) = \{g \in G \mid gD = D\}$ is the smallest possible one. Adding new components if necessary, we may assume that D is given by one equation in projective coordinates.

Applying the Veronese embedding, we may assume that D is a hyperplane section $\mathbb{P}(\langle v^* \rangle^{\perp}) \cap X$, where $v^* \in V^*$ is a highest weight vector. Then $\mathring{X} = X \setminus D$, $P = P(D)$, and each $(B \cap L)$-stable hypersurface in Z is L-stable. Thus each $(B \cap L)$-eigenvector in $\mathbb{C}[Z]$ is an L-eigenvector, whence L-isotypic components of $\mathbb{C}[Z]$ are 1-dimensional, and $[L, L]$ acts trivially on $\mathbb{C}[Z]$ and on Z. $\qquad\square$

Corollary 1.4. *In the notation of Theorem* 1.3 *we have* $c(X) = \dim C$, $r(X) = \dim A$, *and* $\Lambda(X) = \Lambda(A)$.

1.2 Horospherical varieties

The local structure theorems of Brion-Luna-Vust describe the action of a certain parabolic $P \subseteq G$ on a certain open subset of X. There is a remarkable class of G-varieties, which in particular admit a local description of the G-action itself and have a number of other nice properties.

Definition 1.5. A subgroup of G containing a maximal unipotent subgroup is called *horospherical*. A G-variety X is *horospherical* if the stabilizers of all points of X are horospherical.

The terminology, due to Knop [18], is explained by the following example.

Example 1.6. Let L^n be the Lobachevsky space modelled as the upper pole of the hyperboloid $\{x \in \mathbb{R}^{n+1} \mid (x, x) = 1\}$ in an $(n+1)$-dimensional pseudo-euclidean space of signature $(1, n)$. A horosphere in L^n (i.e. a hypersurface perpendicular to a pencil of parallel lines) is defined by the equation $(x, y) = 1$, where $y \in \mathbb{R}^{n+1}$ is a nonzero isotropic vector. The space of horospheres is homogeneous under the connected isometry group $SO_{1,n}^+$ of L^n and is isomorphic to the upper pole of the isotropic cone $\{y \in \mathbb{R}^{n+1} \mid (y, y) = 0\}$. Its complexification is the space of highest weight vectors for $SO_{n+1}(\mathbb{C}) : \mathbb{C}^{n+1}$, which is a horospherical variety in the sense of the above definition.

Horospherical subgroups have an explicit description. Up to conjugacy, we may assume that a horospherical subgroup $S \subseteq G$ contains the 'lower' maximal unipotent subgroup U^-. By the Chevalley theorem, S is the stabilizer of a line $\langle v \rangle$ in a representation $G : V$. Then $v = v_{\lambda_1} + \cdots + v_{\lambda_m}$ is the sum of lowest weight vectors v_{λ_i} of weights λ_i. Let $P^-(\lambda_i)$ be the projective stabilizer of v_{λ_i}, $P^- = \bigcap_i P^-(\lambda_i) = L \cdot U_P^-$

(a Levi decomposition), $T_0 = \bigcap_{i,j} \operatorname{Ker}(\lambda_i - \lambda_j) \subseteq T$, and $L_0 = [L, L]T_0$. Then $S = L_0 \cdot U_P^-$ (a Levi decomposition).

The local structure of horospherical varieties is quite simple.

Theorem 1.7. *Each horospherical G-variety X contains an open G-stable subset $\overset{\circ}{X} \simeq (G/S) \times C$, where $S \subseteq G$ is horospherical and G acts on $\overset{\circ}{X}$ via the first factor.*

Proof. We have $X = GX^{U^-}$. By the structure of horospherical subgroups, for each $x \in X^{U^-}$ there exists a parabolic

$$P^- = L \cdot U_P^- \supseteq G_x \supseteq [L, L]U_P^-.$$

There are finitely many choices for P^-, hence X^{U^-} is covered by finitely many closed subsets of $[L, L]U_P^-$-fixed points. It follows that there exists the smallest P^- and a dense open subset $\overset{\circ}{X}{}^{U^-} \subseteq X^{U^-}$ such that $P^- \supseteq G_x \supseteq [L, L]U_P^-$ for all $x \in \overset{\circ}{X}{}^{U^-}$. Then $\overset{\circ}{X} = G\overset{\circ}{X}{}^{U^-} \simeq G \times^{P^-} \overset{\circ}{X}{}^{U^-}$, and the P^--action on $\overset{\circ}{X}{}^S$ factors through the effective action of the torus $A = L/L_0 = P^-/S$, $L \supseteq L_0 \supseteq [L, L]$, $S = L_0 \cdot U_P^-$. Shrinking $\overset{\circ}{X}$ if necessary, we may assume that $\overset{\circ}{X}{}^{U^-} \simeq (P^-/S) \times C$, whence the desired assertion. $\square$

Affine (or quasiaffine) horospherical varieties are characterized in terms of the multiplication law in their coordinate algebras.

Theorem 1.8 ([34]). *A quasiaffine G-variety X is horospherical if and only if the isotypic decomposition of $\mathbb{C}[X]$ is in fact an algebra grading, i.e. $\mathbb{C}[X]_{(\lambda)} \cdot \mathbb{C}[X]_{(\mu)} \subseteq \mathbb{C}[X]_{(\lambda+\mu)}$ for all $\lambda, \mu \in \Lambda_+(X)$.*

Proof. Assume that X is horospherical. In the notation of Theorem 1.7, $\mathbb{C}[X] \subseteq \mathbb{C}[\overset{\circ}{X}] = \mathbb{C}[G/S] \otimes \mathbb{C}[C]$, hence it suffices to consider $X = G/S$. The torus $A = P^-/S$ acts on G/S by G-automorphisms ('translations from the right'), so that $\mathbb{C}[G/S]_{(\lambda)}$ is the eigenspace of weight $-\lambda$. Indeed, for any highest weight vector $f_\lambda \in \mathbb{C}[G/S]_{(\lambda)}$ which is an eigenvector of A, we have $f_\lambda(eS) \neq 0$ (because the U-orbit of eS is dense in G/S) and $f_\lambda(eS \cdot t) = f_\lambda(tS) = \lambda(t^{-1})f_\lambda(eS)$ for all $t \in T$. Therefore the isotypic decomposition respects the multiplication.

Conversely, suppose that the isotypic decomposition is an algebra grading. We may assume that X is affine. It suffices to show that GX^{U^-} is dense in X, because it is closed being the image of the natural proper morphism $G \times^{B^-} X^{U^-} \hookrightarrow G \times^{B^-} X \simeq (G/B^-) \times X \to X$. In other words, the ideal I of X^{U^-} in $\mathbb{C}[X]$ must not contain nonzero

G-submodules or, equivalently, must not contain highest weight vectors $f_\lambda \in \mathbb{C}[X]$, $\lambda \in \Lambda_+(X)$ (because the orbit $U^- f_\lambda$ spans a G-submodule).

But I is generated by $gf - f$, $g \in U^-$, $f \in \mathbb{C}[X]$. (It even suffices to take for f the restrictions of the coordinate functions in an affine embedding of X.) If $I \ni f_\lambda = \sum_i p_i(g_i f_i - f_i)$, $p_i \in \mathbb{C}[X]_{(\lambda_i)}$, $f_i \in \mathbb{C}[X]_{(\mu_i)}$, $g_i \in U^-$, then $\lambda = \lambda_i + \mu_i$ and p_i, $g_i f_i - f_i$ must be highest weight vectors of weights λ_i, μ_i, which never occurs for $g_i f_i - f_i$, a contradiction. $\quad\square$

The above theorems provide an evidence that horospherical varieties have relatively simple structure. Remarkably, every G-variety degenerates to a horospherical one.

Theorem 1.9 ([34, 18]). *Given a G-variety X, there exists a smooth $(G \times \mathbb{C}^\times)$-variety E and a smooth $(G \times \mathbb{C}^\times)$-equivariant morphism $\pi : E \to \mathbb{A}^1$ (here G acts on $\mathbb{A}^1$ trivially and $\mathbb{C}^\times$ acts by homotheties) such that $X_t = \pi^{-1}(t)$ is G-isomorphic to an open smooth G-stable subset of X whenever $t \neq 0$, X_0 is a smooth horospherical variety, and all fields $\mathbb{C}(X_t)^U$ are B-isomorphic. In particular, all X_t have the same complexity, rank, and weight lattice as X.*

Proof. By the standard techniques of passing to an open G-stable subset and taking the affine cone over a projective variety, the theorem is reduced to the affine case handled by Popov [34]. So we may assume X to be affine.

We define the *height* of any weight λ by decomposing $\lambda = \sum_i c_i \alpha_i + \lambda_0$, where α_i are the simple roots and $\lambda_0 \perp \alpha_i$ for all i, and by putting $\operatorname{ht} \lambda = 2 \sum_i c_i$. (The multiplier 2 forces ht to take integer values. Namely, ht λ is the inner product of λ with the sum of the positive coroots.) It follows from the structure of T-weights of simple G-modules that

$$\mathbb{C}[X]_{(\lambda)} \cdot \mathbb{C}[X]_{(\mu)} \subseteq \mathbb{C}[X]_{(\lambda+\mu)} \oplus \bigoplus_i \mathbb{C}[X]_{(\nu_i)},$$

where $\operatorname{ht} \nu_i < \operatorname{ht} \lambda + \operatorname{ht} \mu$.

Now $R = \bigoplus_{\operatorname{ht} \lambda \leq k} \mathbb{C}[X]_{(\lambda)} t^k$ is a $(G \times \mathbb{C}^\times)$-algebra of finite type generated by $f_1 t^{\operatorname{ht} \lambda_1}, \ldots, f_m t^{\operatorname{ht} \lambda_m}, t$, where $f_i \in \mathbb{C}[X]_{(\lambda_i)}$ are generators of $\mathbb{C}[X]$. Then $R = \mathbb{C}[E]$ is the coordinate algebra of an affine $(G \times \mathbb{C}^\times)$-variety E, and the morphism $\pi : E \to \mathbb{A}^1$ corresponds to the inclusion $R \supseteq \mathbb{C}[t]$.

It is easy to see that all $\mathbb{C}[X_t]$ are canonically isomorphic to $\mathbb{C}[X]$ as G-modules. In fact, all algebras $\mathbb{C}[X_t]^U$ are canonically isomorphic to $\mathbb{C}[X]^U$, and $\mathbb{C}[X_t] \simeq \mathbb{C}[X]$ whenever $t \neq 0$. But the multiplication law in $\mathbb{C}[X_0]$ is obtained from that in $\mathbb{C}[X]$ by 'forgetting' isotypic components of lower height. Hence, by Theorem 1.8, X_0 is horospherical. By [24, III.3] all fibres X_t are reduced and irreducible, hence the smooth locus of E meets X_t. Passing to open subsets completes the proof.   □

Example 1.10. In Example 1.6, the 'horospherical contraction' X_0 of the Lobachevsky space $X = L^n$ is the space of horospheres, the total space of deformation E being given by the 'upper pole' of $\{(x,t) \in \mathbb{R}^{n+1} \times \mathbb{R} \mid (x,x) = t\}$. More precisely, we have to complexify the whole picture, so that X is a sphere in $\mathbb{C}^{n+1}$ and X_0 is the isotropic cone.

1.3 Relation to symplectic geometry

There is a deep connection between the geometry of G/H and the equivariant symplectic geometry of its cotangent bundle.

Recall that the cotangent bundle T^*X of any smooth variety X is equipped with a natural symplectic structure given by the 2-form $\omega = d\ell$, where ℓ is the *action 1-form* defined by $\ell_\alpha(\xi) = \langle \alpha, d\pi(\xi) \rangle$ for all $\alpha \in T^*X$, $\xi \in T_\alpha(T^*X)$, and $\pi : T^*X \to X$ is the canonical projection. In local coordinates $q_1, \ldots, q_n$ on X, which determine the dual coordinates $p_1, \ldots, p_n$ in cotangent spaces, one has $\ell = \sum p_i \, dq_i$ and $\omega = \sum dp_i \wedge dq_i$.

If X is a G-variety, then G acts on T^*X by symplectomorphisms, and the velocity fields $\alpha \mapsto \xi\alpha$ of all $\xi \in \mathfrak{g}$ have global Hamiltonians $H_\xi(\alpha) = \ell_\alpha(\xi\alpha)$. Furthermore, the action $G : T^*X$ is Poisson, i.e. the map $\xi \mapsto H_\xi$ is a homomorphism of $\mathfrak{g}$ to the algebra of functions on T^*X equipped with the Poisson bracket. The dual morphism $\Phi : T^*X \to \mathfrak{g}^*$ given by $\langle \Phi(\alpha), \xi \rangle = H_\xi(\alpha) = \langle \alpha, \xi x \rangle$ for all $\alpha \in T_x^*X$, $\xi \in \mathfrak{g}$, is called the *moment map*.

It is easy to see that the moment map is G-equivariant, and

$$\langle d_\alpha \Phi(\nu), \xi \rangle = \omega_\alpha(\nu, \xi\alpha)$$

for all $\nu \in T_\alpha(T^*X)$, $\xi \in \mathfrak{g}$. It follows that $\operatorname{Ker} d_\alpha\Phi = (\mathfrak{g}\alpha)^\angle$, $\operatorname{Im} d_\alpha\Phi = (\mathfrak{g}_\alpha)^\perp$, where $^\angle$ and $^\perp$ denote the skew-orthocomplement and the annihilator in $\mathfrak{g}^*$, respectively. Let $M_X = \overline{\operatorname{Im} \Phi}$ be the closure of the image of the moment map. It follows that $\dim M_X = \dim G\alpha$ for generic $\alpha \in T^*X$.

For $X = G/H$ we have $T^*(G/H) \simeq G \times^H \mathfrak{h}^\perp$, where $\mathfrak{h}^\perp = (\mathfrak{g}/\mathfrak{h})^*$ is the annihilator of $\mathfrak{h}$ in $\mathfrak{g}^*$. The moment map is given by $\Phi(g * \alpha) = g\alpha$ (with the coadjoint g-action on the r.h.s.). Indeed, the formula is true for $g = e$ since $\langle \Phi(\alpha), \xi \rangle = \langle \alpha, \xi(eH) \rangle$ and $\xi(eH)$ identifies with $\xi \bmod \mathfrak{h}$, and we conclude by G-equivariance. Moreover, for any G-variety X, the moment map of its cotangent bundle restricted to an orbit $Gx \subseteq X$ factors through the moment map of T^*Gx.

Remark 1.11. We may (and will) identify $\mathfrak{g}^*$ with $\mathfrak{g}$ via a G-invariant inner product (e.g. that given by the trace form for any faithful representation of G). Then $\mathfrak{h}^\perp$ identifies with the orthocomplement of $\mathfrak{h}$.

The algebra homomorphism dual to Φ can be defined both in the commutative and in the non-commutative setting. Let $\mathcal{U}(\mathfrak{g})$ denote the universal enveloping algebra of $\mathfrak{g}$, and $\mathcal{D}(X)$ be the algebra of differential operators on X. Each $\xi \in \mathfrak{g}$ determines a vector field on X, i.e. a differential operator of order 1, and this assignment extends to a homomorphism $\Phi^* : \mathcal{U}(\mathfrak{g}) \to \mathcal{D}(X)$. The map Φ^* preserves the natural filtrations, and the associated graded map

$$\operatorname{gr} \Phi^* : \operatorname{gr} \mathcal{U}(\mathfrak{g}) \simeq \mathbb{C}[\mathfrak{g}^*] \longrightarrow \operatorname{gr} \mathcal{D}(X) \subseteq \mathbb{C}[T^*X], \quad \xi \mapsto H_\xi, \quad (\xi \in \mathfrak{g})$$

is the pull-back of functions w.r.t. Φ. Here the isomorphism $\operatorname{gr} \mathcal{U}(\mathfrak{g}) \simeq \mathbb{C}[\mathfrak{g}^*]$ is provided by the Poincaré-Birkhoff-Witt theorem, and the embedding $\operatorname{gr} \mathcal{D}(X) \subseteq \mathbb{C}[T^*X]$ is the symbol map.

We have already seen that the complexity, rank and weight lattice are preserved by the 'horospherical contraction'. The same is true for the closure of the image of the moment map.

Theorem 1.12 ([18]). *In the notation of Theorem 1.9, $M_X = M_{X_0}$.*

Proof. The assertion can be reformulated in algebraic terms: put $I_X = \operatorname{Ker} \operatorname{gr} \Phi^*$, then $I_X = I_{X_0}$. We deduce this equality from its non-commutative analogue: put $\mathcal{I}_X = \operatorname{Ker} \Phi^*$, then $\mathcal{I}_X = \mathcal{I}_{X_0}$.

The latter equality is obvious in the affine case, because $\mathcal{I}_X$ depends only on the G-module structure of $\mathbb{C}[X]$. The general case is reduced to the affine one by standard techniques [18, 5.1].

Put $\mathcal{M}_X = \operatorname{Im} \Phi_X^* \subseteq \mathcal{D}(X)$. By the above, $\mathcal{M}_X \simeq \mathcal{M}_{X_0}$, but the filtrations by the order of differential operators on X and on X_0 are *a priori* different. It suffices to show that in fact they coincide.

There is even a third filtration, the quotient one induced from $\mathcal{U}(\mathfrak{g})$. Let $\operatorname{ord}_X \partial$, $\operatorname{ord} \partial$ denote the order of $\partial \in \mathcal{M}_X$ as a differential operator on X and w.r.t. the quotient filtration, respectively. It is clear that $\operatorname{ord} \geq \operatorname{ord}_X$.

First note that, in the notation of Theorem 1.9, there are obvious isomorphic restriction maps $\mathcal{M}_E \to \mathcal{M}_{X_t}$, which do not raise the order of differential operators and even preserve it whenever $t \neq 0$, because then $E \setminus X_0 \simeq X_t \times (\mathbb{A}^1 \setminus \{0\})$. Thus $\operatorname{ord}_X \geq \operatorname{ord}_{X_0}$.

Secondly, $\operatorname{gr}\mathcal{M}_{X_0}$ is a finite $\mathbb{C}[\mathfrak{g}^*]$-module. To prove this, we may assume by Theorem 1.7 that $X_0 = (G/S) \times C$. Hence $\mathcal{M}_{X_0} \simeq \mathcal{M}_{G/S}$ and $\operatorname{ord}_{X_0} = \operatorname{ord}_{G/S}$. We use the notation of Subsection 1.2. The torus $A = P^-/S$ acts on G/S by G-automorphisms, whence $\mathcal{M}_{G/S} \subseteq \mathcal{D}(G/S)^A$. Therefore $\operatorname{gr}\mathcal{M}_{G/S} \subseteq \mathbb{C}[T^*(G/S)]^A = \mathbb{C}[G \times^{P^-} \mathfrak{s}^\perp]$. But the natural morphism $G \times^{P^-} \mathfrak{s}^\perp \to \mathfrak{g}^*$ is proper with finite generic fibres by Lemma 1.13 below. It follows that $\mathbb{C}[G \times^{P^-} \mathfrak{s}^\perp]$, and hence $\operatorname{gr}\mathcal{M}_{G/S}$, is a finite $\mathbb{C}[\mathfrak{g}^*]$-module.

Now let $\partial_1, \ldots, \partial_m \in \mathcal{M}_{X_0}$ represent generators of $\operatorname{gr}\mathcal{M}_{X_0}$ over $\mathbb{C}[\mathfrak{g}^*]$, $d_i = \operatorname{ord}_{X_0}\partial_i$, and $d = \max_i \operatorname{ord}\partial_i$. If $\operatorname{ord}_{X_0}\partial = n$, then $\partial = \sum_i u_i \partial_i$ for some $u_i \in \mathcal{U}(\mathfrak{g})$, $\operatorname{ord} u_i \leq n - d_i$, hence $\operatorname{ord}\partial \leq n + d$. But if $\operatorname{ord}\partial > \operatorname{ord}_{X_0}\partial$, then $\operatorname{ord}\partial^{d+1} > \operatorname{ord}_{X_0}\partial^{d+1} + d$, a contradiction. Therefore $\operatorname{ord} = \operatorname{ord}_X = \operatorname{ord}_{X_0}$, and we are done.   □

Thus in the study of the image of the moment map, we may assume that X is horospherical and even $X = G/S$, where S is a horospherical subgroup containing U^-. In the sequel we use the notation of Subsection 1.2. The moment map $\Phi : T^*(G/S) \simeq G \times^S \mathfrak{s}^\perp \to \mathfrak{g}^* \simeq \mathfrak{g}$ factors through $\overline{\Phi} : G \times^{P^-} \mathfrak{s}^\perp \to \mathfrak{g}$. We have the decomposition $\mathfrak{g} = \mathfrak{u}_P \oplus \mathfrak{a} \oplus \mathfrak{l}_0 \oplus \mathfrak{u}_P^-$, where $\mathfrak{a}$ embeds into $\mathfrak{l}$ as the orthocomplement of $\mathfrak{l}_0$, so that $\mathfrak{s}^\perp = \mathfrak{a} \oplus \mathfrak{u}_P^-$. The following helpful result is essentially due to Richardson.

Lemma 1.13 ([18, Lemma 4.1]). *The morphism $\overline{\Phi} : G \times^{P^-} \mathfrak{s}^\perp \to \mathfrak{g}$ is proper with finite generic fibres.*

Another nice consequence of 'horospherical contraction' is the conjugacy of the stabilizers of generic points in cotangent bundles [18, §8]. We consider only the quasiaffine case.

Theorem 1.14 ([18]). *In the notation of Theorem 1.3, suppose X is quasiaffine; then the stabilizers in G of generic points in T^*X are all conjugate to L_0.*

Corollary 1.15. *We have $\Lambda(X) = \Lambda(T/(T \cap G_\alpha))$ for some sufficiently general point $\alpha \in T^*X$ such that G_α is an intermediate subgroup between a standard Levi subgroup and its commutator subgroup.*

Remark 1.16. Intermediate subgroups between standard Levi subgroups and their commutator subgroups, as well as embeddings onto such subgroups in G, will be called *standard*. Thus Corollary 1.15 says that a standard embedding of G_α for generic $\alpha \in T^*X$ yields the weight lattice of X. However, in applying this corollary for computing the weight lattice, one should be cautious, because G_α might have different conjugate standard embeddings into G. Some additional argument may be required to specify the weight lattice, see Example 1.22.

Proof. We prove the theorem for horospherical varieties. The general case can be deduced with the aid of 'horospherical contraction' using some additional reasoning [18, Satz 8.1].

We may assume $X = G/S$. As X is quasiaffine, $G/S \simeq Gv$ is an orbit in a representation $G : V$. Then $v = v_{\lambda_1} + \cdots + v_{\lambda_m}$ is the sum of lowest weight vectors, $P^- = \bigcap_i P^-(\lambda_i) = L \cdot U_P^-$, and $S = L_0 \cdot U_P^-$, where $L_0 = [L, L]T_0$, $T_0 = \bigcap_i \operatorname{Ker} \lambda_i \subseteq T$.

Note that $Z(\mathfrak{a}) = L$. Indeed, β is a root of $Z(\mathfrak{a})$ if and only if $\beta|_\mathfrak{a} = 0$ if and only if $\beta \perp \lambda_1, \ldots, \lambda_m$ if and only if β is a root of L.

We have $T^*(G/S) \simeq G \times^S \mathfrak{s}^\perp$, whence the stabilizers in G of generic points in $T^*(G/S)$ are, up to conjugacy, the stabilizers in S of generic points in $\mathfrak{s}^\perp = \mathfrak{a} \oplus \mathfrak{u}_P^-$. If $\xi \in \mathfrak{a}$ is a sufficiently general point (it suffices to have $\beta(\xi) \neq 0$ for all roots β of G that are not roots of L), then $\mathfrak{z}(\mathfrak{a}) = \mathfrak{l}$ yields $[\mathfrak{s}, \xi] = \mathfrak{u}_P^-$. Since the projection map $\pi : \mathfrak{s}^\perp \to \mathfrak{a}$ is S-invariant, $S\xi$ is dense in (in fact, coincides with) $\pi^{-1}(\xi) = \xi + \mathfrak{u}_P^-$. Therefore the stabilizers of generic points in $\mathfrak{s}^\perp$ are conjugate to $S_\xi = L_0$. $\qquad\square$

Remark 1.17. Another proof of Theorem 1.14 using the moment map instead of 'horospherical contraction' is given in [21]. A careful analysis of the moment map together with a refined version of the local structure theorem is applied in [42] to obtain alternative more direct proofs for Theorem 1.12 and for conjugacy of the stabilizers of generic points in cotangent bundles of arbitrary G-varieties which do not use differential operators and deformation arguments.

The following fundamental result of Knop interprets complexity and rank in terms of equivariant symplectic geometry.

Theorem 1.18 ([18]). *Let X be a G-variety with $\dim X = n$, $c(X) = c$, $r(X) = r$. Then*

$$\dim M_X = 2n - 2c - r \tag{1.1}$$
$$d_G(T^*X) = 2c + r \tag{1.2}$$
$$d_G(M_X) = r. \tag{1.3}$$

Proof. We may assume that X is horospherical and even $X = G/S \times C$.

By Lemma 1.13, $d_G(M_X) = d_G(G \times^{P^-} \mathfrak{s}^\perp) = d_{P^-}(\mathfrak{s}^\perp)$ and

$$\dim M_X = \dim(G \times^{P^-} \mathfrak{s}^\perp) = \dim G/P^- + \dim \mathfrak{s}^\perp$$
$$= 2 \dim G/S - \dim A = 2(n - c) - r.$$

The projection map $\pi : \mathfrak{s}^\perp \to \mathfrak{a}$ is P^--invariant, and P^- has a dense orbit in $\pi^{-1}(0) = \mathfrak{u}_P^-$ (the Richardson orbit). By semicontinuity of orbit and fibre dimensions, generic (in fact, all) fibres of π contain dense P^--orbits, whence $d_{P^-}(\mathfrak{s}^\perp) = \dim \mathfrak{a} = r$. Thus we have proved (1.1) and (1.3), and (1.2) stems from (1.1) and from $d_G(T^*X) = 2n - \dim M_X$. $\square$

In particular, for $X = G/H$ we obtain formulæ for complexity and rank in terms of the coisotropy representation $(H : \mathfrak{h}^\perp)$.

Theorem 1.19 (Knop [18], Panyushev [29]).

$$2c(G/H) + r(G/H) = \operatorname{codim}_{\mathfrak{h}^\perp} H\alpha$$
$$= \dim G - 2 \dim H + \dim H\alpha \qquad (1.4)$$
$$r(G/H) = \dim G_\alpha - \dim H\alpha \qquad (1.5)$$

where $\alpha \in \mathfrak{h}^\perp$ is a generic point. For reductive H, formula (1.5) amounts to

$$r(G/H) = \operatorname{rk} G - \operatorname{rk} H_\alpha \qquad (1.6)$$

and also

$$\Lambda(G/H) = \Lambda(T/(T \cap H_\alpha)). \qquad (1.7)$$

Proof. The isomorphism $T^*(G/H) \simeq G \times^H \mathfrak{h}^\perp$ yields $d_G(T^*(G/H)) = d_H(\mathfrak{h}^\perp)$, whence (1.4). Further, $d_G(M_{G/H}) = \dim M_{G/H} - \dim G\alpha = \dim(G * \alpha) - \dim G\alpha = \dim G_\alpha - \dim H_\alpha$ implies (1.5). Finally, if H is reductive, then G/H is affine, and (1.6)–(1.7) stem from Theorem 1.14 and its corollary. $\square$

Example 1.20. Consider the space of quadrics $Q_n = PGL_{n+1}/PO_{n+1}$. Here the coisotropy representation identifies with the natural representation of PO_{n+1} in the space $S_0^2 \mathbb{C}^{n+1}$ of traceless symmetric matrices. The stabilizer of a generic point is $\mathbf{Z}_2^n = \{\operatorname{diag}(\pm 1, \ldots, \pm 1)\}/\{\pm E\}$. The weight lattice of the (standard, diagonal) maximal torus $T \subset PGL_{n+1}$ is the root lattice Λ_{ad} of PGL_{n+1}, whence $\Lambda(Q_n) = 2\Lambda_{\mathrm{ad}}$, and $r(Q_n) = n$. Finally, $2c(Q_n) + r(Q_n) = (n+1)(n+2)/2 - 1 - n(n+1)/2 = n$ yields $c(Q_n) = 0$. The latter equality can be seen directly since $B \cdot PO_{n+1}$ is open in PGL_{n+1}, where $B \subseteq PGL_{n+1}$ is the standard Borel subgroup of upper-triangular matrices. (The Gram-Schmidt orthogonalization.)

Example 1.21. Let $G/H = Sp_n/Sp_{n-2}$. As the adjoint representation $G : \mathfrak{g}$ identifies with $Sp_n : S^2\mathbb{C}^n$, the symmetric square of the standard representation, and similarly for $H : \mathfrak{h}$, we have $\mathfrak{h}^\perp \simeq \mathbb{C}^{n-2} \oplus \mathbb{C}^{n-2} \oplus \mathbb{C}^3$, where $Sp_{n-2} : \mathbb{C}^{n-2}$ is the standard representation, and $Sp_{n-2} : \mathbb{C}^3$ a trivial one. It follows that $H_\alpha = Sp_{n-4}$. There exists a unique standard embedding $Sp_{n-4} \hookrightarrow Sp_n$ as a subgroup generated by all the simple roots except the first two. Therefore $\Lambda = \langle \omega_1, \omega_2 \rangle$, where ω_i are the fundamental weights, and $r = 2$. We also have $2c + r = 2(n-2) + 3 - (n-2)(n-1)/2 + (n-4)(n-3)/2 = 4$, whence $c = 1$.

Example 1.22. Let $G/H = GL_n/(GL_1 \times GL_{n-1})$. Here $\mathfrak{h}^\perp \simeq (\mathbb{C}^1 \otimes (\mathbb{C}^{n-1})^*) \oplus ((\mathbb{C}^1)^* \otimes \mathbb{C}^{n-1})$, where $\mathbb{C}^k$ is the standard representation of GL_k ($k = 1, n-1$). It is easy to find that $H_\alpha = \{\mathrm{diag}(t, A, t) \mid t \in \mathbb{C}^\times, A \in GL_{n-2}\}$. Therefore $r = 1$, and $2c + r = 2(n-1) - 1 - (n-1)^2 + 1 + (n-2)^2 = 1$, whence $c = 0$.

However H_α has three different standard embeddings into G obtained by permuting the diagonal blocks. To choose the right one, note that $\mathfrak{sl}_n$ contains a vector with stabilizer $GL_1 \times GL_{n-1}$. Hence $G/H \hookrightarrow \mathfrak{sl}_n$ and the restriction of the highest weight covector yields a highest weight function in $\mathbb{C}[G/H]$ of highest weight $\omega_1 + \omega_{n-1}$ (the highest root). Thus $\Lambda = \langle \omega_1 + \omega_{n-1} \rangle$, and H_α indeed embeds into G as above. (The simple roots of H_α are the simple roots of G except the first and the last one.)

Example 1.23. The space of twisted (i.e. irreducible non-planar) cubic curves in $\mathbb{P}^2$ is isomorphic to $G/H = PGL_4/PGL_2$, where $GL_2 \hookrightarrow GL_4$ is given by the representation $GL_2 : V_3$. Here V_d denotes the space of binary d-forms. Indeed, each twisted cubic is the image of a Veronese embedding $\mathbb{P}^1 \hookrightarrow \mathbb{P}^3$.

From the H-isomorphisms $\mathfrak{gl}_2 \simeq V_1 \otimes V_1^*$, $\mathfrak{gl}_4 \simeq V_3 \otimes V_3^*$, and the Clebsch-Gordan formula, it is easy to deduce that $\mathfrak{h}^\perp \simeq (V_6 \otimes \det^{-3}) \oplus (V_4 \otimes \det^{-2})$. It follows that $H_\alpha = \{E\}$, $r = 3$, $2c + r = 7 + 5 - 3 = 9$, whence $c = 3$.

If we replace G by PSp_4 in the above computations, then $\mathfrak{h}^\perp \simeq V_6 \otimes \det^{-3}$, still $H_\alpha = \{E\}$, but $r = 2$, $2c + r = 7 - 3 = 4$, whence $c = 1$.

On the other hand, replacing H by PSp_4 yields $\mathfrak{h}^\perp \simeq \bigwedge_0^2 \mathbb{C}^4$, the space of bivectors having zero contraction with the symplectic form. We obtain $H_\alpha = P(SL_2 \times SL_2)$, whence $\Lambda = \langle 2\omega_2 \rangle$, $r = 1$, $2c + r = 5 - 10 + 6 = 1$, and $c = 0$.

In the notation of Theorem 1.3, there is an open embedding $U_P \times A \hookrightarrow PGL_4/PSp_4$. Since $\Lambda(A) \subset \Lambda(T)$ is generated by an indivisible vector, A embeds in T as a subtorus, and $U_P \cdot A \hookrightarrow PGL_4$. This yields an open embedding $PGL_4/PGL_2 \simeq PGL_4 \times^{PSp_4} PSp_4/PGL_2 \hookleftarrow U_P \times A \times PSp_4/PGL_2$. Applying Theorem 1.3 to PSp_4/PGL_2 this time, we obtain open embeddings $PSp_4/PGL_2 \hookleftarrow U_{P_0} \times A_0 \times C$, $PGL_4/PGL_2 \hookleftarrow U_P \times A \times U_{P_0} \times A_0 \times C$, where P_0 is a parabolic in PSp_4, A_0 is its quotient torus, and C is a rational curve by the Lüroth theorem. This proves the theorem of Piene-Schlessinger on rationality of the space of twisted cubics. (See [3, §3] for another proof using homogeneous spaces.)

1.4 Reduction to representations

Theorem 1.19 yields computable formulæ for complexity and rank of affine homogeneous spaces reducing everything to computing stabilizers of general position for representations of reductive groups, which is an accessible problem. Panyushev [32] performed a similar reduction for arbitrary G/H. The idea is to consider a *regular* embedding $H \subseteq Q$ into a parabolic $Q \subseteq G$, i.e. such that there is also the inclusion of the unipotent radicals $U_H \subseteq U_Q$. The existence of a regular embedding into a parabolic subgroup was first proved by Weisfeiler, see e.g. [14, §30.3].

Let $H = K \cdot U_H$, $Q = M \cdot U_Q$ be Levi decompositions. We may assume $K \subseteq M$. The space $G/H \simeq G \times^Q Q/H$ is a homogeneous fibre space with generic fibre $Q/H \simeq M \times^K (U_Q/U_H) \simeq M \times^K (\mathfrak{u}_Q/\mathfrak{u}_H)$ a homogeneous vector bundle with affine base. The K-isomorphism $\mathfrak{u}_Q/\mathfrak{u}_H \overset{\sim}{\to} U_Q/U_H$ is proved in [27] essentially in the same way as in the non-equivariant setting, using a normal K-stable series $U_Q = U_0 \rhd \cdots \rhd U_m = U_H$, considering K-stable decompositions $\mathfrak{u}_{i-1} = \mathfrak{u}_i \oplus \mathfrak{m}_i$, and mapping $x = x_1 + \cdots + x_m \mapsto (\exp x_1) \cdots (\exp x_m)$ for all $x_i \in \mathfrak{m}_i$. Up to conjugacy, we may assume $Q \supseteq B^-$, $M \supseteq T$. Let K_0 denote the stabilizer of a generic point in the coisotropy representation $K : \mathfrak{k}^\perp$, with its standard embedding into M.

Theorem 1.24 ([32]).

$$c(G/H) = c(M/K) + c(\mathfrak{u}_Q/\mathfrak{u}_H) \tag{1.8}$$

$$r(G/H) = r(M/K) + r(\mathfrak{u}_Q/\mathfrak{u}_H) \tag{1.9}$$

and there is an exact sequence of weight lattices

$$0 \longrightarrow \Lambda(M/K) \longrightarrow \Lambda(G/H) \longrightarrow \Lambda(\mathfrak{u}_Q/\mathfrak{u}_H) \longrightarrow 0. \tag{1.10}$$

Here complexities, ranks, and weight lattices are considered for the homogeneous spaces G/H, M/K, and for the linear representation K_0 : $\mathfrak{u}_Q/\mathfrak{u}_H$.

Proof. As $U(eQ)$ is dense in G/Q and $B \cap M$ is a Borel subgroup of M, the complexities and the weight lattices of G/H and of $M : Q/H$ coincide. We may assume that $eK \in M/K$ is a generic point w.r.t the $(B \cap H)$-action. Then by Theorems 1.3 and 1.14, $B \cap K = B \cap K_0$ is a Borel subgroup of K_0, and the stabilizers in $B \cap M$ of generic points in M/K are conjugate to $B \cap K_0$. Now an easy computation of orbit dimensions implies (1.8).

By Theorem 1.3, the stabilizers of generic points for the actions $B : G/H$, $(B \cap M) : Q/H$, $(B \cap K_0) : \mathfrak{u}_Q/\mathfrak{u}_H$ are conjugate to $B \cap L_0$. It follows that $\Lambda(G/H) = \Lambda(T/(T \cap L_0))$, $\Lambda(M/K) = \Lambda(T/(T \cap K_0))$, and $\Lambda(\mathfrak{u}_Q/\mathfrak{u}_H) = \Lambda((T \cap K_0)/(T \cap L_0))$. This yields (1.10), and (1.9) stems from (1.10). $\square$

Example 1.25. Let $G = Sp_n$ and H be the stabilizer of three ordered generic vectors in the symplectic space $\mathbb{C}^n$. Without loss of generality, we may assume that these vectors are e_1, e_{n-1}, e_n, where $e_1, \dots, e_n$ is a symplectic basis of $\mathbb{C}^n$ such that the symplectic form has an antidiagonal matrix in this basis. Take for Q the stabilizer in Sp_n of the isotropic plane $\langle e_{n-1}, e_n \rangle$. Then $M \simeq GL_2 \times Sp_{n-4}$ consists of symplectic operators preserving the decomposition $\mathbb{C}^n = \langle e_1, e_2 \rangle \oplus \langle e_3, \dots, e_{n-2} \rangle \oplus \langle e_{n-1}, e_n \rangle$, $K \simeq Sp_{n-4}$, $\mathfrak{u}_Q$ consists of skew-symmetric (w.r.t. the symplectic form) operators mapping $\mathbb{C}^n \to \langle e_3, \dots, e_n \rangle \to \langle e_{n-1}, e_n \rangle \to 0$, and $\mathfrak{u}_H$ is the annihilator of e_1 in $\mathfrak{u}_Q$.

For M/K we have: $\mathfrak{k}^\perp \simeq \mathbb{C}^4$ is a trivial representation of K, whence $K_0 = K = Sp_{n-4}$, $\Lambda = \langle \omega_1, \omega_2 \rangle$, $r = 2$, $2c + r = 4$, $c = 1$.

Further, $\mathfrak{u}_Q \simeq \mathbb{C}^{n-4} \oplus \mathbb{C}^{n-4} \oplus \mathbb{C}^3$, and $\mathfrak{u}_H \simeq \mathbb{C}^{n-4} \oplus \mathbb{C}^1$, where $\mathbb{C}^{n-4}$ is the standard representation of Sp_{n-4} and $\mathbb{C}^3$, $\mathbb{C}^1$ are trivial ones. Therefore $\mathfrak{u}_Q/\mathfrak{u}_H \simeq \mathbb{C}^{n-4} \oplus \mathbb{C}^2$. One easily finds that the stabilizers of generic points in $T^*(\mathfrak{u}_Q/\mathfrak{u}_H) = \mathfrak{u}_Q/\mathfrak{u}_H \oplus (\mathfrak{u}_Q/\mathfrak{u}_H)^* \simeq \mathfrak{u}_Q/\mathfrak{u}_H \oplus \mathfrak{u}_Q/\mathfrak{u}_H$ (i.e. of generic pairs of vectors) are conjugate to Sp_{n-6}. It follows that $\Lambda = \langle \overline{\omega_3} \rangle$ is generated by the first fundamental weight of Sp_{n-4}, which is the restriction of the 3rd one for Sp_n. Hence $r = 1$,

$$2c + r = 2(n - 3 + 2) - (n - 4)(n - 3)/2 + (n - 6)(n - 5)/2 = 5$$

and $c = 2$.

By Theorem 1.24, we conclude that $c(G/H) = r(G/H) = 3$ and $\Lambda(G/H) = \langle \omega_1, \omega_2, \omega_3 \rangle$.

1.5 Complexity and growth of multiplicities

The complexity of a homogeneous space has a nice representation-theoretic meaning: it provides asymptotics of the growth of multiplicities of simple G-modules in representation spaces of regular functions or global sections of line bundles.

For any G-module M, let $\mathrm{mult}_\lambda\, M$ denote the multiplicity of a simple G-module of highest weight λ in M. Equivalently, $\mathrm{mult}_\lambda\, M = \dim M^U_{(\lambda)}$, where $M_{(\lambda)}$ is the respective isotypic component of M.

Theorem 1.26. *The complexity $c(G/H)$ is the minimal integer c such that $\mathrm{mult}_{n\lambda}\, H^0(G/H, \mathcal{L}^{\otimes n}) = O(n^c)$ over all dominant weights λ and all G-line bundles $\mathcal{L} \to G/H$. If G/H is quasiaffine, then it suffices to consider only $\mathrm{mult}_{n\lambda}\, \mathbb{C}[G/H]$.*

Proof. We may identify $\mathcal{L}$ with $G \times^H \mathbb{C}_\chi$, where H acts on $\mathbb{C}_\chi = \mathbb{C}$ by the character χ. Then $H^0(G/H, \mathcal{L})$ is the H-eigenspace of $\mathbb{C}[G]$ of weight $-\chi$, where H acts on G from the right. From the structure of $\mathbb{C}[G]$ as a $(G \times G)$-module (see Subsection 3.3) we see that $\mathrm{mult}_\lambda\, H^0(G/H, \mathcal{L}) = \dim V^*_{\lambda,-\chi}$, where $V^*_{\lambda,-\chi} \subseteq V^*_\lambda$ is the H-eigenspace of weight $-\chi$.

Put $c = c(G/H)$. Replacing H by a conjugate, we may assume that $\mathrm{codim}\, B(eH) = c$. If $c > 0$, then there exists a minimal parabolic $P_1 \supseteq B$ which does not stabilize $\overline{B(eH)}$. Therefore $\mathrm{codim}\, P_1(eH) = c - 1$. Continuing in the same way, we construct a sequence of minimal parabolics $P_1, \ldots, P_c \supset B$ such that $\overline{P_c \cdots P_1(eH)} = G/H$, i.e. $P_c \cdots P_1 H$ is dense in G. It follows that $\dim P_1 \cdots P_c/B = c$, whence $S_w = \overline{BwB}/B = P_1 \ldots P_c/B \subseteq G/B$ is the Schubert variety corresponding to an element w of the Weyl group W with reduced decomposition $w = s_1 \cdots s_c$, where $s_i \in W$ are the simple reflections corresponding to P_i.

The B-submodule $V_{\lambda,w} \subseteq V_\lambda$ generated by $w v_\lambda$ is called a *Demazure module*. We have $V_{\lambda,w} = \langle P_1 \cdots P_c v_\lambda \rangle = H^0(S_w, \mathcal{L}_{-\lambda})^*$, where $\mathcal{L}_{-\lambda} = G \times^B \mathbb{C}_{-\lambda}$ [15].

Lemma 1.27. *The pairing between V^*_λ and V_λ provides an embedding $V^*_{\lambda,-\chi} \hookrightarrow (V_{\lambda,w})^*$. Consequently $\mathrm{mult}_\lambda\, H^0(G/H, \mathcal{L}) \leq \dim V_{\lambda,w}$.*

Proof of the Lemma. If a nonzero $v^* \in V^*_{\lambda,-\chi}$ vanishes on $V_{\lambda,w}$, then it vanishes on $P_1 \cdots P_c v_\lambda$, whence $\langle P_c \cdots P_1 v^* \rangle = \langle G v^* \rangle = V^*_\lambda$ vanishes on v_λ, a contradiction. $\square$

Replacing λ by $n\lambda$ and $\mathcal{L}$ by $\mathcal{L}^{\otimes n}$ means that we replace $V_{\lambda,w}$ by

$V_{n\lambda,w} = H^0(S_w, \mathcal{L}_{-\lambda}^{\otimes n})^*$. As S_w is a projective variety of dimension c, the dimension of the r.h.s. space of sections grows as $O(n^c)$.

On the other hand, let $f_1, \ldots, f_c$ be a transcendence base of $\mathbb{C}(G/H)^B$. There exists a line bundle $\mathcal{L}$ and B-eigenvectors $s_0, \ldots, s_c \in H^0(G/H, \mathcal{L})$ of the same weight λ such that $f_i = s_i/s_0$ for all $i = 1, \ldots, c$. (Indeed, $\mathcal{L}$ and s_0 may be determined by any B-stable effective divisor dominating the poles of all f_i.) These $s_0, \ldots, s_c$ are algebraically independent in $R = \bigoplus_{n \geq 0} H^0(G/H, \mathcal{L}^{\otimes n})_{(n\lambda)}^U$, hence $\operatorname{mult}_{n\lambda} H^0(G/H, \mathcal{L}^{\otimes n}) = \dim R_n \geq \binom{n+c}{c} \sim n^c$. Therefore the exponent c in the estimate for the multiplicity cannot be made smaller.

Finally, if G/H is quasiaffine, then there even exist $s_0, \ldots, s_c \in \mathbb{C}[G/H]$ with the same properties. $\qquad \square$

For homogeneous spaces of small complexity much more precise information can be obtained.

Theorem 1.28. *In the above notation,*

(i) *If $c(G/H) = 0$, then $\operatorname{mult}_\lambda H^0(G/H, \mathcal{L}) \leq 1$ for all λ and $\mathcal{L}$.*

(ii) *If $c(G/H) = 1$, then there exists a G-line bundle $\mathcal{L}_0$ and a dominant weight λ_0 such that $\operatorname{mult}_\lambda H^0(G/H, \mathcal{L}) = n+1$, where n is the maximal integer such that $\mathcal{L} = \mathcal{L}_0^n \otimes \mathcal{M}$, $\lambda = n\lambda_0 + \mu$, $H^0(G/H, \mathcal{M})_{(\mu)} \neq 0$.*

Proof. In the case $c = 0$, assuming the contrary yields two non-proportional B-eigenvectors $s_0, s_1 \in H^0(G/H, \mathcal{L})$ of the same weight. Hence $f = s_1/s_0 \in \mathbb{C}(G/H)^B$, $f \neq \text{const}$, a contradiction.

In the case $c = 1$, we have $c(G/H)^B \simeq \mathbb{C}(\mathbb{P}^1)$ by the Lüroth theorem. Consider the respective rational map $\pi : G/H \dashrightarrow \mathbb{P}^1$, whose generic fibres are (the closures of) generic B-orbits. In a standard way, π is given by two B-eigenvectors $s_0, s_1 \in H^0(G/H, \mathcal{L}_0)$ of the same weight λ_0 for a certain line bundle $\mathcal{L}_0$. Moreover, s_0, s_1 are algebraically independent, and each $f \in \mathbb{C}(G/H)^B$ can be represented as a homogeneous rational fraction in s_0, s_1 of degree 0.

Now fix $s_\mu \in H^0(G/H, \mathcal{M})_{(\mu)}^U$ and take any $s_\lambda \in H^0(G/H, \mathcal{L})_{(\lambda)}^U$. Then $f = s_\lambda/s_0^n s_\mu \in \mathbb{C}(G/H)^B$, whence $f = F_1/F_0$ for some m-forms F_0, F_1 in s_0, s_1. We may assume the fraction to be reduced and decompose $F_1 = L_1 \ldots L_m$, $F_0 = M_1 \ldots M_m$, as products of linear forms, with all L_i distinct from all M_j. Then $s_\lambda M_1 \ldots M_m = s_\mu s_0^n L_1 \ldots L_m$. Being fibres of π, the divisors of s_0, L_i, M_j on G/H either coincide or have no common components. By the definition of $\mathcal{M}$, the divisor of s_μ does not dominate any one of M_j. Therefore $M_1 = \cdots = M_m = s_0$, $m \leq n$, and s_λ/s_μ is an n-form in s_0, s_1. The assertion follows. $\qquad \square$

Remark 1.29. Theorems 1.26 and 1.28 in full generality were proved in [41]. The algebraic interpretation of complexity in terms of growth of multiplicities is well-known, see versions of Theorem 1.26 in [30] (multiplicities in $\mathbb{C}[G/H]$ for G/H quasiaffine and $\mathbb{C}[G/H]$ finitely generated) and [5, §1.3] (multiplicities in coordinate algebras for affine varieties and in section spaces of line bundles for projective varieties). Part (i) of Theorem 1.28 is due to Vinberg and Kimelfeld [44], and Part (ii) for finitely generated coordinate algebras of quasiaffine homogeneous spaces was handled by Panyushev [30].

1.6 Double flag varieties

We illustrate the method of computing complexity and rank at double flag varieties, which are of importance in representation theory (cf. Subsection 3.6).

Let $P, Q \subseteq G$ be two parabolics. The product $X = G/P \times G/Q$ of the two respective (generalized) flag varieties is called a *double flag variety*. We may assume that P, Q are the projective stabilizers of lowest weight vectors v, w in G-modules V, W, respectively. Consider the Levi decompositions $P = L \cdot U_P$, $Q = M \cdot U_Q$ such that $L, M \supseteq T$. The following theorem is due to Panyushev.

Theorem 1.30 ([31]). *Let S be the stabilizer in $L \cap M$ of a generic point in $(\mathfrak{l} + \mathfrak{m})^\perp \simeq (\mathfrak{u}_P \cap \mathfrak{u}_Q) \oplus (\mathfrak{u}_P \cap \mathfrak{u}_Q)^*$. Then*

$$2c(X) + r(X) = 2\dim(U_P \cap U_Q) - \dim(L \cap M) + \dim S \qquad (1.11)$$
$$= \dim G - \dim L - \dim M + \dim S$$
$$r(X) = \operatorname{rk} G - \operatorname{rk} S \qquad (1.12)$$

and also

$$\Lambda(X) = \Lambda(T/(T \cap S)) \qquad (1.13)$$

provided $S \hookrightarrow L \cap M$ is the standard embedding.

Proof. Let $U_P^+, U_Q^+, U_{P \cap Q}^+$ be the unipotent radicals of the parabolics opposite to $P, Q, P \cap Q$. We have a decomposition

$$U_{P \cap Q}^+ = (U_P^+ \cap U_Q^+) \cdot (L \cap U_Q^+) \cdot (U_P^+ \cap M).$$

Consider the Segre embedding $X \simeq G\langle v \rangle \times G\langle w \rangle \subseteq \mathbb{P}(V) \times \mathbb{P}(W) \hookrightarrow \mathbb{P}(V \otimes W)$. Choose highest weight covectors $v^* \in V^*$, $w^* \in W^*$ such that

$\langle v, v^* \rangle, \langle w, w^* \rangle \neq 0$. By Subsection 1.1, we may restrict our attention to
$\mathring{X} = X \setminus \mathbb{P}((v^* \otimes w^*)^\perp) = U_P^+ \langle v \rangle \times U_Q^+ \langle w \rangle$.

By the above decomposition, $\mathring{X} \simeq U_{P \cap Q}^+ \times (U_P^+ \cap U_Q^+) \langle v \otimes w \rangle$ is an $(L \cap M)$-equivariant isomorphism. (This is nothing but the local structure of X provided by Theorem 1.2.) Therefore the complexity, rank and weight lattice for the actions $G : X$ and $(L \cap M) : (U_P^+ \cap U_Q^+)$ are the same. The latter action is isomorphic to the linear representation of $L \cap M$ in $(\mathfrak{u}_P \cap \mathfrak{u}_Q)^*$, and we may apply Theorems 1.3, 1.14 and their corollaries. This yields (1.12), (1.13) and the first equality in (1.11), whereupon the second equality is derived by a simple dimension count. □

Example 1.31. Let $G = GL_n(\mathbb{C})$, $P = Q = $ the stabilizer of a line in $\mathbb{C}^n$; we may assume this line to be spanned by e_n, the last vector of the standard basis. Then $X = \mathbb{P}^{n-1} \times \mathbb{P}^{n-1}$. Here $L = M = GL_{n-1} \times \mathbb{C}^\times$, and $(\mathfrak{l} + \mathfrak{m})^\perp \simeq (\mathbb{C}^{n-1})^* \oplus \mathbb{C}^{n-1}$, where GL_{n-1} acts on $\mathbb{C}^{n-1}$ in the standard way and $\mathbb{C}^\times$ acts by homotheties.

One easily finds $S = \{\text{diag}(A, t, t) \mid A \in GL_{n-2}, \ t \in \mathbb{C}^\times\}$. (We choose one of the two possible standard embeddings $S \hookrightarrow L \cap M$ by observing the existence of a highest weight linear function on $(\mathfrak{u}_P \cap \mathfrak{u}_Q)^* \simeq \mathbb{C}^{n-1}$ of weight $-\epsilon_{n-1} + \epsilon_n$, where ϵ_i are the T-weights of the standard basic vectors e_i.) It follows that $\Lambda(X) = \langle \epsilon_{n-1} - \epsilon_n \rangle$ is generated by the last simple root, $r(X) = 1$, and $2c(X) + r(X) = n^2 - 2((n-1)^2 + 1) + (n-2)^2 + 1 = 1$, whence $c(X) = 0$.

Example 1.32. Let $G = Sp_n(\mathbb{C})$, P be the stabilizer of a line in $\mathbb{C}^n$, and Q be the stabilizer of a Lagrangian subspace in $\mathbb{C}^n$. Choose a symplectic basis $e_1, \ldots, e_n$ such that $(e_i, e_j) = \text{sgn}(j - i)$ whenever $i + j = n + 1$, and 0, otherwise. We may assume that the above line is $\langle e_n \rangle$, and the Lagrangian subspace is $\langle e_{l+1}, \ldots, e_n \rangle$, $n = 2l$. Then $X = \mathbb{P}^{n-1} \times \text{LGr}(\mathbb{C}^n)$, where LGr denotes the Lagrangian Grassmannian. Here $L \cap M = GL_1 \times GL_{l-1}$, and $(\mathfrak{l} + \mathfrak{m})^\perp \simeq ((\mathbb{C}^1 \otimes \mathbb{C}^{l-1}) \oplus (\mathbb{C}^1)^{\otimes 2})^* \oplus ((\mathbb{C}^1 \otimes \mathbb{C}^{l-1}) \oplus (\mathbb{C}^1)^{\otimes 2})$, where $\mathbb{C}^k$ is the standard representation of GL_k $(k - 1, l - 1)$.

Now the same reasoning as in Example 1.22 shows that

$$S = \{\pm \text{diag}(1, A, 1) \mid A \in GL_{l-2}\} \subset M = GL_l.$$

It follows that $\Lambda(X) = \langle \epsilon_1 + \epsilon_l, \epsilon_1 - \epsilon_l \rangle$, where ϵ_i are the eigenweights of e_i, $i = 1, \ldots, l$, with respect to the standard diagonal maximal torus $T \subset Sp_n$. Therefore $r(X) = 2$, and $2c(X) + r(X) = 2l - (l-1)^2 - 1 + (l-2)^2 = 2$, whence $c(X) = 0$.

2 Embedding theory

The general theory of equivariant embeddings of homogeneous spaces was constructed by Luna and Vust in the seminal paper [26]. It is rather abstract, and we present here only the most important results, required in the sequel, skipping complicated and/or technical proofs. In our exposition, we follow [38], where the Luna-Vust theory is presented in a more compact way (and generalized to non-homogeneous varieties).

In what follows, a $(G$-equivariant$)$ embedding of G/H is a normal algebraic variety X equipped with a G-action and containing an open dense orbit isomorphic to G/H. More precisely, we fix an open embedding $G/H \hookrightarrow X$.

2.1 Uniform study of embeddings

The first thing to do is to patch together all embeddings of G/H in a huge prevariety $\mathbb{X}$. Geometrically, we patch any two embeddings X_1, X_2 of G/H along their largest isomorphic G-stable open subsets $\overset{\circ}{X}_1 \simeq \overset{\circ}{X}_2$. Algebraically, we consider the collection of all local rings $(\mathcal{O}, \mathfrak{m})$ that are localizations at maximal ideals of $\mathfrak{g}$-stable finitely generated subalgebras $R \subset \mathbb{C}(G/H)$ with $\operatorname{Quot} R = \mathbb{C}(G/H)$. We identify these local rings with points of $\mathbb{X}$. The Zariski topology is given by basic affine open subsets formed by all $(\mathcal{O}, \mathfrak{m})$ that are localizations of a given R, with the obvious structure sheaf. From this point of view, an embedding of G/H is just a Noetherian separated G-stable open subset $X \subset \mathbb{X}$.

The next important thing is to observe that an embedding $X \hookleftarrow G/H$ is uniquely determined by the collection of germs of G-stable subvarieties in X. To make this assertion precise, introduce a natural equivalence relation on the set of G-stable subvarieties in $\mathbb{X}$: $Y_1 \sim Y_2$ if $\overline{Y_1} = \overline{Y_2}$. Considering a subvariety up to equivalence means that we are interested only in its generic points. The equivalence classes are called G-germs (of embeddings along subvarieties). G-germs (of embeddings X along subvarieties Y) are determined by the local rings $\mathcal{O}_{X,Y}$, which are just G- and $\mathfrak{g}$-stable local rings of finite type in $\mathbb{C}(G/H)$. Clearly, X is determined by the collection of G-germs along subvarieties intersecting X.

2.2 Invariant valuations and colours

Germs along G-stable prime divisors $D \subset \mathbb{X}$ are of particular importance. The respective local rings $\mathcal{O}_{\mathbb{X},D} = \mathcal{O}_v$ are discrete valuation rings corresponding to G-invariant discrete geometric valuations v of $\mathbb{C}(G/H)$. (A valuation is said to be *geometric* if its valuation ring is the local ring of a prime divisor.) For $v = \mathrm{ord}_D$ the value group is $\mathbb{Z}$, but sometimes it is convenient to multiply v by a positive rational constant. The set of *G-valuations* (= G-invariant discrete $\mathbb{Q}$-valued geometric valuations) of $\mathbb{C}(G/H)$ is denoted by $\mathcal{V}$.

B-stable prime divisors of G/H are also called *colours*. The set of colours is denoted by $\mathcal{D}$. We say that the pair $(\mathcal{V}, \mathcal{D})$ is the *coloured data* of G/H. It is in terms of coloured data that embeddings of G/H are described.

Lemma 2.1. *G-valuations are uniquely determined by restriction to B-semiinvariant functions.*

Proof. We prove it in the quasiaffine case. The general case is more or less reducible to the quasiaffine one, cf. [26, §7.4]. For quasiaffine G/H, any $v \in \mathcal{V}$ is determined by a G-stable decreasing filtration $\mathbb{C}[G/H]_{v \geq c} = \{ f \in \mathbb{C}[G/H] \mid v(f) \geq c \}$, $c \in \mathbb{Q}$, of the coordinate algebra.

Take any $w \in \mathcal{V}$, $w \neq v$. Without loss of generality we may assume that $\mathbb{C}[G/H]_{v \geq c} \not\subseteq \mathbb{C}[G/H]_{w \geq c}$ for a certain c. Consider a G-stable decomposition $\mathbb{C}[G/H]_{v \geq c} = \mathbb{C}[G/H]_{v,w \geq c} \oplus M$, $M \neq 0$, and choose a highest weight vector $f \in M$. Then $v(f) \geq c > w(f)$. $\square$

Clearly the value $v(f)$ of a geometric valuation at a function does not change if we multiply f by a constant. Thus G-valuations are determined by their restrictions to the multiplicative group $\mathcal{A}$ of B-semiinvariant rational functions on G/H regarded up to a scalar multiple. Similarly, colours are mapped (by restriction of the respective valuation) to additive functions on $\mathcal{A}$, but this map is no longer injective in general.

It is natural to think of G-valuations and (the images of) colours as elements of the 'linear dual' of $\mathcal{A}$. We shall see evidence of this principle in Sections 3 and 4, and reflect it in the notation by writing $\langle v, f \rangle = v(f)$, $\langle D, f \rangle = \mathrm{ord}_D(f)$ for all $v \in \mathcal{V}$, $D \in \mathcal{D}$, $f \in \mathcal{A}$.

The following result of Knop is helpful in the study of properties of G-valuations and colours by restriction to $\mathcal{A}$.

Lemma 2.2 ([20]). *Fix $v \in \mathcal{V}$. For any $f \in \mathbb{C}(G/H)$ having B-stable divisor of poles, there exists $\tilde{f} \in \mathcal{A}$ such that:*

$$\begin{cases} \langle v, \tilde{f} \rangle = v(f) \\ \langle w, \tilde{f} \rangle \geq w(f) & \text{for all } w \in \mathcal{V} \\ \langle D, \tilde{f} \rangle \geq \mathrm{ord}_D(f) & \text{for all } D \in \mathcal{D} . \end{cases}$$

2.3 B-charts

In the study of manifolds it is natural to utilize coverings by 'simple' local charts. In our situation, this principle leads to the following

Definition 2.3. A *B-chart* is a B-stable affine open subvariety $\mathring{X} \subset \mathbb{X}$. An embedding $X \hookleftarrow G/H$ is said to be *simple* if $X = G\mathring{X}$.

The ubiquity of B-charts is justified by the following result.

Lemma 2.4. *Given a normal G-variety X and a G-stable subvariety $Y \subseteq X$, there exists a B-stable affine open subvariety $\mathring{X} \subseteq X$ meeting Y.*

Proof. By Sumihiro's theorem (see e.g. [22]), Y intersects a G-stable quasiprojective open subset of X. Shrinking X if necessary, we may assume it to be quasiprojective. Passing to the projective closure, we may assume without loss of generality that $X \subseteq \mathbb{P}(V)$ is a projective variety and $Y = G\langle v \rangle$ is the (closed) projectivized orbit of a lowest weight vector. Now in the notation of Subsection 1.1, it suffices to take $\mathring{X} = X \cap \mathbb{P}(\mathring{V})$. $\qquad\qquad\square$

Theorem 2.5. *Any B-chart $\mathring{X}$ determines a simple embedding $X = G\mathring{X} \subset \mathbb{X}$. Moreover, any embedding is covered by finitely many simple embeddings.*

Proof. For the first part it suffices to verify that X is Noetherian and separated. Being the image of $G \times \mathring{X}$ under the action morphism, X is Noetherian. Assuming X is not separated, i.e. $\operatorname{diag} X$ is not closed in $X \times X$, we take a G-orbit in $Y \subseteq \overline{\operatorname{diag} X} \setminus \operatorname{diag} X$. Then Y intersects the two open subsets $\mathring{X} \times X$ and $X \times \mathring{X}$ of $X \times X$. But $Y \cap (\mathring{X} \times X) \cap (X \times \mathring{X}) = Y \cap (\mathring{X} \times \mathring{X}) = \emptyset$ since $\mathring{X}$ is separated, a contradiction.

For the second part it suffices to note that any G-stable subvariety $Y \subset X$ intersects a certain B-chart, whence X is covered by simple embeddings, and it remains to choose a finite subcover. $\qquad\square$

Being a normal affine variety, a B-chart $\mathring{X}$ is determined by its coordinate algebra $R = \mathbb{C}[\mathring{X}]$, so that

$$R = \bigcap_{D,\ BD \neq D} \mathcal{O}_D \cap \bigcap_{D \in \mathcal{F}} \mathcal{O}_D \cap \bigcap_{w \in \mathcal{W}} \mathcal{O}_w \qquad (2.1)$$

is a finitely generated Krull ring with $\operatorname{Quot} R = \mathbb{C}(G/H)$. Here $\mathcal{W}$ is the set of G-valuations corresponding to G-stable prime divisors intersecting $\mathring{X}$, $\mathcal{F}$ is the set of colours intersecting $\mathring{X}$, and the first intersection runs over all non-B-stable prime divisors in G/H. The pair $(\mathcal{W}, \mathcal{F})$ is said to be the *coloured data of* $\mathring{X}$.

Conversely, consider arbitrary subsets $\mathcal{W} \subseteq \mathcal{V}$, $\mathcal{F} \subseteq \mathcal{D}$, and introduce an equivalence relation on the set of pairs: $(\mathcal{W}, \mathcal{F}) \sim (\mathcal{W}', \mathcal{F}')$ if $\mathcal{W}$ differs from $\mathcal{W}'$ and $\mathcal{F}$ from $\mathcal{F}'$ by finitely many elements. Clearly the coloured data of all B-charts lie in a distinguished equivalence class, denoted by **CD**.

Theorem 2.6. *Suppose* $(\mathcal{W}, \mathcal{F}) \in \mathbf{CD}$; *then*

(i) *the algebra R defined by formula (2.1) is a Krull ring;*

(ii) $\operatorname{Quot} R = \mathbb{C}(G/H)$ *if and only if*

$$\text{for any } \mathcal{W}_0 \subseteq \mathcal{W},\ \mathcal{F}_0 \subseteq \mathcal{F},\ \mathcal{W}_0, \mathcal{F}_0 \text{ finite, there exists } f \in \mathcal{A}$$
$$\text{such that } \langle \mathcal{W}, f \rangle \geq 0,\ \langle \mathcal{F}, f \rangle \geq 0,\ \langle \mathcal{W}_0, f \rangle > 0,\ \langle \mathcal{F}_0, f \rangle > 0\ ; \quad (\mathrm{C})$$

iii) *R is finitely generated if and only if*

$$R^U \text{ is finitely generated}; \qquad (\mathrm{F})$$

iv) *a valuation $v \in \mathcal{W}$ is essential for R if and only if*

$$\text{for some } f \in \mathcal{A},\ \langle \mathcal{W} \setminus \{v\}, f \rangle \geq 0,\ \langle \mathcal{F}, f \rangle \geq 0,\ \langle v, f \rangle < 0\ ; \quad (\mathrm{W})$$

(v) *all the valuations ord_D corresponding to $D \in \mathcal{F}$ are essential for R.*

Corollary 2.7. $(\mathcal{W}, \mathcal{F})$ *is the coloured data of a B-chart if and only if the conditions* (C), (F), (W) *are satisfied.*

Proof. Claim (1) stems from the simple observation that the set of defining valuations for R differs from that of $\mathbb{C}[\mathring{X}]$ by finitely many elements, where $\mathring{X}$ is any B-chart.

(ii) If $\operatorname{Quot} R = \mathbb{C}(G/H)$, then there exists $f \in R$ such that $w(f) > 0$, $\operatorname{ord}_D(f) > 0$ for all $w \in \mathcal{W}_0$, $D \in \mathcal{F}_0$. Replacing f by $\tilde{f}$ from Lemma 2.2 yields (C).

Conversely, suppose that (C) holds, and take any $h \in \mathbb{C}(G/H)$. We have $h = h_1/h_0$ for some $h_i \in \mathbb{C}[\mathring{X}]$, where $\mathring{X}$ is an arbitrary B-chart. Let $\mathcal{W}_0$ be the set of valuations that are negative at h_0, and $\mathcal{F}_0$ given by the poles of h_0. Then $h_0 f^N \in R$ for $N \gg 0$; similarly for h_1. Thus $h \in \operatorname{Quot} R$.

Claim (iii) is well known in the case $\mathcal{F} = \mathcal{D}$, i.e. whenever R is G-stable [24, III.3.1–2]. The general case is reduced to this one by a tricky argument [38, §1.4].

(iv) If v is essential, then there exists $f \in \mathbb{C}(G/H)$ with B-stable poles such that $v(f) < 0$, $w(f) \geq 0$, $\text{ord}_D f \geq 0$ for all $w \in W \setminus \{v\}$, $D \in \mathcal{F}$. Replacing f by $\tilde{f}$ from Lemma 2.2 yields (W).

Conversely, if (W) holds, then obviously v cannot be removed from the left hand side of formula (2.1), i.e. it is essential for R.

(v) Take a G-line bundle $\mathcal{L} \to G/H$ and a section $s \in H^0(G/H, \mathcal{L})$, whose divisor is a multiple of D. Put $f = gs/s$, where $g \in G$, $gD \neq D$. Then $\text{ord}_{D'} f \geq 0$ for all $D' \subset G/H$, $D' \neq D$, and $v(f) = 0$ for all $v \in \mathcal{V}$ (because v can be extended G-invariantly to sections of line bundles [26, §3.2], [20, §3]), but $\text{ord}_D f < 0$. Thus D cannot be removed from formula (2.1). $\qquad\qquad\qquad\qquad\qquad\qquad\qquad\qquad\qquad\qquad\quad\square$

2.4 G-germs

Now we study G-germs of a simple embedding $X = G\mathring{X}$, i.e. G-germs intersecting the B-chart $\mathring{X}$. Let $(\mathcal{W}, \mathcal{F})$ be the coloured data of $\mathring{X}$.

Definition 2.8. The *support* $\mathcal{S}_Y$ of a G-germ along Y is the set of G-valuations having centre Y.

The support is nonempty, which one can see by blowing up Y, normalizing, and taking the valuation corresponding to a component of the exceptional divisor. Each G-subvariety $Y \subset X$ intersects a certain simple embedding X, and any valuation has at most one centre in X by the separation axiom, hence the G-germ along Y is determined by the triple $(\mathcal{W}, \mathcal{F}, \mathcal{S}_Y)$.

There is also an intrinsic way to characterize G-germs regardless of simple embeddings. Let $\mathcal{V}_Y$ be the set of G-valuations corresponding to G-stable divisors containing Y, and $\mathcal{D}_Y = \{D \in \mathcal{D} \mid \overline{D} \supset Y\}$. The pair $(\mathcal{V}_Y, \mathcal{D}_Y)$ is said to be the *coloured data of the G-germ*. Clearly $\mathcal{V}_Y \subseteq W$, $\mathcal{D}_Y \subseteq \mathcal{F}$.

Theorem 2.9. *Let v be a G-valuation.*

(i) *$v \in \mathcal{S}_Y$ for some $Y \subseteq X$ if and only if, for all $f \in \mathcal{A}$,*

$$\langle \mathcal{W}, f \rangle \geq 0, \ \langle \mathcal{F}, f \rangle \geq 0 \implies \langle v, f \rangle \geq 0 . \qquad (\text{V})$$

(ii) *Suppose $v \in \mathcal{S}_Y$, $w \in \mathcal{W}$, $D \in \mathcal{F}$. Then:*

- *$D \in \mathcal{D}_Y$ if and only if, for all $f \in \mathcal{A}$,*

$$\langle \mathcal{W}, f \rangle \geq 0, \ \langle \mathcal{F}, f \rangle \geq 0, \ \langle v, f \rangle = 0 \implies \langle D, f \rangle = 0 ; \qquad (\text{D}')$$

- $w \in \mathcal{V}_Y$ if and only if, for all $f \in \mathcal{A}$,

$$\langle \mathcal{W}, f \rangle \geq 0, \ \langle \mathcal{F}, f \rangle \geq 0, \ \langle v, f \rangle = 0 \implies \langle w, f \rangle = 0 . \qquad \text{(V')}$$

(iii) $v \in \mathcal{S}_Y$ if and only if, for all $f \in \mathcal{A}$,

$$\langle \mathcal{V}_Y, f \rangle \geq 0, \ \langle \mathcal{D}_Y, f \rangle \geq 0 \implies \langle v, f \rangle \geq 0, \qquad \text{where } \langle v, f \rangle > 0$$

if either inequality on the left hand side is strict. $\qquad \text{(S)}$

(iv) *G-germs are uniquely determined by their coloured data.*

Proof. (i) A G-valuation v has a centre in X if and only if it has a centre in $\mathring{X}$ if and only if it is nonnegative on $\mathbb{C}[\mathring{X}]$, which implies (V). Conversely, if there exists $f \in \mathbb{C}[\mathring{X}]$, $v(f) < 0$, then replacing f by $\tilde{f}$ from Lemma 2.2 we see that (V) fails.

(ii) By assumption, $\mathcal{O}_v$ dominates $\mathcal{O}_Y$. Assume $\overline{D} \supset Y$, and take $f \in \mathcal{A}$ satisfying the l.h.s. of (D'). Then f is invertible in $\mathcal{O}_v$, whence in $\mathcal{O}_Y$, and in $\mathcal{O}_D$ as well. This implies (D').

On the other hand, if $\overline{D} \not\supset Y$, then for some $f \in \mathbb{C}[\mathring{X}]$, $f = 0|_D$, $f \neq 0|_Y$, hence $v(f) = 0$. Applying Lemma 2.2, we see that (D') fails.

A similar reasoning proves the second equivalence.

(iii) Assume $v \in \mathcal{S}_Y$. If the left hand side inequalities hold, then the poles of f do not contain Y, whence $f \in \mathcal{O}_Y$ and $\langle v, f \rangle \geq 0$. If one of these inequalities is strict, then the zeroes of f contain Y, whence $\langle v, f \rangle > 0$. This implies (S).

Conversely, if $v \notin \mathcal{S}_Y$, then there exists $f \in \mathcal{O}_Y$ such that either $v(f) < 0$ or $f|_Y = 0$, $v(f) = 0$. Applying Lemma 2.2 again, we see that (S) fails.

(iv) Consider the algebra R defined by formula (2.1) with $\mathcal{W} = \mathcal{V}_Y$, $\mathcal{F} = \mathcal{D}_Y$. Then $\mathcal{O}_Y$ is the localization of R at the ideal given by the conditions $v > 0$ for all $v \in \mathcal{S}_Y$. But $\mathcal{S}_Y$ is determined by $(\mathcal{V}_Y, \mathcal{D}_Y)$. $\quad \square$

2.5 Résumé

Summing up, we can construct all embeddings $X \hookleftarrow G/H$ in the following way:

- Take a finite collection of coloured data $(\mathcal{W}_i, \mathcal{F}_i)$ satisfying (C), (F), (W). These coloured data determine B-charts $\mathring{X}_i$ and simple embeddings $X_i = G\mathring{X}_i$.

- Compute the coloured data $(\mathcal{V}_Y, \mathcal{D}_Y)$ of G-germs $Y \subseteq X_i$ using the conditions (V), (V'), (D').
- Compute the supports $\mathcal{S}_Y$ using (S).
- Finally, simple embeddings X_i can be pasted together in an embedding X if and only if the supports $\mathcal{S}_Y$ are all disjoint, which stems from the following version of the valuative criterion of separation.

Theorem 2.10. *An open G-stable subset $X \subset \mathbb{X}$ is separated if and only if each G-valuation has at most one centre in X.*

Proof. If X is not separated, and $Y \subseteq \overline{\mathrm{diag}\, X} \setminus \mathrm{diag}\, X$ is a G-orbit, then the projections Y_i of Y to the copies of X ($i = 1, 2$) are disjoint. Now any G-valuation having centre Y in $\overline{\mathrm{diag}\, X}$ has at least two centres Y_1, Y_2 in X. The converse implication stems from the usual valuative criterion of separation (involving all valuations). $\qquad\square$

The above 'combinatorial' description of embeddings looks rather cumbersome and inaccessible for practical use. However, we shall see in the sequel, that for homogeneous spaces of small complexity, this theory looks much nicer.

3 Spherical varieties

3.1 Spherical homogeneous spaces

The most elegant and deep theory can be developed for *spherical* homogeneous spaces, namely those of complexity 0. A homogeneous space G/H is spherical if and only if B has an open orbit in G/H. It should be noted that a number of classical varieties are in fact spherical: e.g. all examples in the introduction (except the first one), flag varieties, varieties of matrices of given rank, of complexes, symmetric spaces etc. Also the class of spherical homogeneous spaces is stable under degeneration.

The importance of this class of homogeneous spaces is also justified by a number of particularly nice properties characterizing them. Some of these properties are listed in

Theorem 3.1. *The following conditions are equivalent:*

(i) *B acts on G/H with an open orbit;*
(ii) *$\mathbb{C}(G/H)^B = \mathbb{C}$;*
(iii) *$\mathfrak{g} = \mathfrak{b} + \mathrm{Ad}(g)\mathfrak{h}$ for some $g \in G$;*
(iv) *For any G-line bundle $\mathcal{L} \to G/H$, the representation $G : H^0(G/H, \mathcal{L})$ is multiplicity free;*

(v) *(For quasiaffine G/H) The representation $G : \mathbb{C}[G/H]$ is multiplicity free.*

Proof. (i) $\Longleftrightarrow$ (ii) This holds by Rosenlicht's theorem.

(i) $\Longleftrightarrow$ (iii) $\mathfrak{b} + \mathrm{Ad}(g)\mathfrak{h}$ is the tangent space at e of $BgHg^{-1} \subseteq G$, the latter being a translate of the preimage of $B(gH) \subseteq G/H$.

(ii) $\Longleftrightarrow$ (iv) $\Longleftrightarrow$ (v) This follows from Theorems 1.26, 1.28(i). $\square$

3.2 Embedding theory

See [26, §8.10], [19], [5] and [38, §1.7].

Definition 3.2. A *spherical variety* is an algebraic variety G-isomorphic to an embedding of a spherical homogeneous space G/H, i.e. a normal algebraic G-variety X containing an open orbit isomorphic to G/H.

We are going to apply the theory of Section 2 to spherical varieties. As $\mathbb{C}(G/H)^B = \mathbb{C}$, any B-semiinvariant rational function of G/H is determined by its weight uniquely up to a scalar multiple. Therefore $\mathcal{A} = \Lambda(G/H)$, and G-valuations $v \in \mathcal{V}$ may be regarded as vectors in $\Lambda_{\mathbb{Q}}^* = \mathrm{Hom}(\Lambda, \mathbb{Q})$ given by $\langle v, \lambda \rangle = v(f_\lambda)$ for all $\lambda \in \Lambda$, where f_λ is a function of weight λ. Colours $D \in \mathcal{D}$ are also mapped to vectors $v_D \in \Lambda^* = \mathrm{Hom}(\Lambda, \mathbb{Z})$ given by $\langle v_D, \lambda \rangle = \mathrm{ord}_D(f_\lambda)$. Colours are just the components of the complement of the open B-orbit in G/H, whence $\mathcal{D}$ is finite.

Theorem 3.3. *G-valuations form a solid convex polyhedral cone $\mathcal{V} \subseteq \Lambda_{\mathbb{Q}}^*$ (the valuation cone).*

Proof. We consider the quasiaffine case, the general case being reducible to this one. Since the G-module $\mathbb{C}[G/H]$ is multiplicity free, there is a unique G-stable complement of each G-stable subspace. Thus for any $v \in \mathcal{V}$, the filtration $\mathbb{C}[G/H]_{v \geq c}$ comes from a unique G-stable grading of $\mathbb{C}[G/H]$, the latter being given by the vector $v \in \Lambda_{\mathbb{Q}}^*$, so that $v(\mathbb{C}[G/H]_{(\lambda)}) = \langle v, \lambda \rangle$ for all $\lambda \in \Lambda_+$.

Conversely, each $v \in \Lambda_{\mathbb{Q}}^*$ determines a G-stable grading and a decreasing filtration of $\mathbb{C}[G/H]$, and $v \in \mathcal{V}$ if and only if this filtration respects the multiplication. We have $\mathbb{C}[G/H]_{(\lambda)} \cdot \mathbb{C}[G/H]_{(\mu)} = \mathbb{C}[G/H]_{(\lambda+\mu)} \oplus \bigoplus_i \mathbb{C}[G/H]_{(\lambda+\mu-\beta_i)}$ for all $\lambda, \mu \in \Lambda_+(G/H)$, where β_i are positive linear combinations of positive roots. Thus $v \in \mathcal{V}$ if and only if $\langle v, \beta_i \rangle \leq 0$ for all λ, μ, β_i.

These inequalities define a convex cone containing the image of the antidominant Weyl chamber. Brion and Pauer proved that $\mathcal{V}$ is polyhedral by constructing a projective 'colourless' embedding, i.e. $X \hookleftarrow G/H$ such that $\mathcal{D}_Y = \emptyset$ for all $Y \subset X$, see e.g. [19, §5], [5, §2.4]. (Then $\mathcal{V}$ is generated by finitely many vectors corresponding to G-stable divisors in X by Theorem 3.8(iii) below.) Brion [2] proved that $\mathcal{V}$ is even cosimplicial and is in fact a fundamental chamber of a certain crystallographic reflection group, called the *little Weyl group* of G/H. A nice geometric interpretation for this group in the spirit of Subsection 1.3 was found by Knop [21].   □

Example 3.4. If G/H is horospherical, then $\mathcal{V} = \Lambda_{\mathbb{Q}}^*$. In particular, this is the case if $G = T$ is a torus. In the toric case, there are no colours, and we may also assume $H = \{e\}$ without loss of generality.

Now we reorganize coloured data in a more convenient way.

The class **CD** consists of the pairs of finite subsets. Take $(\mathcal{W}, \mathcal{F}) \in$ **CD** and consider the polyhedral cone $\mathcal{C}$ generated by $\mathcal{W}$ and (the image of) $\mathcal{F}$.

Condition (C) means that $\mathcal{C}$ is strictly convex, and no $D \in \mathcal{F}$ maps to 0.

Condition (F) is automatically satisfied, because R^U is just the semigroup algebra of $\mathcal{C}^{\vee} \cap \Lambda$, the semigroup of lattice points in the dual cone, which is finitely generated by Gordan's lemma.

Condition (W) says that $\mathcal{W}$ is recovered from $(\mathcal{C}, \mathcal{F})$ as the set of generators of those edges of $\mathcal{C}$ which do not intersect $\mathcal{F}$.

Definition 3.5. A *coloured cone* is a pair $(\mathcal{C}, \mathcal{F})$, where $\mathcal{C}$ is a strictly convex cone generated by $\mathcal{F} \subseteq \mathcal{D}$ and by finitely many vectors of $\mathcal{V}$, and $\mathcal{F} \not\ni 0$. The coloured cone is said to be *supported* if $(\operatorname{int} \mathcal{C}) \cap \mathcal{V} \neq \emptyset$.

Thus B-charts are in bijection with coloured cones. Let us consider G-germs of the simple embedding X spanned by the B-chart $\mathring{X}$ given by a coloured cone $(\mathcal{C}, \mathcal{F})$.

Condition (V) means simply that $v \in \mathcal{C}$.

Conditions (V′) and (D′) say that $\mathcal{V}_Y$, $\mathcal{D}_Y$ consist of those elements of $\mathcal{W}$, $\mathcal{F}$, respectively, which lie in the face $\mathcal{C}_Y \subseteq \mathcal{C}$ such that $v \in \operatorname{int} \mathcal{C}_Y$.

Condition (S) means that $v \in \mathcal{V} \cap \operatorname{int} \mathcal{C}_Y$.

Thus G-germs are in bijection with supported coloured cones.

Definition 3.6. A *face* of a coloured cone $(\mathcal{C}, \mathcal{F})$ is a coloured cone $(\mathcal{C}', \mathcal{F}')$ such that $\mathcal{C}'$ is a face of $\mathcal{C}$, and $\mathcal{F}' = \mathcal{F} \cap \mathcal{C}'$.

A *coloured fan* is a finite collection of supported coloured cones which is closed under passing to supported faces and such that different cones intersect along faces inside $\mathcal{V}$.

The arguments of Subsection 2.5 yield

Theorem 3.7. *Spherical embeddings are in bijection with coloured fans.*

Amazingly, a lot of geometry of a spherical variety can be read off its coloured fan. We illustrate this principle by the following result.

Theorem 3.8. *Let X be a spherical variety.*

(i) *The G-orbits $Y \subseteq X$ are in bijection with the coloured cones in the respective coloured fan. Moreover, $Y \subset \overline{Y'}$ if and only if $(\mathcal{C}_{Y'}, \mathcal{D}_{Y'})$ is a face of $(\mathcal{C}_Y, \mathcal{D}_Y)$.*

(ii) *X is affine if and only if its fan is formed by all supported faces of a coloured cone $(\mathcal{C}, \mathcal{D})$.*

(iii) *X is complete if and only if its fan covers the valuation cone.*

Proof. (i) It follows from the above that there are finitely many germs along G-subvarieties in X, whence each G-subvariety contains a dense orbit. If $Y \subset \overline{Y'}$, then $\mathcal{V}_Y \supseteq \mathcal{V}_{Y'}$, $\mathcal{D}_Y \supseteq \mathcal{D}_{Y'}$, hence $(\mathcal{C}_{Y'}, \mathcal{D}_{Y'})$ is a face of $(\mathcal{C}_Y, \mathcal{D}_Y)$.

Conversely, suppose $Y \not\subset \overline{Y'}$, and take $v \in \mathcal{S}_Y = (\operatorname{int}\mathcal{C}_Y) \cap \mathcal{V}$. There exists $f \in \mathbb{C}[\mathring{X}]$ such that $f|_{Y'} = 0$, $f|_Y \neq 0$, whence $v(f) = 0$. Applying Lemma 2.2, we replace f by a B-eigenfunction f_λ, and obtain $\langle v, \lambda \rangle = 0$, whence $\langle \mathcal{C}_Y, \lambda \rangle = 0$, but $\langle v', \lambda \rangle > 0$ for all $v' \in (\operatorname{int}\mathcal{C}_{Y'}) \cap \mathcal{V}$. Therefore $\mathcal{C}_{Y'}$ is not a face of $\mathcal{C}_Y$.

(ii) X is affine if and only if X is a G-stable B-chart, i.e. $\mathcal{D}$ is the set of colours of X.

(iii) If the fan of X does not cover $\mathcal{V}$, then it is easy to construct an open embedding $X \hookrightarrow \overline{X}$ by adding more coloured cones (e.g. one ray in $\mathcal{V}$) to the fan. Conversely, if X is non-complete, we choose a G-equivariant completion $X \hookrightarrow \overline{X}$ and take any orbit $Y \subseteq \overline{X} \setminus X$. Then $\mathcal{S}_Y$ is not covered by the fan of X. $\qquad\square$

Corollary 3.9 (Servedio). *Any spherical variety has finitely many orbits.*

It is instructive to deduce this assertion directly from the multiplicity-free property, see e.g. [5, §2.1].

Example 3.10.

- The (well-known) toric varieties [7, 11] are nothing but spherical embeddings of algebraic tori. Since there are no colours in this case, toric varieties are classified by usual fans, i.e. collections of strictly convex rational polyhedral cones intersecting along faces, which are closed under passing to faces.

- Complete symmetric varieties [9, 10] are certain compact embeddings of homogeneous symmetric spaces.

- Determinantal varieties are affine embeddings of spaces of matrices with given rank.

Example 3.11. Consider the space of plane conics Q_2 acted on by $G = PGL_3$. The smooth conics in $\mathbb{P}^2$ are represented by non-degenerate symmetric (3×3)-matrices of the respective quadratic forms: a matrix q determines a conic by the equation $x^\top qx = 0$ (x is a vector of projective coordinates). Let $\Delta_i(q)$ be the upper-left corner i-minor of q ($i = 1, 2, 3$).

We have seen in Example 1.20 that Q_2 is spherical and $\Lambda = 2\Lambda_{\mathrm{ad}} = \langle 2\alpha_1, 2\alpha_2 \rangle$, where α_i are the simple roots. We may take $f_{2\alpha_1} = \Delta_1^2/\Delta_2$, $f_{2\alpha_2} = \Delta_2^2/\Delta_1\Delta_3$. There are the two colours: D_1 consists of conics passing through the B-fixed point, and D_2 of those tangent to the B-stable line, D_i being given by the equation $\Delta_i = 0$, whence $v_{D_i} = \alpha_i^\vee/2$, where $\alpha_i^\vee$ are the simple coroots.

Consider the embedding $Q_2 \hookrightarrow \mathbb{P}^5 = \{\text{all conics in } \mathbb{P}^2\}$. The boundary is the G-stable prime divisor D of singular conics, given by the equation $\Delta_3 = 0$, whence $v_D = -\omega_2^\vee/2$, where $\omega_i^\vee$ are the fundamental coweights. There are 3 orbits: the open one Q_2, the closed one $Y = \{\text{double lines}\}$, and $D \setminus Y = \{\text{pairs of distinct lines}\}$. We have $\mathcal{V}_Y = \{v_D\}$, $\mathcal{D}_Y = \{D_2\}$, hence $\mathcal{C}_Y$ is generated by $-\omega_2^\vee/2, \alpha_2^\vee/2$.

The dual embedding $Q_2 \hookrightarrow (\mathbb{P}^5)^* = \{\text{all conics in } (\mathbb{P}^2)^*\}$ is given by mapping each smooth conic in $\mathbb{P}^2$ to the dual one in $(\mathbb{P}^2)^*$ consisting of all lines tangent to the given conic. In coordinates, $q \mapsto q^\vee$, the adjoint matrix formed by the cofactors of the entries in q. All the above considerations can be repeated, but the indices 1, 2 are interchanged. In particular, there is a unique G-stable divisor $D' \subset (\mathbb{P}^5)^*$ with $v_{D'} = -\omega_1^\vee/2$, and a unique closed orbit Y' with $\mathcal{C}_{Y'}$ generated by $-\omega_1^\vee/2, \alpha_1^\vee/2$.

By Theorem 3.8(iii), $\mathcal{C}_Y, \mathcal{C}_{Y'} \supseteq \mathcal{V}$, whence $\mathcal{V} = \mathcal{C}_Y \cap \mathcal{C}_{Y'}$ is generated by $-\omega_1^\vee/2, -\omega_2^\vee/2$, i.e. $\mathcal{V}$ is the antidominant Weyl chamber.

Now consider the diagonal embedding $Q_2 \hookrightarrow \mathbb{P}^5 \times (\mathbb{P}^5)^*$ and let $X = \overline{Q_2}$ be the closure of its image. It is given by the equation $q \cdot q^* = \lambda E$ ($\lambda \in \mathbb{C}$), where q, q^* are nonzero symmetric (3×3)-matrices. It is easy to see that there are four orbits $Y_{ij} \subset X$ given by $(\mathrm{rk}\, q, \mathrm{rk}\, q^*) = (i,j) = (3,3),\ (2,1),\ (1,2),\ (1,1)$, respectively. Differentiating the equation at a point of the unique closed orbit Y_{11}, one verifies that X is smooth. Since Y_{11} projects onto Y, Y', we have $\mathcal{C}_{Y_{11}} \subseteq \mathcal{C}_Y \cap \mathcal{C}_{Y'} = \mathcal{V}$, $\mathcal{D}_{Y_{11}} \subseteq \mathcal{D}_Y \cap \mathcal{D}_{Y'} = \emptyset$. But X is a complete simple embedding of Q_2, whence $(\mathcal{C}_{Y_{11}}, \mathcal{D}_{Y_{11}}) = (\mathcal{V}, \emptyset)$ by Theorem 3.8(iii). The space X, called the space of *complete conics*, was first considered by Chasles (1864).

3.3 Algebraic semigroups

A nice application of the embedding theory in Subsection 3.2 is the classification of reductive algebraic monoids, i.e. linear algebraic semigroups with unity whose groups of invertibles are reductive. The general study of algebraic semigroups was undertaken by Putcha and Renner; particular cases were classified by them. A complete classification of normal reductive monoids was developed by Vinberg [43]. It soon became clear that this classification can be easily derived from the embedding theory of spherical varieties. Rittatore [36] made this last step.

The point is that a reductive monoid X with unit group $G \subseteq X$ can be considered as a $(G \times G)$-variety, where the factors act by left/right multiplication. From this viewpoint, X is a $(G \times G)$-equivariant embedding of $G = (G \times G)/\operatorname{diag} G$.

Theorem 3.12 ([43, 36]). *X is an affine embedding of G. Conversely, any affine $(G \times G)$-embedding of G carries a structure of algebraic monoid with unit group G.*

Proof. The actions of the left and right copy of $G \times G$ on X define coactions $\mathbb{C}[X] \to \mathbb{C}[G] \otimes \mathbb{C}[X]$ and $\mathbb{C}[X] \to \mathbb{C}[X] \otimes \mathbb{C}[G]$, which are the restrictions to $\mathbb{C}[X] \subseteq \mathbb{C}[G]$ of the comultiplication $\mathbb{C}[G] \to \mathbb{C}[G] \otimes \mathbb{C}[G]$. Hence the image of $\mathbb{C}[X]$ lies in $(\mathbb{C}[G] \otimes \mathbb{C}[X]) \cap (\mathbb{C}[X] \otimes \mathbb{C}[G]) = \mathbb{C}[X] \otimes \mathbb{C}[X]$, and we have a comultiplication in $\mathbb{C}[X]$. Now G is open in X and consists of invertibles. For any invertible $x \in X$, we have $xG \cap G \neq \emptyset$, hence $x \in G$. □

To apply Subsection 3.2 we have to determine the coloured data for $(G \times G)/\operatorname{diag} G$. This was done by Vust [45] in the more general context of symmetric spaces.

First, the isotypic decomposition of the coordinate algebra has the form $\mathbb{C}[G] = \bigoplus_{\lambda \in \Lambda_+} \mathbb{C}[G]_{(\lambda)}$, where Λ_+ denotes the set of dominant weights, and $\mathbb{C}[G]_{(\lambda)} \cong V_\lambda^* \otimes V_\lambda$ is the linear span of the matrix entries of the representation $G : V_\lambda$. It is convenient to choose the Borel subgroup $B^- \times B$ in $G \times G$. Thus Λ is naturally identified with $\Lambda(B)$.

Secondly, the valuation cone $\mathcal{V} \subseteq \Lambda_{\mathbb{Q}}^*$ is identified with the antidominant Weyl chamber. To see it, we recall the proof of Theorem 3.3. A vector $v \in \Lambda_{\mathbb{Q}}^*$ determines a G-valuation if and only if $\langle v, \beta_i \rangle \leq 0$ for all β_i which occur in the decompositions $\mathbb{C}[G]_{(\lambda)} \cdot \mathbb{C}[G]_{(\mu)} = \mathbb{C}[G]_{(\lambda+\mu)} \oplus \bigoplus_i \mathbb{C}[G]_{(\lambda+\mu-\beta_i)}$ for all $\lambda, \mu \in \Lambda_+$. But $\mathbb{C}[G]_{(\lambda)} \cdot \mathbb{C}[G]_{(\mu)}$ is the linear span of the matrix entries of $G : V_\lambda \otimes V_\mu = V_{\lambda+\mu} \oplus \bigoplus_i V_{\lambda+\mu-\beta_i}$, and all simple roots occur among β_i for generic λ, μ.

The colours are the Schubert subvarieties $D_j = \overline{B^- s_j B} \subset G$ of codimension 1, where s_j is the reflection along the simple root α_j in the Weyl group W. It is easy to see (e.g. from [5, §3.1]) that $v_{D_j} = \alpha_j^\vee$ are the simple coroots.

From Theorems 3.8(ii), 3.12 and other results of 3.2, we deduce this consequence.

Theorem 3.13. *Normal reductive monoids X are in bijection with strictly convex cones $\mathcal{C}(X) \subset \Lambda_{\mathbb{Q}}^*$ generated by all simple coroots and finitely many antidominant vectors. The set $\mathcal{C}(X)^\vee \cap \Lambda$ of lattice points in the dual cone consists of all highest weights of $\mathbb{C}[X]$, and determines $\mathbb{C}[X] \subseteq \mathbb{C}[G]$ completely.*

Remark 3.14. This is in terms of highest weights of the coordinate algebra that the classification of Vinberg was initially presented. The semigroup $\mathcal{C}(X)^\vee \cap \Lambda$ is formed by the highest weights of the representations $G \to GL(V_\lambda)$ extendible to X. If we are interested in non-normal reductive monoids, then we have to replace $\mathcal{C}(X)^\vee \cap \Lambda$ by any finitely generated subsemigroup $S \subseteq \Lambda_+$ such that $\mathbb{Z}S = \Lambda$ and $\bigoplus_{\lambda \in S} \mathbb{C}[G]_{(\lambda)} \subseteq \mathbb{C}[G]$ is closed under multiplication, i.e. all highest weights $\lambda + \mu - \beta$ of $V_\lambda \otimes V_\mu$ belong to S whenever $\lambda, \mu \in S$. X is normal if and only if S is the semigroup of all lattice vectors in a polyhedral cone.

Definition 3.15. We say that $\lambda_1, \ldots, \lambda_m$ G-*generate* S if S consists of all highest weights $k_1 \lambda_1 + \cdots + k_m \lambda_m - \beta$ of G-modules $V(\lambda_1)^{\otimes k_1} \otimes \cdots \otimes V(\lambda_m)^{\otimes k_m}$, $k_1, \ldots, k_m \in \mathbb{Z}_+$. (In particular any generating set G-generates S.)

It is easy to see that $X \hookrightarrow \operatorname{End} V$ if and only if the highest weights $\lambda_1, \ldots, \lambda_m$ of $G : V$ G-generate S.

Lemma 3.16 ([40, §2]). $\mathbb{Q}_+ S = (\mathbb{Q}_+ W\{\lambda_1, \ldots, \lambda_m\}) \cap C$, *where* $C = \mathbb{Q}_+ \Lambda_+$ *is the dominant Weyl chamber. (In other words, a multiple of* *each dominant vector in the weight polytope eventually occurs as a highest weight in a tensor power of V.)*

If $V = V_\lambda$ is irreducible, then the centre of G acts by homotheties, whence $G = \mathbb{C}^\times \cdot G_0$, where G_0 is semisimple, $\Lambda \subseteq \mathbb{Z} \oplus \Lambda_0$ is a cofinite sublattice, Λ_0 being the weight lattice of G_0, and $\lambda = (1, \lambda_0)$. By Lemma 3.16, $\mathbb{Q}_+ S$ is the intersection of $\mathbb{Q}_+(W\lambda)$ with the dominant Weyl chamber. Recently de Concini showed that $\mathbb{Q}_+(W\lambda) \cap \Lambda_+$ is G-generated by $(\operatorname{conv} W\lambda) \cap \Lambda_+$ [8]. It follows that X is normal if and only if λ_0 is a minuscule weight [8], [40, §12].

Example 3.17. Let $G = GL_n$, and $X = \operatorname{Mat}_n$ be the full matrix algebra. For B take the standard Borel subgroup of upper-triangular matrices. We have $\Lambda = \langle \epsilon_1, \dots, \epsilon_n \rangle$, where the ϵ_i are the diagonal matrix entries of B. We identify Λ with Λ^* via the inner product such that the ϵ_i form an orthonormal basis. Let $(k_1, \dots, k_n)$ denote the coordinates of $\lambda \in \Lambda_\mathbb{Q}$ with respect to this basis.

The upper-left corner i-minors Δ_i are highest weight vectors in $\mathbb{C}[X]$, and their weights $\epsilon_1 + \dots + \epsilon_i$ generate Λ. Put $D_i = \{x \in X \mid \Delta_i(x) = 0\}$. Then $\mathcal{D} = \{D_1, \dots, D_{n-1}\}$, $v_{D_i} = \epsilon_i - \epsilon_{i+1}$ for all $i < n$, and D_n is the unique G-stable prime divisor, $v_{D_n} = \epsilon_n$. Therefore $\mathcal{C}(X) = \mathbb{Q}_+ v_{D_1} + \dots + \mathbb{Q}_+ v_{D_n} = \{k_1 + \dots + k_i \geq 0, \ i = 1, \dots, n\}$, and $\mathcal{C}(X)^\vee = \{k_1 \geq \dots \geq k_n \geq 0\}$. The lattice vectors of $\mathcal{C}(X)^\vee$ are exactly the dominant weights of polynomial representations, and $S = \mathcal{C}(X)^\vee \cap \Lambda$ is generated by $\epsilon_1 + \dots + \epsilon_i$, $i = 1, \dots, n$, and G-generated by ϵ_1.

3.4 Projective group compactifications

Given a faithful representation $G : V$, we obtain a reductive monoid $X = \overline{G} \subseteq \operatorname{End} V$, whose weight semigroup S is G-generated by the highest weights of V. The projective counterpart of this situation is studied in [40]: given a faithful projective representation $G : \mathbb{P}(V)$ with highest weights $\lambda_0, \dots, \lambda_m$, we examine the geometry of $X = \overline{G} \subseteq \mathbb{P}(\operatorname{End} V)$ in terms of the weight polytope $\mathcal{P} = \operatorname{conv} W\{\lambda_0, \dots, \lambda_m\}$ of V. Without loss of generality we may assume $V = V_{\lambda_0} \oplus \dots \oplus V_{\lambda_m}$. The affine situation can be regarded as a particular case of the projective one, since $\operatorname{End} V \hookrightarrow \mathbb{P}(\operatorname{End}(V \oplus \mathbb{C}))$ is an affine chart. To a certain extent, the projective case reduces to the affine case by taking the affine cone.

Theorem 3.18 ([16, 40]). *$(G \times G)$-orbits $Y \subset X$ are in bijection with the faces $\Gamma \subseteq \mathcal{P}$ such that $(\operatorname{int} \Gamma) \cap \mathcal{C} \neq \emptyset$. They are represented by*

$y = \langle e_\Gamma \rangle$, where e_Γ is the projector of V onto the sum of T-eigenspaces of weights in Γ. The cone $\mathcal{C}_Y$ is dual to the cone of $\mathcal{P} \cap C$ at the face $\Gamma \cap C$, and $\mathcal{D}_Y$ consists of simple coroots orthogonal to $\langle \Gamma \rangle$.

Remark 3.19. One can also describe the stabilizers $(G \times G)_y$ [40, §9].

Proof. It is easy to see that the points $y = \langle e_\Gamma \rangle$ are limits of 1-parameter subgroups in T, whence $y \in \overline{T}$. Moreover, one deduces from elementary toric geometry that wy ($w \in W$) represent all T-orbits in $\overline{T}$, because $w\Gamma$ run over all faces of $\mathcal{P}$.

Recall the Cartan decomposition $G = KTK$, where $K \subset G$ is a maximal compact subgroup. Hence $X = K\overline{T}K$, and therefore y represent all $(G \times G)$-orbits $Y \subset X$. In particular, closed $(G \times G)$-orbits $Y_i \subset X$ correspond to the dominant vertices $\lambda_i \in \mathcal{P}$, and the representatives are $y_i = \langle v_{\lambda_i} \otimes v^*_{-\lambda_i} \rangle$, where $v_{\lambda_i} \in V$ is a highest weight vector, and $v^*_{-\lambda_i} \in V^*$ the dual lowest weight vector.

Take one of these vertices, say λ_0, and consider the parabolic $P = P(\lambda_0) = L \cdot U_P$. There is an L-stable decomposition $V = \langle v_{\lambda_0} \rangle \oplus V_0$. Let $\mathring{X} = X \cap \mathbb{P}((\operatorname{End} V) \setminus \langle v^*_{-\lambda_0} \otimes v_{\lambda_0} \rangle^\perp)$. Here is a (projectivized) version of Theorem 1.2.

Lemma 3.20. $\mathring{X} \simeq U_P^- \times U_P \times Z$, where $Z \simeq \overline{L} \subseteq \operatorname{End}(V_0 \otimes \mathbb{C}_{-\lambda_0})$, and $y_0 \in \mathring{X}$ corresponds to $0 \in Z$.

Proof. By Theorem 1.2, the affine chart $\mathring{X}$ has the above structure with $Z = X \cap \mathbb{P}\big(\mathbb{C}^\times (v_{\lambda_0} \otimes v^*_{-\lambda_0}) \oplus W_0\big)$, where $W_0 = (\mathfrak{g} \times \mathfrak{g})(v^*_{-\lambda_0} \otimes v_{\lambda_0})^\perp = \big(\mathfrak{g}v^*_{-\lambda_0} \otimes v_{\lambda_0} + v^*_{-\lambda_0} \otimes \mathfrak{g}v_{\lambda_0}\big)^\perp \supseteq V_0 \otimes V_0^* = \operatorname{End} V_0$. Hence $Z = \overline{(L \times L)e} = \overline{L} \subseteq \mathbb{P}\big(\mathbb{C}^\times (v_{\lambda_0} \otimes v^*_{-\lambda_0}) \oplus \operatorname{End} V_0\big) \simeq \operatorname{End}(V_0 \otimes \mathbb{C}_{-\lambda_0})$. $\square$

By Lemma 3.16, $\mathbb{Q}_+ S = (C \cap P)_{\lambda_0}$ is the cone of $C \cap \mathcal{P}$ at λ_0, and $\mathcal{C}_{Y_0} = (\mathbb{Q}_+ S)^\vee$ by 3.2. It is also clear that $D_j \ni y_0$ if and only if $\alpha_j \perp \lambda_0$. Thus Theorem 3.18 is proven for closed orbits, and the assertion for other orbits is deduced by passage to coloured faces; see details in [40, §9]. $\square$

Example 3.21. $X = \mathbb{P}(\operatorname{Mat}_n)$ is a projective embedding of $G = PGL_n$. In the notation of Example 3.17, we have $\mathcal{P} = \operatorname{conv}\{\epsilon_1, \ldots, \epsilon_n\}$, $\mathcal{P} \cap C = \{k_1 \geq \cdots \geq k_n \geq 0,\ k_1 + \cdots + k_n = 1\} = \operatorname{conv}\{(\epsilon_1 + \cdots + \epsilon_i)/i \mid i = 1, \ldots, n\}$, $\Gamma = \operatorname{conv}\{\epsilon_1, \ldots, \epsilon_i\}$ ($i = 1, \ldots, n$), e_Γ is the projector onto the span of the first i basic vectors of $V = \mathbb{C}^n$, and $Y = \mathbb{P}(\text{matrices of rank } i)$ are the $(G \times G)$-orbits in X.

Finally, we give criteria of normality and smoothness of X. It clearly suffices to look at singularities at points of closed orbits.

Theorem 3.22. *In the above notation,*

(i) X *is normal at points of* Y_0 *if and only if the weights* $\lambda_1 - \lambda_0, \ldots, \lambda_m - \lambda_0$ *and negative simple roots* $-\alpha_j \not\perp \lambda_0$ *L-generate* $\Lambda \cap (\mathcal{P} \cap \mathcal{C})_{\lambda_0}$.

(ii) X *is smooth at points of* Y_0 *if and only if* $L \simeq GL_{n_1} \times \cdots \times GL_{n_p}$, *the representation* $(L : V_0 \otimes \mathbb{C}_{-\lambda_0})$ *is polynomial and contains the minimal representations* $(GL_{n_i} : \mathbb{C}^{n_i})$ *of factors of* L.

Proof. (i) X is normal along Y_0 if and only if Z is normal at 0 if and only if $\Lambda \cap (\mathcal{P} \cap \mathcal{C})_{\lambda_0}$ is L-generated by the highest weights $\mu_1, \ldots, \mu_s$ of $(L : V_0 \otimes \mathbb{C}_{-\lambda_0})$. The weights $\lambda_1 - \lambda_0, \ldots, \lambda_m - \lambda_0, -\alpha_j$ occur among them, being the highest weights of $v_{\lambda_1}, \ldots, v_{\lambda_m}, e_{-\alpha_j} v_{\lambda_0} \in V_0$, where $e_{-\alpha_j} \in \mathfrak{g}$ are root vectors. But

$$V = \sum_{k,i} \underbrace{\mathfrak{p}^- \cdots \mathfrak{p}^-}_{k} v_{\lambda_i} = \sum_{n,i,j_1,\ldots,j_n} \mathfrak{g}_{L,-\alpha_{j_1}} \cdots \mathfrak{g}_{L,-\alpha_{j_n}} \cdot V_{L,\lambda_i}$$

where $V_{L,\lambda_i} \subseteq V$, $\mathfrak{g}_{L,-\alpha_j} \subseteq \mathfrak{g}$ are simple L-modules generated by v_{λ_i}, $e_{-\alpha_j}$, respectively. The summands on the r.h.s. are quotients of $\mathfrak{g}_{L,-\alpha_{j_1}} \otimes \cdots \otimes \mathfrak{g}_{L,-\alpha_{j_n}} \otimes V_{L,\lambda_i}$. Hence $\lambda_i - \lambda_0, -\alpha_j$ L-generate all remaining μ_k.

(ii) Again it suffices to consider the smoothness of Z at 0. Z naturally embeds into $\bigoplus_{i=1}^{s} \operatorname{End} V_{L,\mu_i}$ and $T_0 Z = \bigoplus_{i=1}^{p} \operatorname{End} V_{L,\mu_i}$, $p \leq s$, after reordering μ_i. If Z is smooth, then the L-equivariant projection $Z \to T_0 Z$ is étale at 0 and in fact isomorphic by a weak version of Luna's fundamental lemma from the étale slice theory, see [40, §3]. Now it is easy to conclude that $L \simeq GL_{n_1} \times \cdots \times GL_{n_p}$, $Z \simeq \operatorname{Mat}_{n_1} \times \cdots \times \operatorname{Mat}_{n_p}$, and μ_i ($i \leq p$) are the highest weights of $(GL_{n_i} : \mathbb{C}^{n_i})$, whence all the required conditions hold. The converse implication is obvious. $\square$

Example 3.23. Take $G = Sp_4$, with simple roots $\alpha_1 = \epsilon_1 - \epsilon_2$, $\alpha_2 = 2\epsilon_2$, $\pm\epsilon_i$ being the weights of the minimal representation $Sp_4 : \mathbb{C}^4$. Let $\lambda_0 = 3\epsilon_1$, $\lambda_1 = 2(\epsilon_1 + \epsilon_2)$ be the highest weights of V. We have $\alpha_1 \not\perp \lambda_0 \perp \alpha_2$ and $L \simeq SL_2 \times \mathbb{C}^*$, so that α_2 is the simple root of SL_2, and ϵ_1 is a generator of $\Lambda(\mathbb{C}^*)$. The Clebsch-Gordan formula implies that $\lambda_1 - \lambda_0 = 2\epsilon_2 - \epsilon_1$, $-\alpha_1 = \epsilon_2 - \epsilon_1$ L-generate all lattice points in the cone $\mathbb{Q}_+\{2\epsilon_2 - \epsilon_1, -\epsilon_1\}$ except $-\epsilon_1$. Thus X is non-normal along Y_0. But if we increase V by adding V_{λ_2}, $\lambda_2 = 2\epsilon_1$, then X becomes normal.

Example 3.24. Suppose $G = SO_{2l+1}$, and $V = V_{\omega_i}$ is a fundamental representation. We have a unique closed orbit $Y_0 \subset X$. If $i < l$, then $L \not\simeq GL_{n_1} \times \cdots \times GL_{n_p}$, hence X is singular. But for $i = l$, $L \simeq GL_l$ is the common stabilizer of two transversal maximal isotropic subspaces in

$\mathbb{C}^{2l+1}$. It follows e.g. from the realization of the spinor representation in the Clifford algebra that $V_{\omega_l} \otimes \mathbb{C}_{-\omega_l}$ is L-isomorphic to $\bigwedge^{\bullet} \mathbb{C}^l$. Here all the conditions of Theorem 3.22(ii) are satisfied, whence X is smooth.

3.5 Divisors and line bundles

The theory of divisors on spherical varieties is due to Brion [1]. The starting point is to show that each divisor on a spherical variety is rationally equivalent to a combination of colours and of G-stable prime divisors.

Theorem 3.25. *Each Weil divisor δ on a spherical variety X is rationally equivalent to a B-stable Weil divisor δ'.*

Proof. Let $\mathring{X}$ be the B-chart, corresponding to the coloured cone $(0, \emptyset)$, i.e. just the open B-orbit in G/H. Since $\mathring{X}$ is a factorial variety, $\delta|_{\mathring{X}} = \operatorname{div}_{\mathring{X}} f$ for some $f \in \mathbb{C}(\mathring{X})$. Now take $\delta' = \delta - \operatorname{div}_X f$. $\qquad\square$

Remark 3.26. This assertion is a particular case of a more general result [12] stating that each effective algebraic cycle on a B-variety is rationally equivalent to a B-stable effective one. The idea here is to apply Borel's fixed point theorem to Chow varieties of cycles.

Next, we describe the relations between the B-stable generators of the divisor class group $\operatorname{Cl} X$, i.e. between colours and G-stable divisors on X.

Theorem 3.27. *There is a finite presentation*

$$\operatorname{Cl} X = \left\langle D_1, \dots, D_n \right\rangle \bigg/ \left\langle \sum_{i=1}^{n} \langle v_i, \lambda \rangle D_i \,\bigg|\, \lambda \in \Lambda \right\rangle$$

where D_i are all the B-stable divisors on X, represented by indivisible vectors $v_i \in \Lambda^*$. (Of course, it suffices to take λ from a basis of Λ.)

Proof. Just note that B-stable principal divisors are of the form $\operatorname{div} f_\lambda$, and $\operatorname{ord}_{D_i} f_\lambda = \langle v_i, \lambda \rangle$. $\qquad\square$

There are transparent combinatorial criteria in terms of coloured data for a B-stable divisor to be Cartier, base point free, or ample.

Theorem 3.28. *Let $\delta = \sum m_i D_i$ be a B-stable divisor on X.*

(i) *δ is Cartier if and only if for any G-orbit $Y \subseteq X$, there exists $\lambda_Y \in \Lambda^*$ such that $m_i = \langle v_i, \lambda_Y \rangle$ whenever $\overline{D_i} \supseteq Y$.*

(ii) δ *is base point free if and only if these* λ_Y *can be chosen in such a way that* $\lambda_Y \geq \lambda_{Y'}|_{\mathcal{C}_Y}$ *and* $m_i \geq \langle v_i, \lambda_Y \rangle$ *for all* $Y, Y' \subseteq X$ *and all* $D_i \in \mathcal{D} \setminus \bigcup_{Y \subseteq X} \mathcal{D}_Y$.

(iii) δ *is ample if and only if these* λ_Y *can be chosen in such a way that* $\lambda_Y > \lambda_{Y'}|_{\mathcal{C}_Y \setminus \mathcal{C}'_Y}$ *and* $m_i > \langle v_i, \lambda_Y \rangle$ *for all* $Y, Y' \subseteq X$ *and all* $D_i \in \mathcal{D} \setminus \bigcup_{Y \subseteq X} \mathcal{D}_Y$.

Remark 3.29. Theorem 3.28 says that a Cartier divisor is determined by a piecewise linear function on the fan, and it is base point free, resp. ample, if and only if this function is convex, resp. strictly convex with respect to the fan, with some additional positivity condition on the coefficients at the colours which do not contain G-orbits in their closures.

Proof. Note that δ is Cartier outside a G-stable subvariety in $\operatorname{supp} \delta$ [21, §2.2], because $g\delta \sim \delta$ for all $g \in G$.

(i) If δ satisfies the condition, then $\operatorname{supp}(\delta - \operatorname{div} f_{\lambda_Y}) \not\supseteq Y$, whence δ is Cartier on an open subset $\mathring{X} \subseteq X$, $\mathring{X} \cap Y \neq \emptyset$. By the above remark, δ is Cartier on X.

Conversely, suppose δ is Cartier. By Sumihiro's theorem, we may assume that X is quasiprojective and δ is very ample, since each Cartier divisor on a quasiprojective variety is the difference of two very ample divisors. Then there exists a B-eigenvector $s_Y \in H^0(X, \mathcal{O}(\delta))$, $s_Y \neq 0|_Y$, and $\delta = \operatorname{div}(f_{\lambda_Y} s_Y)$ for some $\lambda_Y \in \Lambda$, which obviously satisfies the required condition.

(ii) δ is base point free if and only if for any G-orbit $Y \subseteq X$, there exists $s_Y \in H^0(X, \mathcal{O}(\delta))$, $s_Y \neq 0|_Y$. We may assume s_Y to be a B-eigenvector. Then $\delta = \operatorname{div}(f_{\lambda_Y} s_Y)$ for some $\lambda_Y \in \Lambda$ satisfying the required condition.

(iii) If δ is ample, then, replacing δ by a multiple, we may assume that $\delta' = \delta - \sum_{\overline{D_i} \not\supseteq Y} D_i$ is base point free for a given $Y \subseteq X$ and apply the argument from the previous paragraph to δ' in order to obtain the required λ_Y.

Conversely, assume that the condition on λ_Y is satisfied. Then we have $\delta = \operatorname{div}(f_{\lambda_Y} s_Y)$, where $s_Y \in H^0(X, \mathcal{O}(\delta))$ has the zero locus $X \setminus \mathring{X}$, $\mathring{X}$ being the B-chart given by $(\mathcal{C}_Y, \mathcal{D}_Y)$. Then clearly

$$\mathbb{C}[\mathring{X}] = \bigcup_{m \geq 0} H^0(X, \mathcal{O}(m\delta))/s_Y^m.$$

Replacing δ by a multiple, we may assume that $H^0(X, \mathcal{O}(\delta))/s_Y$ contains generators of $\mathbb{C}[\mathring{X}]$ for all $Y \subseteq X$. Furthermore, we may replace $H^0(X, \mathcal{O}(\delta))$ here by a finite-dimensional G-submodule M containing all s_Y. Then the natural map $\phi : X \to \mathbb{P}(M^*)$ is well defined on $\mathring{X}$, whence on the whole X, $\mathring{X} = \phi^{-1}(\mathbb{P}(M^* \setminus \langle s_Y \rangle^\perp))$, and $\phi|_{\mathring{X}}$ is a closed embedding into $\mathbb{P}(M^* \setminus \langle s_Y \rangle^\perp)$ for all $Y \subseteq X$. It follows that ϕ is a closed embedding, and δ is ample. $\square$

Now we describe the G-module structure of $H^0(X, \mathcal{O}(\delta))$ for a Cartier divisor δ.

Theorem 3.30. *In the notation of Theorem 3.28,*

$$H^0(X, \mathcal{O}(\delta)) \simeq \bigoplus_{\lambda \in \mathcal{P}(\delta) \cap \Lambda} V_{\lambda + \pi(\delta)}$$

where $\pi(\delta)$ is the B-weight of the canonical rational section s_δ of $\mathcal{O}(\delta)$ with div $s_\delta = \delta$, and

$$\mathcal{P}(\delta) = \{\lambda \mid \langle v_i, \lambda \rangle \geq -m_i \text{ for all } i = 1, \dots, n\}$$

$$= \bigcap_{Y \subseteq X} (-\lambda_Y + \mathcal{C}_Y^\vee) \cap \left\{ \lambda \;\middle|\; \begin{array}{l} \langle v_i, \lambda \rangle \geq -m_i \\ \text{for all } D_i \in \mathcal{D} \setminus \bigcap_{Y \subseteq X} \mathcal{D}_Y \end{array} \right\}$$

is the weight polytope *of δ.*

Proof. Since all simple G-modules occur in $H^0(X, \mathcal{O}(\delta))$ with multiplicities ≤ 1 by Theorem 3.1(iv), it suffices to describe the set of highest weights. But $s = f_\lambda s_\delta$ is a highest weight section if and only if div $f_\lambda \geq -\delta$ if and only if $\lambda \in \mathcal{P}(\delta) \cap \Lambda$. $\square$

Remark 3.31. In order to find $\pi(\delta)$, we may identify $\mathcal{O}(\delta)|_{G/H}$ with $G \times^H \mathbb{C}_\chi$, where H acts on $\mathbb{C}_\chi = \mathbb{C}$ by a character χ. Then rational sections of $\mathcal{O}(\delta)$ are identified with rational functions on G that are H-semiinvariant from the right with character $-\chi$, and $\pi(\delta)$ is the weight of the equation of the pull-back of δ to G, up to a shift by a character of G.

3.6 Application: tensor product decompositions

If $P, Q \subset G$ are two parabolics and $X = G/P \times G/Q$ is a spherical variety, then the geometry of X can be applied to finding decompositions of certain tensor products of simple modules. Namely, by the Borel-Weil-Bott theorem, the space of global sections of any line bundle on G/P or G/Q is a simple G-module (maybe zero). The tensor product of pull-backs to X of line bundles $\mathcal{L} \to G/P$, $\mathcal{M} \to G/Q$ equals $\mathcal{O}(\delta)$ for some

B-stable Cartier divisor δ. Computing $\mathcal{P}(\delta)$ leads to a decomposition of $H^0(\mathcal{L}) \otimes H^0(\mathcal{M})$ into simple G-modules.

If P, Q stabilize the lines generated by lowest weight vectors $v_{-\lambda}, v_{-\mu}$ in two G-modules, respectively, and $\mathcal{L} = G \times^P \mathbb{C}_\lambda$, $\mathcal{M} = G \times^Q \mathbb{C}_\mu$ are pull-backs of ample line bundles on $G\langle v_{-\lambda}\rangle$, $G\langle v_{-\mu}\rangle$, then $H^0(\mathcal{L}) = V_\lambda$, $H^0(\mathcal{M}) = V_\mu$. All pairs of fundamental weights (λ, μ) such that X is spherical were classified by Littelmann [25] and the respective decompositions were computed. Recently all pairs of weights with spherical X were classified by Stembridge [37] and decompositions of $V_\lambda \otimes V_\mu$ were found in all cases.

Example 3.32. Consider the double flag variety $X = \mathbb{P}^{n-1} \times \mathbb{P}^{n-1}$ of Example 1.31. We have seen that X is spherical and $\Lambda = \langle \epsilon_{n-1} - \epsilon_n \rangle \simeq \mathbb{Z}$, where ϵ_i are the diagonal matrix entries of B, the standard Borel subgroup of upper-triangular matrices. There are three B-stable divisors D, D', D'' given by equations

$$\Delta = \begin{vmatrix} x_{n-1} & y_{n-1} \\ x_n & y_n \end{vmatrix} = 0, \qquad x_n = 0, \qquad y_n = 0$$

in homogeneous coordinates. Any B-eigenfunction is (up to a scalar multiple) an integer power of $f_{\epsilon_{n-1}-\epsilon_n}(x,y) = x_n y_n / \Delta$, whence D, D', D'' are represented by the vectors $v = -1$, $v' = v'' = 1$ in $\Lambda^* \simeq \mathbb{Z}$.

There are the two orbits in X: the closed one $Y = \operatorname{diag} \mathbb{P}^{n-1}$, and the open orbit $X \setminus Y$. We have $\mathcal{D}_Y = \{D\}$, $\mathcal{V}_Y = \emptyset$ (or vice versa for $n = 2$), hence $\mathcal{C}_Y = \mathbb{Q}_-$.

There is a relation $D = D' + D''$ in $\operatorname{Pic} X$, hence any divisor on X is equivalent to $\delta = pD' + qD''$. We have

$$H^0(X, \mathcal{O}(pD')) = H^0(\mathbb{P}^{n-1}, \mathcal{O}(p)) = \mathbb{C}[\mathbb{A}^n]_p \simeq V_{-p\epsilon_n},$$

and similarly $H^0(X, \mathcal{O}(qD'')) = \mathbb{C}[\mathbb{A}^n]_q \simeq V_{-q\epsilon_n}$. On the other hand, it is easy to compute $\mathcal{P}(\delta) = \{k(\epsilon_{n-1} - \epsilon_n) \mid 0 \geq k \geq -p, -q\}$. Shifting by the highest weight $\pi(\delta) = -(p+q)\epsilon_n$ of the canonical section $s_\delta = x_n^p \otimes y_n^q$ yields a decomposition

$$\mathbb{C}[\mathbb{A}^n]_p \otimes \mathbb{C}[\mathbb{A}^n]_q = \bigoplus_{k=0}^{\min(p,q)} V_{(k-p-q)\epsilon_n - k\epsilon_{n-1}}$$

generalizing the Clebsch-Gordan formula.

Example 3.33. Consider another spherical double flag variety $X = \mathbb{P}^{n-1} \times \operatorname{LGr}(\mathbb{C}^n)$ of Example 1.32. In the notation of that example, $\Lambda = \langle \epsilon_1 + \epsilon_l, \epsilon_1 - \epsilon_l \rangle$ with respect to the standard Borel subgroup of upper-triangular matrices in Sp_n. There are the two orbits in X: the closed one $Y = \{(\ell, F) \in X \mid \ell \subseteq F\}$, and the open orbit $X \setminus Y$. There are four B-stable divisors $D_1, \ldots, D_4$ given by the conditions $\ell \perp \langle e_1 \rangle$, $F \cap \langle e_1, \ldots, e_l \rangle \neq 0$, $(F+\ell) \cap \langle e_1, \ldots, e_{l-1} \rangle \neq 0$, $(F+\ell) \cap \ell^\perp \cap \langle e_1, \ldots, e_l \rangle \neq 0$, respectively. (One verifies it by proving that the complement of the union of the D_i is a single B-orbit.) Clearly $\mathcal{D}_Y = \{D_3, D_4\}$.

It is easy to see from the above description that the D_i can be determined by bihomogeneous equations F_i in projective coordinates of $\mathbb{P}^{n-1}$ and Plücker coordinates of $\mathrm{LGr}(\mathbb{C}^n)$ of bidegrees $(1,0)$, $(0,1)$, $(1,1)$, $(2,1)$, and B-eigenweights $\omega_1 = \epsilon_1$, $\omega_l = \epsilon_1 + \cdots + \epsilon_l$, $\omega_{l-1} = \epsilon_1 + \cdots + \epsilon_{l-1}$, ω_l, respectively. We have $f_{\epsilon_1 + \epsilon_l} = F_1 F_2 / F_3$, $f_{\epsilon_1 - \epsilon_l} = F_1 F_3 / F_4$, whence D_i are represented by the vectors $v_i \in \Lambda_{\mathbb{Q}}^*$, where $v_1 = \epsilon_1$, $v_2 = (\epsilon_1 + \epsilon_l)/2$, $v_3 = -\epsilon_l$, $v_4 = (\epsilon_l - \epsilon_1)/2$, under the identification of $\Lambda_{\mathbb{Q}}$ with $\Lambda_{\mathbb{Q}}^*$ via the inner product such that the ϵ_1, ϵ_l form an orthonormal basis. In particular, $\mathcal{C}_Y$ is generated by $-\epsilon_l, (\epsilon_l - \epsilon_1)/2$.

Every divisor on X is rationally equivalent to $\delta = pD_1 + qD_2$. We have $H^0(X, \mathcal{O}(pD_1)) = V_{p\omega_1}$, $H^0(X, \mathcal{O}(qD_2)) = V_{q\omega_l}$. Computing $\mathcal{P}(\delta) = \{\lambda = a\epsilon_1 + b\epsilon_n \mid 0 \geq b \geq a \geq -p, \ a + b \geq -2q\}$ and shifting by $\pi(\delta) = p\omega_1 + q\omega_l$ finally yields a decomposition

$$V_{p\omega_1} \otimes V_{q\omega_l} = \bigoplus_{\substack{0 \leq b \leq a \leq p \\ a+b \leq 2q \\ a \equiv b \pmod{2}}} V_{(p+q-a)\epsilon_1 + q\epsilon_2 + \cdots + q\epsilon_{l-1} + (q-b)\epsilon_l}$$

3.7 Intersection theory

The approach to enumerative problems on homogeneous spaces mentioned in the introduction leads to the definition of the *intersection ring* $C^*(G/H)$ [10]. It may be defined without use of compactifications, but one proves that $C^*(G/H) = \varinjlim H^*(X)$ over all smooth completions $X \supseteq G/H$.

In the simplest case, we have to compute the intersection number of divisors on G/H. Everything reduces to computing the self-intersection number (δ^d) for an effective divisor $\delta \subset G/H$, $d = \dim G/H$.

Translating δ by a generic element of G, we may assume that no colours are among the components of δ. Since the open B-orbit $\mathring{X} \subseteq G/H$ is a factorial variety, we may consider the equation $f \in \mathbb{C}[\mathring{X}]$ of $\delta|_{\mathring{X}}$.

Definition 3.34. The *Newton polytope* of δ is

$$\mathcal{N}(\delta) = \{\lambda \mid \langle v, \lambda \rangle \geq v(f), \ \langle v_D, \lambda \rangle \geq \operatorname{ord}_D(f) \text{ for all } v \in \mathcal{V}, \ D \in \mathcal{D}\}$$

Example 3.35. Suppose G/H is quasiaffine and, for simplicity, $\delta = \operatorname{div} f$ is a principal divisor, $f = f_1 + \cdots + f_m$, $f_j \in \mathbb{C}[G/H]_{(\lambda_j)}$, $f_j \neq 0$. Then $v(f) = \min_j \langle v, \lambda_j \rangle$ for all $v \in \mathcal{V}$, $\operatorname{ord}_D f = 0$ for all $D \in \mathcal{D}$, and

$$\mathcal{N}(\delta) = (\operatorname{conv}\{\lambda_1, \ldots, \lambda_m\} + \mathcal{V}^\vee) \cap \{\lambda \mid \langle v_D, \lambda \rangle \geq 0 \text{ for all } D \in \mathcal{D}\}.$$

In particular, if G is a torus, then $\mathcal{N}(\delta) = \operatorname{conv}\{\lambda_1, \ldots, \lambda_m\}$ is the usual Newton polytope of a Laurent polynomial f.

Theorem 3.36 ([4]).

$$(\delta^d) = d! \int_{\mathcal{N}(\delta)} \prod_{\alpha \not\perp \Lambda + \langle \pi(\delta) \rangle} \frac{(\lambda + \pi(\delta), \alpha)}{(\rho, \alpha)} \, d\lambda \qquad (3.1)$$

where α runs over positive roots, ρ is half the sum of positive roots, $\pi(\delta) = -\sum_{D \in \mathcal{D}} (\operatorname{ord}_D f) \pi(D)$, and the Lebesgue measure $d\lambda$ is normalized in such a way that the fundamental parallelepiped of Λ has volume 1.

Proof. Consider a smooth projective embedding $X \hookleftarrow G/H$. The divisor $\delta_X = \delta - \operatorname{div}_X f = -\sum_{i=1}^n (\operatorname{ord}_{D_i} f) D_i$ is B-stable, and

$$\mathcal{P}(\delta_X) = \{\lambda \mid \langle v_i, \lambda \rangle \geq \operatorname{ord}_{D_i} f \text{ for all } i\}.$$

It is clear that $\mathcal{N}(\delta) = \bigcap_{X \hookleftarrow G/H} \mathcal{P}(\delta_X)$.

There exists X such that the closure of δ contains no G-orbits [10]. Then δ is base point free, $(\delta^d) = [\delta_X]^d \in H^{2d}(X)$, and $\mathcal{N}(\delta) = \mathcal{P}(\delta_X)$. Indeed, take any $\lambda \in \mathcal{P}(\delta_X)$ and $v \in \mathcal{V}$. Consider an embedding $\hat{X}$ obtained by subdividing the fan of X by v, and let $D \subset \hat{X}$ be the divisor corresponding to v. It is easy to see that there is a map $\hat{X} \to X$ contracting D to the centre of v in X. For $k \gg 0$ we have

$$s = f_{k\lambda} s_{\delta_X}^k = f_{k\lambda} s_\delta^k / f^k \in H^0(X, \mathcal{O}(\delta)) \subseteq H^0(\hat{X}, \mathcal{O}(\delta)),$$

whence $\operatorname{ord}_D s = \langle v, k\lambda \rangle + \operatorname{ord}_D s_\delta^k - v(f^k) \geq 0$. But $\operatorname{ord}_D s_\delta = 0$, hence $\langle v, \lambda \rangle \geq v(f)$, which yields $\lambda \in \mathcal{N}(\delta)$.

It remains to compute $[\delta_X]^d$. By [13, Exer. II.7.5] base point free divisors lie in the closure of the ample cone in $(\operatorname{Pic} X) \otimes \mathbb{Q}$ (this is also visible from Theorem 3.28), and both sides of (3.1) depend continuously on δ_X. Therefore we may assume δ_X to be ample. Then $[\delta_X]^d = d! \cdot I$, where $\dim H^0(X, \mathcal{O}(k\delta_X)) = I \cdot k^d + \text{lower terms}$.

Recall Weyl's dimension formula: $\dim V_\lambda = \prod_\alpha (\lambda + \rho, \alpha)/(\rho, \alpha)$ (over all positive roots α). By Theorem 3.30,

$$
\begin{aligned}
\dim H^0(X, \mathcal{O}(k\delta_X)) &= \sum_{\lambda \in \mathcal{P}(k\delta_X) \cap \Lambda} \prod_\alpha \frac{(\lambda + \pi(k\delta_X) + \rho, \alpha)}{(\rho, \alpha)} \\
&= \sum_{\lambda \in \mathcal{P}(\delta_X) \cap \Lambda/k} \prod_\alpha \frac{(k\lambda + k\pi(\delta) + \rho, \alpha)}{(\rho, \alpha)}.
\end{aligned}
$$

The leading coefficient I equals the integral on the r.h.s. of (3.1). $\qquad\square$

Theorem 3.36 can be regarded as a generalization of the classical Bézout theorem.

Example 3.37. If G is a torus, then $(\delta^d) = d!\operatorname{vol}\mathcal{N}(\delta)$. Polarization yields $(\delta_1, \dots, \delta_d) = d!\operatorname{vol}(\mathcal{N}(\delta_1), \dots, \mathcal{N}(\delta_d))$, with the mixed volume of $\mathcal{N}(\delta_1), \dots, \mathcal{N}(\delta_d)$ on the right hand side, giving the number of solutions for a system of d equations in general position on a d-dimensional torus (Bernstein-Kouchnirenko [23]).

Example 3.38. More generally, consider $G = (G \times G)/\operatorname{diag} G$ as a homogeneous space under the doubled group, cf. Subsections 3.3–3.4. Suppose $\delta = \operatorname{div} f$, $f \in \mathbb{C}[G]$. (There is no essential loss of generality, because a finite cover of G is a factorial variety.) From Example 3.35 and results of Subsection 3.3 we see that $\mathcal{N}(\delta) = (\operatorname{conv}\{\lambda_1, \dots, \lambda_m\} - C^\vee) \cap C = (\operatorname{conv} W\{\lambda_1, \dots, \lambda_m\}) \cap C$ if f is expressed as the sum of matrix entries of $G : V_{\lambda_i}$, $i = 1, \dots, m$, and $\pi(\delta) = 0$. We have $\Lambda = \{(-\lambda, \lambda) \mid \lambda \in \Lambda(B)\}$, the positive roots of $G \times G$ are $(-\alpha, 0)$, $(0, \alpha)$, where α is a positive root of G, and $(-\rho, \rho)$ is half the sum of positive roots for $G \times G$. Now Theorem 3.36 yields Kazarnovskii's 'Bézout theorem' on any reductive group [17]:

$$
(\delta^d) = d! \int_{\mathcal{N}(\delta)} \prod_\alpha \frac{(\lambda, \alpha)^2}{(\rho, \alpha)^2} \, d\lambda
$$

Example 3.39. Consider the Grassmannian $\operatorname{Gr}_k(\mathbb{P}^n)$, acted on by $G = GL_{n+1}$. Let δ be a hyperplane section of its Plücker embedding into $\mathbb{P}(\bigwedge^{k+1} \mathbb{C}^{n+1})$. We have $\delta \sim D$, where D is the unique colour which generates $\operatorname{Pic} \operatorname{Gr}_k(\mathbb{P}^n)$. Here $\Lambda = 0$, whence $\mathcal{N}(\delta) = \{0\}$, and $\pi(\delta) = \pi(D) = -\epsilon_{k+2} - \cdots - \epsilon_{n+1}$. Positive roots are of the form $\alpha = \epsilon_i - \epsilon_j$, $i < j$, and $\rho = (n/2)\epsilon_1 + (n/2 - 1)\epsilon_2 + \cdots + (-n/2)\epsilon_{n+1}$. The degree of the Plücker embedding equals

$$(\delta^d) = d! \prod_{\alpha \not\perp \pi(\delta)} \frac{(\pi(\delta), \alpha)}{(\rho, \alpha)} = [(k+1)(n-k)]! \prod_{i \leq k+1 < j} \frac{1}{j-i}$$

$$= [(k+1)(n-k)]! \frac{0! \dots k!}{n! \dots (n-k)!}$$

This is a classical result of Schubert.

Example 3.40. Now we come back to the classical enumerative problem mentioned in the introduction. In the notation of Example 3.11, all conics tangent to a given one fill the divisor δ given by the equation $f(q) = \mathrm{Dis}\det(sq - tq_0) = 0$, where q_0 is the matrix of the given conic, s, t are indeterminates, and Dis denotes the discriminant of a binary form. Note that $f \in \mathbb{C}[Q_2]$, whence $\delta = \mathrm{div}\, f$ is principal.

From the expression for the discriminant of a binary cubic form and from Example 3.11, it is easy to see that $f = f_{(4\omega_1 + 4\omega_2)} + f_{(6\omega_1)} + f_{(6\omega_2)} + f_{(2\omega_1 + 2\omega_2)} + f_{(0)}$, where $f_{(\lambda)}$ is the projection to $\mathbb{C}[Q_2]_{(\lambda)}$. It follows by Examples 3.35, 3.11 that $\mathcal{N}(\delta) = \mathrm{conv}\{4\omega_1 + 4\omega_2, 6\omega_1, 6\omega_2, 0\}$ and $\pi(\delta) = 0$. (Actually, it suffices to know the highest weight $4\omega_1 + 4\omega_2$ occurring in f.) We subdivide $\mathcal{N}(\delta)$ into 2 triangles $\mathcal{N}_i = \mathrm{conv}\{4\omega_1 + 4\omega_2, 6\omega_i, 0\}$ $(i = 1, 2)$.

The positive roots are $\alpha_1, \alpha_2, \rho = \alpha_1 + \alpha_2$. For any $\lambda \in \Lambda \otimes \mathbb{Q}$ write $\lambda = 2x_1\alpha_1 + 2x_2\alpha_2$. The number of plane conics tangent to 5 given conics in general position equals

$$(\delta^5) = 5! \int_{\mathcal{N}(\delta)} \frac{(\lambda, \alpha_1)(\lambda, \alpha_2)(\lambda, \rho)}{(\rho, \alpha_1)(\rho, \alpha_2)(\rho, \rho)} \, d\lambda$$

$$= 5! \int_{\mathcal{N}_1} (4x_1 - 2x_2)(4x_2 - 2x_1)(2x_1 + 2x_2) \, dx_1 \, dx_2$$

$$= 5! \int_0^2 dx_1 \int_{x_1/2}^{x_1} dx_2 \, (4x_1 - 2x_2)(4x_2 - 2x_1)(2x_1 + 2x_2) = 3264$$

(Chasles, 1864).

4 Spaces of complexity one

The embedding theory of homogeneous spaces of complexity one is developed in [38] from the general Luna-Vust theory of embeddings in a way similar to the theory of spherical varieties. In this survey, we will only give a brief exposition of this theory, skipping most proofs and attracting reader's attention to common points and distinctions from the spherical case.

4.1 Coloured data

In contrast with the spherical case, a B-semiinvariant rational function on a homogeneous space G/H of complexity 1 is not uniquely determined (up to a constant) by its weight. Observe that, by the Lüroth theorem, $\mathbb{C}(G/H)^B \simeq \mathbb{C}(\mathbb{P}^1)$ is the field of rational functions in one variable, and a B-eigenfunction f_λ is determined by its weight $\lambda \in \Lambda$ only up to a multiple in $\mathbb{C}(\mathbb{P}^1)^\times$. We have a short exact sequence

$$0 \longrightarrow \mathbb{C}(\mathbb{P}^1)^\times/\mathbb{C}^\times \longrightarrow \mathcal{A} \longrightarrow \Lambda \longrightarrow 0$$

recalling $\mathcal{A} = \mathbb{C}(G/H)^{(B)}/\mathbb{C}^\times$ from Subsection 2.2. It is convenient to fix a (non-canonical) splitting $\mathcal{A} \simeq \Lambda \times (\mathbb{C}(\mathbb{P}^1)^\times/\mathbb{C}^\times)$, so that each B-semiinvariant function is represented as $f = f_\lambda q$, where f_λ is a fixed function of weight λ, and $q \in \mathbb{C}(\mathbb{P}^1)$.

Geometrically, the identification $\mathbb{C}(G/H)^B \simeq \mathbb{C}(\mathbb{P}^1)$ gives rise to a surjective rational map $\pi : G/H \dashrightarrow \mathbb{P}^1$, whose generic fibres are (the closures of) generic B-orbits in G/H. Thus the set of colours depends on one continuous parameter. We may fix a cofinite subset $\overset{\circ}{\mathcal{D}} \subseteq \mathcal{D}$ consisting of $D_z = \pi^{-1}(z)$, $z \in \mathbb{P}^1$, a cofinite subset of $\mathbb{P}^1$.

To any colour $D \in \mathcal{D}$ we associate a vector $v_D \in \Lambda^*$ by restriction of ord_D to $\{f_\lambda \mid \lambda \in \Lambda\}$. The restriction of ord_D to $\mathbb{C}(G/H)^B$ yields a valuation of $\mathbb{C}(\mathbb{P}^1)$ with centre $z_D \in \mathbb{P}^1$ and the order $h_D \in \mathbb{Z}_+$ of a local coordinate at z_D. We have $\operatorname{ord}_D f = \langle v_D, \lambda \rangle + h_D(\operatorname{ord}_{z_D} q)$. (If ord_D vanishes on $\mathbb{C}(\mathbb{P}^1)$, then we put $h_D = 0$ and take any point of $\mathbb{P}^1$ for z_D.) Similarly, G-valuations are determined by triples (v, h, z), where $v \in \Lambda^*_{\mathbb{Q}}$, $h \in \mathbb{Q}_+$, $z \in \mathbb{P}^1$.

Consider the union $\Lambda^+_{\mathbb{Q}} = \bigcup_{z \in \mathbb{P}^1} \Lambda^+_{\mathbb{Q}}(z)$, where $\Lambda^+_{\mathbb{Q}}(z) = \Lambda^*_{\mathbb{Q}} \times \mathbb{Q}_+$ are half-spaces naturally attached together along their common boundary hyperplane $\Lambda^*_{\mathbb{Q}}$. We say that $\Lambda^+_{\mathbb{Q}}$ is the *hyperspace* associated with G/H. By the above, colours and G-valuations are represented by points of the hyperspace. Reducing $\overset{\circ}{\mathcal{D}}$ if necessary, we may assume that $\operatorname{ord}_D f_\lambda = 0$ for all $D \in \overset{\circ}{\mathcal{D}}$, $\lambda \in \Lambda$. Hence D_z is represented by the vector $(0, 1) \in \Lambda^+_{\mathbb{Q}}(z)$ for all $z \in \mathbb{P}^1$.

The following result generalizes Theorem 3.3:

Theorem 4.1 ([20]). *G-valuations form a subset $\mathcal{V} \subseteq \Lambda_{\mathbb{Q}}^+$, called the valuation hypercone, such that the $\mathcal{V}(z) = \mathcal{V} \cap \Lambda_{\mathbb{Q}}^+(z)$ are solid convex polyhedral (in fact, cosimplicial) cones.*

4.2 Equivariant embeddings

Now we reorganize coloured data of B-charts and G-germs in a way similar to the spherical case.

The class **CD** consists of the pairs $(\mathcal{W}, \mathcal{F})$ such that $\mathcal{W}$ is finite and $\mathcal{F}$ differs from $\mathring{\mathcal{D}}$ by finitely many elements. Take $(\mathcal{W}, \mathcal{F}) \in$ **CD**.

Condition (F) is always satisfied, but in this case it is non-trivial, see [38, §3.1].

Let $\mathcal{C}(z)$ be the cone generated by those elements of $\mathcal{W}$ and $\mathcal{F}$ which map to $\Lambda_{\mathbb{Q}}^+(z)$ and by

$$\mathcal{Z} = \sum_{z \in \mathbb{P}^1} \mathcal{Z}(z) \subseteq \Lambda_{\mathbb{Q}}^* \qquad \text{(Minkowski sum), where}$$

$$\mathcal{Z}(z) = \text{conv} \left\{ v/h, \ v_D/h_D \ \middle| \ \begin{array}{l} (v, h) \in \mathcal{W} \cap \Lambda_{\mathbb{Q}}^+(z) \\ (v_D, h_D) \in \mathcal{F} \cap \Lambda_{\mathbb{Q}}^+(z) \\ h, h_D \neq 0 \end{array} \right\}. \qquad (4.1)$$

Put $\mathcal{C} = \bigcup_{z \in \mathbb{P}^1} \mathcal{C}(z)$. Condition (C) means that $(\mathcal{C}, \mathcal{F})$ is a coloured hypercone in the sense of the following

Definition 4.2. A *coloured hypercone* is a pair $(\mathcal{C}, \mathcal{F})$, where $\mathcal{C} \subseteq \Lambda_{\mathbb{Q}}^+$, $\mathcal{F} \subseteq \mathcal{D}$, and there exists a finite subset $\mathcal{W} \subset \mathcal{V}$ such that:

- $\mathcal{F}$ differs from $\mathring{\mathcal{D}}$ by finitely many elements, and $\mathcal{F} \not\ni 0$.
- $\mathcal{Z} \not\ni 0$, where $\mathcal{Z}$ is defined by formula (4.1).
- $\mathcal{C}(z) = \mathcal{C} \cap \Lambda_{\mathbb{Q}}^+(z)$ are strictly convex cones generated by $\mathcal{W} \cap \Lambda_{\mathbb{Q}}^+(z)$, $\mathcal{F} \cap \Lambda_{\mathbb{Q}}^+(z)$, and by $\mathcal{Z}$.

The *interior* of $(\mathcal{C}, \mathcal{F})$ is $\text{int}\,\mathcal{C} = \left(\bigcup_{z \in \mathbb{P}^1} \text{int}\,\mathcal{C}(z) \right) \cup \text{int}(\mathcal{C} \cap \Lambda_{\mathbb{Q}}^*)$ whenever $\mathcal{C}(z) \not\subseteq \Lambda_{\mathbb{Q}}^*$ for all $z \in \mathbb{P}^1$, and $\emptyset$ otherwise. The coloured hypercone is said to be *supported* if $(\text{int}\,\mathcal{C}) \cap \mathcal{V} \neq \emptyset$.

A *face* of $(\mathcal{C}, \mathcal{F})$ is either a coloured cone $(\mathcal{C}', \mathcal{F}')$ in some $\Lambda_{\mathbb{Q}}^+(z)$ such that $\mathcal{C}'$ is a face of $\mathcal{C}(z)$ and $\mathcal{C}' \cap \mathcal{Z} = \emptyset$, or a coloured hypercone $(\mathcal{C}', \mathcal{F}')$ such that $\mathcal{C}'(z)$ are faces of $\mathcal{C}(z)$ and $\mathcal{C}' \cap \mathcal{Z} \neq \emptyset$, and $\mathcal{F}' = \mathcal{F} \cap \mathcal{C}'$ in both cases.

A *coloured hyperfan* is a collection of supported coloured cones and hypercones which is obtained from finitely many coloured hypercones by taking all the supported faces, and has the property that different cones and hypercones intersect along faces inside $\mathcal{V}$.

Condition (W) says that $\mathcal{W}$ is recovered from $(\mathcal{C}, \mathcal{F})$ as the set of generators of those edges of $\mathcal{C}$ which do not intersect $\mathcal{F}$ and $\mathcal{Z}$. Conditions (V), (V'), (D'), (S) are reformulated exactly as in the spherical case.

The following theorem is a counterpart of Theorem 3.7.

Theorem 4.3. *B-charts are in bijection with coloured hypercones, G-germs with supported coloured cones and hypercones, and embeddings of G/H are in bijection with coloured hyperfans.*

Theorem 3.8 transfers verbatim to the case of complexity 1 if we only replace 'G-orbits' by 'closed G-subvarieties', 'cones' by 'cones and hypercones', and 'fan' by 'hyperfan'.

4.3 Divisors and intersection theory

Results of 3.5–3.7 are generalized in [39] to the complexity one case (and even, to some extent, to arbitrary complexity).

Theorem 3.25 generalizes together with the proof if we take $\mathring{X} = U_P \times A \times C$ from Theorem 1.3 and observe that C is a smooth rational curve, hence $\mathring{X}$ is factorial. There is a description of B-stable Cartier, base point free, and ample divisors similar to Theorem 3.28; see [39, §4].

However, the G-module structure of global sections for a B-stable Cartier divisor $\delta = \sum m_i D_i$ on an embedding $X \hookleftarrow G/H$ is more complicated. We may assume that the sum ranges over all B-stable prime divisors $D_i \subset X$ (with only finitely many $m_i \neq 0$), and let $(v_i, h_i) \in \Lambda_{\mathbb{Q}}^+(z_i)$ be the respective vectors of the hyperspace. Put

$$\mathcal{P}(\delta) = \{\lambda \in \Lambda_{\mathbb{Q}} \mid \langle v_i, \lambda \rangle \geq -m_i \text{ whenever } h_i = 0\}$$

$$m_z = \min_{\substack{z_i = z \\ h_i \neq 0}} \frac{\langle v_i, \lambda \rangle + m_i}{h_i} \qquad \text{for all } z \in \mathbb{P}^1$$

$$m(\delta, \lambda) = \max\left(1 + \sum_{z \in \mathbb{P}^1} m_z, \ 0\right).$$

Theorem 4.4. *Let $\pi(\delta)$ be the B-weight of the canonical section s_δ of $\mathcal{O}(\delta)$ with div $s_\delta = \delta$. Then the multiplicity of $V_{\lambda + \pi(\delta)}$ in $H^0(X, \mathcal{O}(\delta))$ equals $m(\delta, \lambda)$ if $\lambda \in \mathcal{P}(\delta)$ and 0 otherwise.*

Remark 4.5. Note that the multiplicity function $m(\delta, \lambda)$ is a piecewise affine concave function of λ on its support.

Proof. It suffices to examine the space of highest weight vectors of a given weight in $H^0(X, \mathcal{O}(\delta))$. A section $s = f_\lambda q s_\delta$ ($\lambda \in \Lambda$, $q \in \mathbb{C}(\mathbb{P}^1)$) is a highest weight vector if and only if $\operatorname{div} f_\lambda q \geq -\delta$ if and only if $\langle v_i, \lambda \rangle + h_i(\operatorname{ord}_{z_i} q) \geq -m_i$ for all i. The latter condition is equivalent to $\lambda \in \mathcal{P}(\delta)$ and $\operatorname{ord}_z q \geq -m_z$ for all $z \in \mathbb{P}^1$. It follows that the dimension of the space of highest weight vectors equals $\dim H^0\big(\mathbb{P}^1, \mathcal{O}(\sum_z m_z z)\big) = m(\delta, \lambda)$. $\qquad\square$

Unfortunately, the intersection theory on homogeneous spaces of complexity one is not as nice as for spherical spaces. The reason is that embeddings of G/H generally have infinitely many G-orbits, and there might exist no compactification $X \hookleftarrow G/H$ with finitely many orbits such that the closures $\overline{Z_i}$ of given subvarieties $Z_1, \ldots, Z_s \subset G/H$ intersect $X \setminus (G/H)$ properly. Then $\overline{Z_1} \cap \cdots \cap \overline{Z_s}$ may have points 'at infinity', and the intersection product of $[\overline{Z_i}]$ in $H^*(X)$ has no relation with $|Z_1 \cap \cdots \cap Z_s|$. In particular, there is generally no 'Bézout theorem' for the intersection number of hypersurfaces in G/H. However, there is a weaker version of Theorem 3.36.

Theorem 4.6 ([39]). *Let δ be a base point free divisor on a projective embedding $X \hookleftarrow G/H$, $\dim G/H = d$. Then*

$$(\delta^d) = d! \int_{\mathcal{P}(\delta)} m(\delta, \lambda) \prod_{\alpha \not\perp \Lambda + \langle \pi(\delta) \rangle} \frac{(\lambda + \pi(\delta), \alpha)}{(\rho, \alpha)} \, d\lambda. \qquad (4.2)$$

The proof is essentially the same as for Theorem 3.36 using Theorem 4.4 instead of Theorem 3.30. Details are left to the reader.

Consider the problem of finding the intersection number of divisors on G/H. Suppose we managed to construct a compactification $X \supset G/H$ with finitely many orbits such that all divisors, whose intersection number we are looking for, intersect each orbit properly. Then Theorem 4.6 leads to a 'Bézout theorem' on G/H. Another application of Theorem 4.6 is the computation of the degree of any orbit in any $SL_2(\mathbb{C})$-module or projective representation [39]. (For irreducible representations this degree was computed in [28] using the description of Chow rings for smooth embeddings of $SL_2/\{e\}$.)

Bibliography

[1] M. Brion. Groupe de Picard et nombres charactéristiques des variétés sphériques. *Duke Math. J.*, 58(2):397–424, 1989.

[2] M. Brion. Vers une généralisation des espaces symétriques. *J. Algebra*, 134:115–143, 1990.

[3] M. Brion. Parametrization and embeddings of a class of homogeneous spaces. In *Proceedings of the International Conference on Algebra, Part 3, Novosibirsk, 1989*, volume 131 of *Contemp. Math.*, pages 353–360. Amer. Math. Society, Providence, 1992.

[4] M. Brion. Piecewise polynomial functions, convex polytopes and enumerative geometry. In *Parameter spaces*, volume 36 of *Banach Centre Publ.*, pages 25–44. Inst. of Math., Polish Acad. Sci., Warszawa, 1996.

[5] M. Brion. Variétés sphériques. Notes de la session de la S.M.F. 'Opérations hamiltoniennes et opérations de groupes algébriques', Grenoble, 1997. Available at
http://www-fourier.ujf-grenoble.fr/~mbrion/spheriques.ps.

[6] M. Brion, D. Luna, and Th. Vust. Espaces homogènes sphériques. *Invent. Math.*, 84:617–632, 1986.

[7] V. I. Danilov. The geometry of toric varieties. *Russian Math. Surveys*, 33(2):97–154, 1978.

[8] C. de Concini. Normality and non-normality of certain semigroups and orbit closures. In *Invariant theory and algebraic transformation groups*, III, volume 132 of *Encyclopædia of Mathematical Sciences*, pages 15–35. Springer-Verlag, Berlin, 2004.

[9] C. de Concini and C. Procesi. Complete symmetric varieties. In *Invariant theory (Montecatini, 1982)*, volume 996 of *Lecture Notes in Math.*, pages 1–44. Spinger-Verlag, 1983.

[10] C. de Concini and C. Procesi. Complete symmetric varieties, II. In R. Hotta, editor, *Algebraic groups and related topics (Kyoto/Nagoya, 1983)*, volume 6 of *Adv. Studies in Pure Math.*, pages 481–513. North-Holland, Amsterdam, 1985.

[11] W. Fulton. *Introduction to toric varieties*, volume 131 of *Ann. Math. Studies*. Princeton Univ. Press, Princeton, 1993.

[12] W. Fulton, R. MacPherson, F. Sottile, and B. Sturmfels. Intersection theory on spherical varieties. *J. Algebraic Geom.*, 4:181–193, 1995.

[13] R. Hartshorne. *Algebraic Geometry*, volume 52 of *Graduate Texts in Mathematics*. Springer-Verlag, New York-Heidelberg-Berlin, 1977.

[14] J. E. Humphreys. *Linear algebraic groups*, volume 21 of *Graduate Texts in Mathematics*. Springer-Verlag, New York, 1975.

[15] J. C. Jantzen. *Representations of algebraic groups*. Academic Press, New York, 1987.

[16] M. M. Kapranov. Hypergeometric functions on reductive groups. In M.-H. Saito, editor, *Integrable systems and algebraic geometry*, pages 236–281. World Scientific, Singapore, 1998.

[17] B. Y. Kazarnovskii. Newton polyhedra and the Bézout theorem for matrix-valued functions of finite-dimensional representations. *Funct. Anal. Appl.*, 21:319–321, 1987.

[18] F. Knop. Weylgruppe und Momentabbildung. *Invent. Math.*, 99:1–23, 1990.

[19] F. Knop. The Luna-Vust theory of spherical embeddings. In S. Ramanan, editor, *Proc. Hyderabad Conf. on Algebraic Groups*, pages 225–249. Manoj Prakashan, Madras, 1991.

[20] F. Knop. Über Bewertungen, welche unter einer reduktiven Gruppe invariant sind. *Math. Ann.*, 295:333–363, 1993.

[21] F. Knop. The asymptotic behavior of invariant collective motion. *Invent. Math.*, 116:309–328, 1994.

[22] F. Knop, H. Kraft, D. Luna, and Th. Vust. Local properties of algebraic group actions. In H. Kraft, P. Slodowy, and T. A. Springer, editors, *Algebraische Transformationsgruppen und Invariantentheorie*, volume 13 of *DMV Seminar*, pages 63–76. Birkhäuser, Basel-Boston-Berlin, 1989.

[23] A. G. Kouchnirenko. Polyèdres de Newton et nombres de Milnor. *Invent. Math.*, 32(1):1–31, 1976.

[24] H. Kraft. *Geometrische Methoden in der Invariantentheorie*. Vieweg Verlag, Braunschweig-Wiesbaden, 1985.

[25] P. Littelmann. On spherical double cones. *J. Algebra*, 166:142–157, 1994.

[26] D. Luna and Th. Vust. Plongements d'espaces homogènes. *Comment. Math. Helv.*, 58:186–245, 1983.

[27] P.-L. Montagard. Une nouvelle propriété de stabilité du pléthysme. *Comment. Math. Helv.*, 71:475–505, 1996.

[28] L. Moser-Jauslin. The Chow ring of smooth complete $SL(2)$-embeddings. *Compositio Math.*, 82(1):67–106, 1992.

[29] D. I. Panyushev. Complexity and rank of homogeneous spaces. *Geom. Dedicata*, 34:249–269, 1990.

[30] D. I. Panyushev. Complexity of quasiaffine homogeneous varieties, t-decompositions, and affine homogeneous spaces of complexity 1. In E. B. Vinberg, editor, *Lie groups, their discrete subgroups and invariant theory*, volume 8 of *Adv. Sov. Math.*, pages 151–166. AMS, Providence, 1992.

[31] D. I. Panyushev. Complexity and rank of double cones and tensor product decompositions. *Comment. Math. Helv.*, 68:455–468, 1993.

[32] D. I. Panyushev. Complexity and nilpotent orbits. *Manuscripta Math.*, 83:223–237, 1994.

[33] D. I. Panyushev. Complexity and rank of actions in invariant theory. *J. Math. Sci. (New York)*, 95(1):1925–1985, 1999.

[34] V. L. Popov. Contractions of the actions of reductive algebraic groups. *Math. USSR-Sb.*, 58(2):311–335, 1987.

[35] V. L. Popov and E. B. Vinberg. Invariant theory. In *Algebraic geometry (IV)*, volume 55 of *Encyclopædia of Mathematical Sciences*, pages 123–278. Springer-Verlag, Berlin-Heidelberg, 1994.

[36] A. Rittatore. Algebraic monoids and group embeddings. *Transformation Groups*, 3(4):375–396, 1998.

[37] J. Stembridge. Multiplicity-free products and restrictions of Weyl characters. *Represent. Theory*, 7:404–439, 2003.

[38] D. A. Timashev. Classification of G-varieties of complexity 1. *Math. USSR-Izv.*, 61(2):363–397, 1997.

[39] D. A. Timashev. Cartier divisors and geometry of normal G-varieties. *Transformation Groups*, 5(2):181–204, 2000.

[40] D. A. Timashev. Equivariant compactifications of reductive groups. *Sbornik: Mathematics*, 194(4):589–616, 2003.

[41] D. A. Timashev. Complexity of homogeneous spaces and growth of multiplicities. *Transformation Groups*, 9(1):65–72, 2004.

[42] D. A. Timashev. Equivariant symplectic geometry of cotangent bundles, II. Preprint, `arXiv:math.AG/0502284`; *Moscow Math. J.*, to appear, 2006.

[43] E. B. Vinberg. On reductive algebraic semigroups. In S. Gindikin and E. Vinberg, editors, *Lie Groups and Lie Algebras: E. B. Dynkin Seminar*, volume 169 of *AMS Transl.*, pages 145–182. Amer. Math. Soc., 1995.

[44] E. B. Vinberg and B. N. Kimelfeld. Homogeneous domains on flag manifolds and spherical subsets of semisimple Lie groups. *Funct. Anal. Appl.*, 12(3):168–174, 1978.

[45] Th. Vust. Plongements d'espaces symétriques algébriques: une classification. *Ann. Scuola Norm. Sup. Pisa Cl. Sci. (4)*, XVII(2):165–194, 1990.

Department of Algebra

Faculty of Mechanics and Mathematics

Moscow State University

Leninskie Gory, 119992 Moscow

Russia

timashev@mech.math.msu.su

http://mech.math.msu.su/department/algebra/staff/timashev

Geometric quantization and
algebraic Lagrangian geometry

Nikolai A. Tyurin

Introduction

The main theme of this survey is the quantization of classical mechanical systems in terms of algebraic geometry. The aim is thus to relate questions of mathematics and theoretical physics. We first recall briefly the main problems and methods which turn us to study the new subject.

Quantization itself is the main problem of theoretical physics. The need to introduce and develop it was dictated by the creators of quantum theory. According to the 'Copenhagen philosophy', the physical predictions of a quantum theory must be formulated in terms of classical concepts (the first sentence of Woodhouse [27]; here we quote the beginning of this survey). Thus in addition to the usual structures (Hilbert space, unitary transformations, selfadjoint operators ..., see, for example, Landau and Lifschitz [13]) any reasonable quantum theory has to admit an appropriate passage to a classical limit under which quantum observables are transferred to their classical analogues. However, as Dirac pointed out at the beginning of the quantum age, the correspondence between quantum theory and classical theory should be based not just on numerical coincidences taking place in the limit $h \to \infty$, but on an analogy between their mathematical structures. Classical theory does approximate the quantum theory, but it does more – it supplies a framework for some interpretation of quantum theory. Using this idea, we can in general understand a quantization procedure as a correspondence between classical and quantum theories. In this sense, quantization of the classical mechanical systems is a movement in one direction, while taking the quasiclassical limit goes in the opposite direction. More abstractly: the moduli space of quantum theories is an n-sheeted cover of the moduli space of classical theories (one supposes that n equals 2), and quantization is the structure of this cover.

Quantization itself is a very popular subject. There are a number of different approaches to this problem. But one of them is honoured as the first one in theoretical physics, and is named *canonical quantization*. In simple cases, the correspondence comes with some choice of fixed coordinates. If a classical observable is represented by a function $f(p_a, q^b)$ in these coordinates, the corresponding quantum observable is the operator

$$f\left(-\imath h \frac{\partial}{\partial q^a}, q^a\right).$$

The canonical quantization of the harmonic oscillator is a standard computation in theoretical physics: any alternative approach should be compared with it, and rejected if it gives essentially different answers from the classical one. However this formal substitution (replacing the coordinates p_a by differential operators) introduces many problems. Indeed, beyond the simplest cases, in this process, the result of the quantization depends on the order of p and q in the expression for the classical observable f, and moreover, the result depends strongly on the choice of coordinates, and is not invariant under generic canonical transformations. Nevertheless this canonical quantization supplied by some physical intuition together with its various generalizations plays a central role in modern theoretical physics.

Geometric quantization provides one way of developing the canonical method while avoiding the difficulty. As a term, geometric quantization has two slightly different meanings. One can understand it either as a specific construction, well-known as Souriau-Kostant quantization (see, for example, Hurt [10], Kostant [12], Souriau [17], Woodhouse [27], etc.), or as a general approach to the problem based on the underlying geometry. Nowadays the problem of quantization is treated by quite different methods: the algebraic approach includes deformation quantization, formal geometry, noncommutative geometry, quantum groups; analytical approaches include the theory of integral Fourier operators, Toeplitz structures and others. All the methods discussed above have one feature in common – one almost completely forgets about the structure of the given system (and Dirac's suggestion mentioned above) and the 'homecoming' turns to be quite impossible. At the same time, in the direction of geometric quantization, one at least attempts to keep the original system in mind. The corresponding symplectic manifold remains the basis of all the constructions, and genuinely takes part in the definition of all the auxiliary geometric objects giving the result of quantization. At the same time, geometric quantization does not need any choice of coordinates, and this basic feature makes it possible to deal with complicated systems not admitting any global coordinates. But starting from a given classical phase space the geometric quantization should give a result that for simple systems is comparable with the canonical one. Thus in any case, geometric quantization is a generalization of canonical quantization. To keep the relation one usually pays the price of losing generality in the construction: out of the whole space of classical observables we keep only the comparatively small subclass of 'quantizable' functions. These quantizable objects are selected in terms of a choice of 'polarization' of the given symplectic manifold (see Śniatycki [16], Woodhouse [27]), and distinguished by the condition that their Hamiltonian vector fields preserve the polarization.

The known geometric quantization schemes are unified by the fact that they usually take some spaces of regular sections of a prequantization bundle as their Hilbert spaces (and again, one imposes some additional conditions on these sections to be regular in our sense). The original Souriau-Kostant construction takes all smooth sections with bounded L^2-norm (with respect to a given Hermitian structure on the fibres of the prequantization bundle, weighted by the Liouville form). Further specializations come in different ways: the Rawnsley-Berezin method (see Rawnsley, Cahen and Gutt [14]) uses only the sections which are holomorphic with respect to a complex polarization (= a fixed complex structure on M) as does the Toeplitz-Berezin approach (see Bordemann, Meinrenken and Schlichenmaier [4]) while in the case of real polarization one collects only such sections (weighted by half-weights) that are invariant under infinitesimal transformations tangent to the fibres of a real polarization (= Lagrangian fibration).

The introduction of an additional structure – the complex polarization – related geometric quantization to the most highly developed subject in modern mathematics, namely, algebraic geometry. As we mentioned above, several of the methods use a complex polarization. This imposes the additional condition that our symplectic manifold (M, ω) admits a Kähler structure: there exists an integrable complex structure J compatible with ω. Together, these two structures ω, J give a corresponding Riemannian metric g such that the complex manifold M, J carries a Hermitian metric; since ω is closed by the definition it provides a Kähler structure on M. Moreover, it is a common requirement of all methods of quantization that ω should have integral cohomology class:

$$[\omega] \in H^2(M, \mathbb{Z}) \subset H^2(M, \mathbb{R})$$

(the 'integrality of charge' condition). This implies that the Kähler metric is of Hodge type, so that the Kähler manifold is an algebraic variety (see, for example, Griffiths and Harris [9]). Thus a symplectic manifold can be quantized if it admits an algebro-geometric structure!

This is not very surprising in view of the so-called geometric formulation of quantum mechanics, the basic idea of which is to replace the algebraic methods of quantum mechanics by algebro-geometric ones. The author learned these ideas from Ashtekar and Schilling [3] and Schilling [15], but of course, as one can imagine, the original sources go back to the birth of quantum theory itself. In any case, the history of the question is discussed in Schilling [15]. The starting point, roughly, is that a state in quantum mechanics is given by a ray in a Hilbert space, with two vectors ψ_1, ψ_2 representing the same state if they are proportional. Thus it is natural to consider the projectivization $\mathbb{P}(\mathcal{H})$ as the space of quantum states, rather than $\mathcal{H}$ itself. This (finite or infinite-dimensional) complex manifold automatically carries a Hermitian metric (the Fubini-Study metric), so we can view it as a real manifold with Kähler structure. This (finite or infinite-dimensional) real manifold comes with a symplectic structure and Riemannian metric. Quantum states are represented simply by points of this manifold. Quantum observables are represented by smooth real functions of a special type called Berezin symbols. These ideas allow us to generalize the problem of quantization in a nonlinear way: namely, rather than a Hilbert space, one could try to find (or to construct) some finite or infinite-dimensional Kähler manifold $\mathcal{K}$ together with a correspondence between the smooth functions on a given symplectic manifold (= the classical observables on a given phase space, see, for example, Abraham and Marsden [1]) and Berezin symbols on this Kähler manifold. This nonlinear generalization is called algebro-geometric quantization. Following the suggestion of Ashtekar and Schilling [3], we should construct this Kähler manifold without introducing the intermediate Hilbert spaces of the usual methods of geometric quantization.

Our main aim in this text is to present an example of a successful algebro-geometric quantization for compact simply-connected symplectic manifolds. We call this method ALG(a)-quantization. To decode the acronym we need to recall some basic facts belonging to a new subject created right on the border between algebraic and symplectic geometries (if such a border exists).

Modern mathematics mixes up different subjects. For example, the mirror symmetry conjecture proposes the idea that the algebraic geometry of a manifold X corresponds to the symplectic geometry of its mirror partner X'. The ingredients of algebraic geometry over X (bundles, sheaves, divisors ...) then correspond to some objects of symplectic geometry (Lagrangian submanifolds of special types). The so-called homological mirror symmetry compares two categories coming from algebraic and symplectic geometry respectively; this approach gives the desired results in some particular cases (for example, elliptic curves). On the other hand, the framework of algebraic geometry over X generates a number of moduli spaces, and a different approach is to look for the moduli spaces corresponding to these within the framework of symplectic geometry. These ideas have developed in different ways, and we could report about a number of promising results and new ideas clarifying and extending the original program (see, for example, Kapustin and Orlov [11], A. Tyurin [20]). But these results are nowhere near complete, and also far from covering all the problems. But the main idea, proclaiming the creation of a new synthethis (or at least synergy) unifying algebraic geometry and symplectic geometry remains very attractive, and seems to be the right approach.

One step in this direction was taken in 1999 when the moduli space of half-weighted Bohr-Sommerfeld Lagrangian cycles of specified topological type and fixed volume was proposed by A. Tyurin [19] and constructed by Gorodentsev and Tyurin [8]. Starting from a simply-connected compact symplectic manifold with an integral symplectic form (read 'classical mechanical system with compact simply-connected phase space satisfying the Dirac condition'), the authors construct a set of infinite-dimensional moduli spaces that are infinite-dimensional algebraic manifolds depending on the choice of some specified topological invariants and a real number – the volume of the half-weighted cycles. Lagrangian geometry is mixed in the construction with algebraic geometry and this construction itself belongs to some new synthetic geometry. The authors called it ALAG – *Abelian Lagrangian algebraic geometry* (not a garbled version of their initials!). It was created as a step in a new approach to the mirror symmetry conjecture generalizing some notions from standard geometric quantization (prequantization data, Bohr-Sommerfeld condition, etc.) so it is not really surprising that this construction plays an important role in geometric quantization. Namely,

I showed in [22] and [23] that these moduli spaces of half-weighted Bohr-Sommerfeld Lagrangian subcycles of fixed volume solve the problem of algebro-geometric quantization stated above for simply-connected compact symplectic manifolds. This method, proved in [22] and [23], was called ALG(a)-*quantization*; it gives new results which are nevertheless entirely consistent with the old ones for an appropriate choice of polarization on (M, ω) (see [23]).

Acknowledgments My work on this problem started in 1999 at the Max-Planck-Institute for Mathematics (Bonn) when I happened to come across the two preprints of Ashtekar and Schilling [3] and Gorodentsev and Tyurin [7]. The main results of the text were established at MPI (Bonn), the Korean Institute of Advanced Study (Seoul) and the Joint Institute for Nuclear Research (Dubna); I would like first to thank all the people from the institutions for hospitality, friendly attention and very good working conditions. Personally I would like to thank Prof. I. R. Shafarevich as the leader of the scientific school to which the author belongs and as an attentive listener whose remarks were extremely important and useful for the work. This publication is supported by a grant of the London Mathematical Society and I would like to thank the LMS for support, the Mathematical Institute of the Warwick University for hospitality and personally Prof. Miles Reid for friendly attention and help in the preparation of the final version of this text.

1 Geometric quantization and its geometric formulation

We first recall what the quantization problem is.

Let (M, ω) be a symplectic manifold with symplectic form ω. We view M as the phase space of a classical mechanical system (see, for example, Abraham and Marsden [1] or Arnol'd and Givental [2]). The space $C^\infty(M, \mathbb{R})$ consists of classical observables. Any distinguished Hamiltonian function $H \in C^\infty(M, \mathbb{R})$ gives rise to a corresponding Hamiltonian dynamical system, and the infinitesimal deformation of an observable f under the Hamiltonian transformation is given by the Poisson bracket $\{f, H\}$. The equation of motion (Hamilton's equations) says that every point $x \in M$ moves in the direction of the Hamiltonian vector field X_H.

Now what is a quantization of the system? This attaches to (M, ω) a corresponding Hilbert space $\mathcal{H}$, together with a map

$$q \colon C^\infty(M, \mathbb{R}) \to \mathrm{Op}(\mathcal{H}) \subset \{\text{selfadjoint operators on } \mathcal{H}\};$$

here $\mathrm{Op}(\mathcal{H})$ is some algebra consisting of selfadjoint operators (quantum observables), and q satisfies a number of conditions listed by Dirac (see, for example, Hurt [10], Śniatycki [16], Woodhouse [27]). We can summarise these conditions briefly as follows: q should be a homomorphism of Lie algebras (where $\mathrm{Op}(\mathcal{H})$ is a Lie algebra under the commutator $i[\,\cdot\,;\,\cdot\,]$), taking the constant function $f \equiv 1$ to the identity operator, and $\mathcal{H}$ should be an irreducible representation of $\mathrm{Op}(\mathcal{H})$.

Geometric quantization arises if one supposes in addition that the symplectic structure on M is integral: the cohomology class $[\omega]$ belongs to $H^2(M,\mathbb{Z})$. This condition provides a complex line bundle $L \to M$, uniquely determined by the condition $c_1(L) = [\omega]$, called the *prequantization line bundle*. At the same time, fixing a Hermitian structure on L gives a prequantization connection $a \in \mathcal{A}_h(L)$ that satisfies $F_a = 2\pi i\omega$. If M is simply connected then a is unique up to gauge transformation. We usually consider this case in what follows.

Once the prequantization data is fixed, the Hilbert space of geometric quantization is given by the space of smooth measurable sections of L or an appropriate subspace.

Example 1.1. The Souriau-Kostant approach takes the space

$$\mathcal{H} = \Gamma(M,L) \cap L_2(M,L),$$

where the L_2-norm is given by the Hermitian structure on L together with the Liouville volume form:

$$\langle s_1, s_2 \rangle = \int_M \langle s_1, s_2 \rangle_h \, d\mu_L.$$

Then the operator $\widehat{Q}_f \colon \mathcal{H} \to \mathcal{H}$ corresponding to $f \in C^\infty(M,\mathbb{R})$ is given by

$$\widehat{Q}_f s = i\nabla_{X_f} s + 2\pi \cdot s.$$

It is not hard to establish that this correspondence is a Lie algebra homomorphism. The problem with this method arises when one considers the simplest case: for $M = \mathbb{R}^{2n}$ this correspondence is reducible (see Hurt [10], Kostant [12], Woodhouse [27]).

Example 1.2. To get away from the reducibility of $\mathcal{H}$ one fixes additional data – a (real or complex) polarization. A *complex polarization* is an integrable complex structure I compatible with ω. Together, I and ω give a Kähler structure on M. Since the curvature form F_a is proportional to ω it follows that a defines a holomorphic structure on L, and one takes

$$\mathcal{H} = H^0(M_I, L).$$

In the compact case this space is finite-dimensional. Then we have the following results.

A) In the Rawnsley-Berezin method, for some special 'quantizable' functions f, the corresponding operator Q_f is defined as in the Souriau-Kostant method; but here Q_f preserves $H^0(M_I, L)$, since f is quantizable (see Rawnsley, Cahen and Gutt [14]).

B) In the Toeplitz-Berezin method one takes the composite

$$s \mapsto f \cdot s \mapsto A_f s \in H^0(M_I, L),$$

where the original s lies in $H^0(M_I, L)$ and the final map is the orthogonal projection from $\Gamma(M, L)$ to a finite-dimensional subspace which is our $H^0(M_I, L)$. In this method the correspondence principle only holds asymptotically (see Bordemann, Meinrenken and Schlichenmaier [4]). At the same time, the correspondence has a very large kernel.

   Tuynman [18] discusses some relations between A) and B).

C) In the case of a real polarization the situation is specified as a completely integrable system. Thus there is a Lagrangian fibration

$$\pi \colon M \to B,$$

and one takes some subset in B of Bohr-Sommerfeld fibres. In this case one can quantize only the first integrals of the system, and the corresponding operators are diagonal (see Śniatycki [16]).

Now to generalize the quantization problem we need some natural translation of standard quantum mechanics into the language of projective geometry. In Ashtekar and Schilling [3] and Schilling [15], this was called the *geometric formulation* of quantum mechanics. It is based on the well-known fact that in ordinary quantum mechanics, physical states are given not by vectors of the Hilbert space but by complex rays, so that vectors ψ_1 and ψ_2 represent the same physical state if and only if they are proportional. It shows that 'real' quantum states are given by points of the projective space $\mathbb{P}\mathcal{H}$. So the question is, can we translate all the notions of ordinary quantum mechanics to the projective language?

   The answer is yes. Indeed, the projectivization $\mathbb{P}\mathcal{H}$ of a Hilbert space is a finite or infinite-dimensional Kähler manifold; it carries the Fubini-Study metric, and is thus a symplectic manifold with compatible integrable complex structure. It also carries a corresponding Riemannian metric. We have already taken the first step in the translation: rather than vectors up to proportionality, we just have points.

Any selfadjoint operator $\widehat{F} \in \mathrm{Op}(\mathcal{H})$ defines a special smooth function $f \in C^\infty(\mathbb{P}\mathcal{H}, \mathbb{R})$. Starting from the operator $\widehat{F}$ and its expectation value $F = \langle \widehat{F}\psi; \psi \rangle$, one restricts F to the unit sphere $S \subset \mathcal{H}$. It is not hard to see that this restriction is invariant under canonical phase rotations, and therefore $F|_S$ can be pushed down under the Hopf fibration

$$S \to \mathbb{P}\mathcal{H}.$$

This gives a smooth function f on $\mathbb{P}\mathcal{H}$. This correspondence is clearly linear and nondegenerate; thus we can consider the functions f instead of selfadjoint operators. The following proposition allows us to distinguish these special smooth functions without reference to the Hilbert space:

Proposition 1.3 (Ashtekar and Schilling [3]). *A smooth function on $\mathbb{P}\mathcal{H}$ $f \in C^\infty(\mathbb{P}\mathcal{H}, \mathbb{R})$ is induced by a selfadjoint operator on $\mathcal{H}$ if and only if its Hamiltonian vector field X_f preserves the Kähler structure, that is,*

$$\mathrm{Lie}_{X_f}\, g \equiv 0,$$

where g is the Riemannian metric.

We write $C_q(\mathbb{P}, \mathbb{R})$ for the space of such smooth functions, and call the functions *symbols* following Berezin. Moreover, it is not hard to check that if $\widehat{K} = i[\widehat{F}_1, \widehat{F}_2]$ then, on the level of symbols,

$$k = \{f_1, f_2\}_\Omega,$$

where the Poisson bracket is defined by the canonical symplectic form Ω on $\mathbb{P}\mathcal{H}$. If ψ is an eigenvector of $\widehat{F}$ with the eigenvalue λ then $\mathbb{P}(\psi) = p \in \mathbb{P}\mathcal{H}$ is a critical point of f with critical value λ. The Schrödinger equation for a distinguished Hamiltonian $\widehat{H}$ corresponds simply to Hamilton's equations for the Hamiltonian h. The orthogonal decomposition with respect to an eigenbasis turns into a simple trigonometric function on geodesic distances to the corresponding critical points.

To sum up, we have the picture of a projective space $\mathcal{P}$, a Kähler manifold equipped with the corresponding Kähler structure. $\mathcal{P}$ thus has a fixed symplectic structure, defining a Lie algebra structure on the function space, that governs the evolution of the system. However, there are two major differences between our picture and the case of classical mechanics. First, as a projective space, the quantum phase space has the very special nature of a Kähler manifold; we discuss possible generalizations later. Second, as a Kähler manifold, it is equipped with a Riemannian metric, and it is this that governs the measurement process. This ingredient was absent in the classical theory – in quantum theory, it is responsible for such notions as uncertainty, state reduction and so on.

We summarize the translation in a short glossary.

Physical states Physical states of a quantum system correspond to points of an appropriate Kähler manifold (a projective space in the basic example).

Kähler evolution The time evolution of the physical system is defined by a flow on $\mathcal{P}$ which preserves the whole Kähler structure. This flow is generated by a vector field which is dense everywhere on $\mathcal{P}$.

Observables Physical observables are given by special real smooth functions on $\mathcal{P}$ whose Hamiltonian vector fields preserve the Kähler structure. In other words physical observables are represented by symbols.

Probability aspects Let $\Lambda \subset \mathbb{R}$ be a closed subset of the spectrum $\mathrm{sp}(f)$, and suppose that the state of the system is represented by $p \in \mathcal{P}$. Then, as the result of measuring f, the probability of getting a result contained in Λ is given by the formula

$$\delta_p(\Lambda) = \cos^2\Big(\sigma\big(p, P_{f,\Lambda}(p)\big)\Big),$$

where $P_{f,\Lambda}$ is the projection taking p to the nearest point of $\mathcal{E}_{f,\Lambda}$, and σ is the geodesic distance.

Reduction The ideal measurements that can be performed correspond to choosing an arbitrary closed subset $\Lambda \subset \mathrm{sp}(f)$. The measurement determines whether or not the critical value f belongs to Λ. After performance of the measurement, the state of the system is represented by either $P_{f,\Lambda}(p)$ or $P_{f,\Lambda^c}(p)$, depending on the result of the measurement.

We now give two direct quotations from Ashtekar and Schilling [3] to clarify possible generalizations. First, the postulates of quantum mechanics can be stated purely geometrically, without reference to Hilbert spaces. Of course, standard Hilbert space considerations and its related algebraic machinery provide a good set-up for concrete computations. But mathematically, the situation is analogous to the usual geometric consideration of compact manifolds with nontrivial topology: a practical starting point to study our manifold is to view it as embedded in an appropriate ambient space (Euclidean, projective, ...); but the embedding is only a convenience; one could derive everything one needs directly from the geometry of the manifold.

Second, in quantum mechanics, the assumption of linearity is an analogue of the inertial systems of special relativity, and the geometric formulation of quantum mechanics could be though of as an analogue of Minkowski's formulation of special relativity; in the same way that the latter paved the way to general relativity, the geometric formulation of quantum mechanics should lead us to a new theory.

The geometric formulation of quantum mechanics leads immediately to one way of generalizing it. Namely, we could suppose that there exist some Kähler manifolds that carry quantum mechanical systems, other than projective spaces. The dynamic properties are easy to satisfy (after all, classical mechanics allows us to consider symplectic manifolds other than projective spaces). The first question here is that of observables: indeed, there exist Kähler manifolds not admitting any real functions whose Hamiltonian vector fields preserve the Kähler structure, so this question is really of primary importance.

For example, one might require the Kähler manifold to admit the maximal possible such functions, corresponding to Kähler manifolds of constant holomorphic sectional curvature (see Ashtekar and Schilling [3]). In finite dimensions it is well known that only projective spaces satisfy this condition. In the infinite-dimensional case this problem is still open, and one could hope for some infinite-dimensional Kähler manifolds other than projective spaces satisfying the condition.

We take the view here that the above requirement is too strong: we only need to impose the condition that the space of quantum observables allowed over a tested Kähler manifold is *sufficiently large*. That is, the Kähler structure of the manifold should provide us with a good supply of observables to use in our investigations. We will keep this view in mind during what follows, and continue with the following natural definition.

Definition 1.4. A real smooth function f on a Kähler manifold $\mathcal{K}$ is a *quasisymbol* if its Hamiltonian vector field preserves the Kähler structure. We write $C_q^\infty(\mathcal{K}, \mathbb{R})$ for the space of all quasisymbols.

A manifest property of such functions follows immediately:

Proposition 1.5. *For any Kähler manifold $\mathcal{K}$ the space $C_q^\infty(\mathcal{K}, \mathbb{R})$ is a Lie subalgebra of the Poisson algebra.*

To prove this, take the Poisson bracket of any two quasisymbols, ensuring that the corresponding Hamiltonian vector field is proportional to the commutator of the Hamiltonian vector field of the given functions and then differentiate the given Riemannian metric in the commutator direction. The answer is obvious.

We are now ready to formulate what we mean by algebro-geometric quantization.

Definition 1.6 ([23]). The *algebro-geometric quantization* of a symplectic manifold M is a procedure which results in a Kähler manifold $\mathcal{P}$ together with a map

$$q \colon C^\infty(M, \mathbb{R}) \to C_q^\infty(\mathcal{P}, \mathbb{R}),$$

satisfying the following conditions:

Linearity: $q(af + b) = aq(f) + b$ for any $f \in C^\infty(M, \mathbb{R})$, and $a, b \in \mathbb{R}$;

Correspondence principle: $q(\{f_1, f_2\}_\omega) = \{q(f_1), q(f_2)\}_\Omega$;

Irreducibility: An irreducibility condition, of which the strongest form is: $\ker q = 0$, and for each $p \in \mathcal{P}$ and each tangent vector $v \in T_p\mathcal{P}$ there exists a function $f \in C^\infty(M, \mathbb{R})$ such that $X_{q(f)}(p) = v$, where $X_{q(f)}$ is the Hamiltonian vector field of the quasisymbol $q(f) \in C_q^\infty(\mathcal{P}, \mathbb{R})$.

Example 1.7. In this new framework, the Rawnsley-Berezin method seems the most natural. Indeed, we start with M_I, a Kähler manifold built over M itself. Then we quantize exactly the functions from $C_q^\infty(M_I, \mathbb{R})$. Thus this is in some sense a tautology.

Example 1.8. The Toeplitz-Berezin method translates to the projective language in the following style. Consider the projectivization $\mathbb{P}H^0(M_I, L)$ together with its natural Kähler structure. Then there is a universal kernel on the direct product $M \times \mathbb{P}H^0$, namely

$$u(x, p) = \langle s(x), s(x) \rangle_h,$$

where $s \in H^0(M_I, L)$ is a holomorphic section with unit norm representing the point $p \in \mathbb{P}H^0$. Then it is not hard to see that the symbol, given by the Toeplitz operator $\widehat{A}_f$, is defined by the Fourier-Berezin transform:

$$a_f(p) = \int_M f \cdot u(x, p) d\mu_L.$$

At the same time $u(x, p)$ is a universal object which gives, for example, the η function of Rawnsley, Cahen and Gutt [14].

2 The correspondence principle in algebraic Lagrangian geometry

Once more, let (M, ω) be a compact simply-connected symplectic manifold of dimension $2n$ with integral symplectic form, and consider the prequantization data (L, a), where L is defined by the condition

$$c_1(L) = [\omega] \quad \text{and} \quad F_a = 2\pi i \omega.$$

For an appropriate smooth oriented connected n-dimensional manifold S, consider the space of smooth Lagrangian embeddings of fixed topological type, that is, smooth maps

$$\phi \colon S \to M \quad \text{such that} \quad \phi^* \omega \equiv 0,$$

with image representing a specified homology class $[S] \in H_n(M, \mathbb{Z})$. As we mentioned in the previous section, the choice of the prequantization data allows us to impose an extra condition on the Lagrangian embeddings: we say that an embedding ϕ is *Bohr-Sommerfeld* if the restriction of the prequantization data to the image admits covariant constant sections. In other words, the flat connection $\phi^* a$ on the trivial line bundle $\phi^* L$ has trivial periods with respect to the fundamental group of S. If one takes the corresponding $U(1)$-principal bundle with the corresponding connection 1-form A then it is an example of contact manifold (for which see, for example, Arnol'd and Givental [2]). The connection A, multiplied by i, satisfies the standard condition

$$\alpha \wedge (d\alpha)^n = d\mu,$$

where $d\mu$ is a volume form on P. The Bohr-Sommerfeld condition can easily be reformulated in terms of the principal bundle. A Lagrangian submanifold satisfies the Bohr-Sommerfeld condition if and only if it can be lifted to P along the fibres of the canonical projection $P \to M$. Our connection A decomposes the tangent to P space at every point into the direct sum of the horizontal and the vertical parts and a map

$$\widetilde{\phi} \colon S \to P$$

is called *Planckian* if $T(\widetilde{\phi})$ is horizontal at every point of $\widetilde{\phi}$, and $\widetilde{\phi}^* \pi^* \omega \equiv 0$ where π is the standard projection.

We now define Bohr-Sommerfeld and Planckian cycles. Let $\widetilde{\mathcal{B}}_S$ be the space of all Bohr-Sommerfeld Lagrangian embeddings of fixed topological type. Then the moduli space of Bohr-Sommerfeld Lagrangian cycles is given by the factorization

$$\mathcal{B}_S = \tilde{\mathcal{B}}_S / \operatorname{Diff}_0 S,$$

where $\operatorname{Diff}_0 S$ is the identity component in the diffeomorphism group of S. Recall that S is oriented, so that $\operatorname{Diff}_0 S$ can be understood as the parameterization group of S. Points of the moduli space are called Bohr-Sommerfeld cycles of fixed topological type. The moduli space of Planckian cycles has almost the same definition: one just starts with the space of all Planckian embeddings of S to P described above. We denote it $\mathcal{P}_S$, following Gorodentsev and Tyurin [8] and [19]. Every Planckian cycle is represented by a covariant constant lifting of a Bohr-Sommerfeld cycle, so the natural map

$$\pi \colon \mathcal{P}_S \to \mathcal{B}_S$$

gives a principal $U(1)$-bundle structure on $\mathcal{P}_S$ such that the canonical $U(1)$-action is generated by the canonical $U(1)$-action on P. This principal bundle is called the *Berry bundle*.

The picture includes an integer parameter – the level k, so that if one takes the corresponding tensor power (L^k, a_k) then one can define the moduli spaces with respect to this power in the same way. This gives us a set of moduli spaces parameterized by k:

$$\mathcal{B}_S = \mathcal{B}_{S,1}, \ldots, \mathcal{B}_{S,k}, \ldots \quad \text{and} \quad \mathcal{P}_S = \mathcal{P}_{S,1}, \ldots, \mathcal{P}_{S,k}, \ldots$$

We need only bear in mind that if we start with the symplectic manifold $(M, k\omega)$ then the pair (L^k, a_k) is precisely the prequantization data for it. Passing from L to L^k has two major effects: the Poisson bracket for $k\omega$ is slightly different from the original one and the Liouville volume form for $k\omega$ is slightly different too. Berezin proposed a natural relation of this integer parameter with the Planck constant, so we will exploit this relation and these remarks to construct an appropriate quasiclassical limit of our algebro-geometric quantization.

Our first aim is to describe smooth structures on the moduli spaces $\mathcal{B}_S, \mathcal{P}_S$. The first is given by the following result.

Proposition 2.1 ([8, 19]). *The tangent space $T_S \mathcal{B}_S$ at any point $S \in \mathcal{B}_S$ is isomorphic to $C^\infty(M, \mathbb{R})$ modulo constant functions.*

The proof can be found in Gorodentsev and Tyurin [8]. Briefly, let S be a regular point of the moduli space $\mathcal{B}_S$. We identify it in our discussion with the image of the corresponding class of maps so we understand S as an oriented smooth Bohr-Sommerfeld Lagrangian submanifold of M. By the Darboux-Weinstein theorem (see Weinstein [26]) there exists a tubular neighbourhood $N(S)$ symplectomorphic to an ε-neighbourhood of the zero section of the cotangent bundle:

$$\psi \colon N(S) \to N_\varepsilon(T^*S),$$

where the latter is equipped with the restriction of the canonical symplectic form. Thus the question reduces to the canonical case. Recall that T^*S is endowed with a natural 1-form η called the canonical 1-form. Its differential $d\eta$ is an everywhere nondegenerate closed 2-form that defines the canonical symplectic structure over T^*S. Therefore one can understand submanifolds of M contained in $N(S)$ as submanifolds of T^*S sufficiently close to S. When we discuss Lagrangian submanifolds, all of them are described by sufficiently 'small' closed 1-forms over S, with Bohr-Sommerfeld Lagrangian submanifolds represented by exact 1-forms. This gives us

$$T_S \mathcal{B}_S = B^1(S) = \{df\}.$$

Turning to the principal bundle $\mathcal{P} \to \mathcal{B}$, it is easy to see that the lifting corresponds to use of the constant functions, so that one gets

$$T_{\widetilde{S}} \mathcal{P}_S = C^\infty(M, \mathbb{R}),$$

where $\widetilde{S}$ is the corresponding Planckian cycle. Moreover, the Darboux-Weinstein theorem ensures that the representations for the tangent spaces are integrable, so that one has distinct local coordinate systems on both the moduli spaces. The set of the Darboux-Weinstein neighbourhoods gives atlases of the smooth structures. This fact shows that Lagrangian submanifolds of symplectic manifolds look like points of a symplectic manifold. And according to the generalization of the old classical result the same is almost true for Lagrangian submanifolds: they differ only by the topological type. Therefore one sees that 'points' and 'Lagrangian submanifolds' have quite similar behaviour from the kinematic point of view. We will see that the same holds for the dynamic behaviour.

The description of the tangent bundles for $\mathcal{B}_S$ and $\mathcal{P}_S$ in terms of smooth functions on S has quite important consequences. We make four remarks before proceeding with the Kähler set-up:

(i) If S has trivial fundamental group (or even trivial first homology group) then each Lagrangian cycle is Bohr-Sommerfeld.

(ii) The linearization of the Bohr-Sommerfeld condition is exactly the same as the so-called isodractic (or Hamiltonian) deformations. Indeed, any smooth function f on S can be extended to a smooth function $\widetilde{f}$ over M. Then the Hamiltonian vector field $X_{\widetilde{f}}$ generates some deformation of S. This infinitesimal deformation preserves the Lagrangian condition. Moreover, it preserves the Bohr-Sommerfeld condition, and the linearly deformed cycle is exactly the cycle given in the neighbourhood of S by df. This means that the Bohr-Sommerfeld condition is a classical dynamical condition over symplectic manifolds. Therefore we can introduce a kind of Bohr-Sommerfeld condition even in the case when ω is not integral at all. Namely this analogy is given by the flows of all strictly Hamiltonian vector fields over M. These flows induce a fibration on the space of all Lagrangian cycles of a fixed topological type. Then any leaf can be taken as a component $\mathcal{B}_S$. But in the integral case, we can first of all avoid the questions about the completeness of the Hamiltonian vector fields and define the fibration on the 'kinematic' level. The same remark also applies to the Planckian cycles.

(iii) We present some formulas to illustrate how we work in the present set-up. If we fix a smooth structure on $\mathcal{B}_S$, we choose any function f on $S \in \mathcal{B}_S$ and extend it arbitrarily to M, getting a smooth function $\widetilde{f}$. Then one can decompose the corresponding Hamiltonian vector field $X_{\widetilde{f}}$ on the horizontal and vertical components at each point of S and this decomposition is absolutely canonical. This fact is presented in Tyurin [25]. Thus we have

$$X_{\widetilde{f}} = X_{\mathrm{ver}} + X_{\mathrm{hor}},$$

where X_{hor} belongs to TS while X_{ver} can be identified with a section of the normal to S bundle

$$N_S = TM|_S / TS.$$

It is clear that X_{hor} corresponds to the part of deformation which preserves the cycle S (its flow generates some motion on S). Thus the deformation of S depends only on X_{ver}. We use the isomorphism

$$\omega \colon TM \to T^*M,$$

to get the formula

$$X_{\mathrm{ver}} = \omega^{-1}(d\widetilde{f}|_S) = \omega^{-1}(d(\widetilde{f}|_S)) = \omega^{-1}(df).$$

Therefore the deformation depends only on the restriction to S.

(iv) As we have seen, Hamiltonian vector fields have a natural infinitesimal action on the moduli space of Bohr-Sommerfeld Lagrangian cycles. Indeed, every Hamiltonian vector field gives an infinitesimal deformation of the base manifold, so generates a vector field on the moduli space of Bohr-Sommerfeld Lagrangian cycles (that is, a dynamic correspondence) since the definition is stated in invariant terms. The point is that the dynamic vector field A_f for any (global) smooth function f on whole $\mathcal{L}_S$ is given by the following simple formula:

$$A_f(S) = d(f|_S) \in T_S \mathcal{L}_S.$$

Thus this 'quantum' vector field preserves the leaves of the foliation defined on $\mathcal{L}$.

Following Gorodentsev and Tyurin [8, 19], the next step is to complexify the moduli space $\mathcal{B}_S$. We first take the moduli space of Planckian cycles $\mathcal{P}_S$. The source manifold S carries a space of half-weights (see [8, 19]). Since S is orientable, the determinant line bundle

$$\det T^*S = \Lambda^n T^*S$$

is trivial. Roughly speaking, a half-weight is almost the same thing as a half form without zeros (at least we can understand it in this way in our case, when S carries a fixed orientation). For any pair of half-weights there are two derivations:

$$\int_S \theta_1 \cdot \theta_2 \in \mathbb{R} \quad \text{and} \quad \frac{\theta_1}{\theta_2} \in C^\infty(S, \mathbb{R}),$$

where the latter is a nowhere vanishing smooth function. Moreover, the space of half weights admits a canonical involution that is simply multiplication by -1 in the half form representation. The tangent space to the set of half-weights over each point is modeled by $C^\infty(S, \mathbb{R})$ ([8, 19]), and we consider the moduli space of half-weighted Planckian cycles ([8, 19]) consisting of pairs

$$(\widetilde{S}, \theta) \in \mathcal{P}_S^{\text{hw}},$$

where $\widetilde{S}$ is a Planckian cycle and θ a half-weight, the first element of which one understands as the image of the corresponding half weight on the source manifold. The volume of this pair is given by

$$\int_S \theta^2 \in \mathbb{R}.$$

By definition, the moduli space of half-weighted moduli space is fibred over the old one

$$\pi_{un}\colon \mathcal{P}_S^{\mathrm{hw}} \to \mathcal{P}_S \quad \text{defined by} \quad \pi_{un}\colon (\widetilde{S}, \theta) \mapsto \widetilde{S}.$$

Moreover, there is another natural fibration

$$\pi_c\colon \mathcal{P}_S^{\mathrm{hw}} \to \mathcal{B}_S$$

equal to the composite of the Berry bundle and the forgetful map, removing the half-weight. Therefore the moduli space of half-weighted Planckian cycles inherits a $U(1)$-principal bundle structure coming from the structure of the Berry bundle.

We have already mentioned that $\mathcal{P}_S^{\mathrm{hw}}$ carries the canonical volume function

$$\mu\colon \mathcal{P}_S^{\mathrm{hw}} \to \mathbb{R}, \quad \text{defined by} \quad \mu(\widetilde{S}, \theta) = \int_{\widetilde{S}} \theta^2,$$

which is obviously invariant under the $U(1)$-action. The last remark will be very important after the following fact is established.

Proposition 2.2 ([8, 19]). *The moduli space of half-weighted Planckian cycles $\mathcal{P}_S^{\mathrm{hw}}$ admits a Kähler structure invariant under the $U(1)$-action.*

The idea of the proof is to exploit the specialty of the tangent spaces to the moduli space $\mathcal{P}_S^{\mathrm{hw}}$. Over a point it is the direct sum

$$T_{(\widetilde{S},\theta)} \mathcal{P}_S^{\mathrm{hw}} = C^\infty(S, \mathbb{R}) \oplus C^\infty(S, \mathbb{R}),$$

and the summands are identified canonically. Moreover, as one has canonical Darboux-Weinstein local coordinates for the Planckian 'unweighted' cycles as well there are canonical *complex* Darboux-Weinstein local coordinates for the moduli space of half-weighted Planckian cycles (see Gorodentsev and Tyurin [8]). These coordinates were introduced in [8]. Thus at an arbitrary point $(\widetilde{S}_0, \theta_0)$ belonging to the moduli space the canonical local coordinates are given by a pair of real smooth functions

$$(\psi_1, \psi_2), \quad \text{with} \quad \psi_i \in C^\infty(S, \mathbb{R}),$$

where the first function parameterizes deformations of the Planckian cycle while the second parametrizes deformations of the half-weight part. In these coordinates one can easily express two natural tensors 'living on' the moduli space. The first one, of type $(1, 1)$, is the linear operator:

$$I|_{(\widetilde{S}_0, \theta_0)}(\psi_1, \psi_2) = (-\psi_2, \psi_1).$$

The second, of type (2,0), is the skew-symmetric 2-form:

$$\Omega_{(\widetilde{S}_0, \theta_0)}(v_1, v_2) = \int_{\widetilde{S}_0} [\psi_1 \phi_2 - \psi_2 \phi_1] \theta_0^2,$$

where $v_1 = (\psi_1, \psi_2), v_2 = (\phi_1, \phi_2)$ are tangent vectors. One checks that this 2-form is everywhere nondegenerate and that Ω is compatible with I. The corresponding Riemannian metric has the form

$$G_{(\widetilde{S}_0, \theta_0)}(v_1, v_2) = \int_{\widetilde{S}_0} [\phi_1 \psi_1 + \phi_2 \psi_2] \theta_0^2.$$

Gorodentsev and Tyurin [8, 19] showed that the form is closed and the complex structure is integrable. Hence $\mathcal{P}_S^{\mathrm{hw}}$ is an infinite-dimensional Kähler manifold; we stress that the Kähler structure was constructed canonically without any additional choices.

One checks moreover that this Kähler structure is invariant under the action of $U(1)$ described above. The function μ, defined as the volume function, is the moment map for this action (see [8]). Thus one can produce a new Kähler manifold using the standard mechanism of Kähler reduction. To get this new manifold we choose a regular value of the moment map function

$$\mu(\widetilde{S}, \theta) = \int_{\widetilde{S}} \theta^2 = r \in \mathbb{R}.$$

The reduced Kähler manifold is called the moduli space of Bohr-Sommerfeld Lagrangian cycles of fixed volume and denoted by $\mathcal{B}_S^{\mathrm{hw}, r}$. Thus the real parameter r measures the volume of the weighted cycles. This moduli space is fibred over $\mathcal{B}_S$, so that as a symplectic manifold it admits a canonical real polarization. At the same time it admits a canonical complex polarization, since it is a Kähler manifold. Moreover, it is algebraic since the Kähler metric is of so-called Hodge type (the Berry bundle is related to the Kähler class, see [8]).

It remains to discuss some basic question which arises here. What we have described is the local theory of the moduli space $\mathcal{B}_S^{\mathrm{hw}, r}$. This theory works provided that we can prove that just a single point of the moduli space exists. Here we can claim the following fact.

Proposition 2.3 (Existence Theorem). *Let (M, ω) be an integral symplectic manifold and $S \subset M$ a smooth oriented Lagrangian submanifold representing a homology class $[S] \in H_n(M, \mathbb{Z})$. Then there exists a level k such that for the prequantization data (L^k, a_k) the moduli space $\mathcal{B}_S^{\mathrm{hw}, r}$ is nonempty.*

The proof is obvious: take the character χ_S of S defined by the original prequantization data (L, a) and the 'radius' of the Darboux-Weinstein neighbourhood of S in M. Any closed 1-form $\beta \in \Omega_S^1$ with character χ_S with respect to S gives us a Bohr-Sommerfeld Lagrangian cycle if the graph Γ_β of this form is contained within the Darboux-Weinstein radius of S. Therefore it can be done increasing the level k since the character of S on (L^k, a_k) equals

$$\chi_S \cdot e^{-i\pi l / k},$$

where l is an appropriate integer. Since the character becomes smaller when $k \to \infty$, it is easy to see that there exists a k such that the graph of the closed 1-form β is contained in the Darboux-Weinstein neighbourhood, and the proposition is proved.

One sees that for any integral compact simply-connected symplectic manifold one can construct an induced infinite-dimensional algebraic manifold $\mathcal{B}_S^{\mathrm{hw},r}$; this subject is called algebraic Lagrangian geometry (see Gorodentsev and Tyurin [8]).

Digression: mirror symmetry

The main idea underlying this construction was the following. At the present time the mirror conjecture is understood in the most broad context as a relation between algebraic geometry and symplectic geometry (linguistically, the words 'complex' and 'simplex' have the same meaning, but in Latin and Greek respectively). So in some sense if M, W are mirror partners then the algebraic geometry of M should be equivalent to the symplectic geometry of W and vice versa. From this point of view it looks extremely meaningful if one can construct canonically some algebraic variety starting with a symplectic manifold. In any case, the behaviour of a symplectic manifold with an integral symplectic form is in many respects sufficiently close to the standard set-up of algebraic geometry – as a striking example one can mention the symplectic analogue of the Kodaira embedding result obtained by Donaldson [6]. In the same integral symplectic set-up one has an algebraic manifold $\mathcal{B}_S^{\mathrm{hw},r}$, constructed in [8]. Although the moduli space $\mathcal{B}_S^{\mathrm{hw},r}$ is infinite-dimensional, Gorodentsev and Tyurin proposed the following way to make the picture finite-dimensional. Namely, the group $\mathrm{Symp}_0(M, \omega)$ of symplectomorphisms isotopic to the identity acts naturally on the moduli space $\mathcal{B}_S^{\mathrm{hw},r}$ preserving its Kähler structure. Therefore one can try to factorize $\mathcal{B}_S^{\mathrm{hw},r}$ by this action, in the hope of producing some finite-dimensional Kähler manifold. This idea is attractive, but the problem is that, even on the virtual level, this quotient space should be zero-dimensional (this follows from the irreducibility of ALG(a)-quantization, see Tyurin [24]). Nevertheless, it might be possible to correct this approach to mirror symmetry, for example if one could constructs some natural holomorphic vector bundle on $\mathcal{B}_S^{\text{hw},r}$, equivariant under $\text{Symp}_0(M, \omega)$ action. Then on passing to the quotient, this vector bundle should give a positive finite-dimensional algebro-geometric object.

Mirror symmetry relates to geometric quantization in some cases, so it is a natural question to ask whether algebraic Lagrangian geometry can solve the problem of algebro-geometric quantization. To do this one would first have to find a correspondence between smooth functions on a given symplectic manifold and some smooth functions on the moduli space $\mathcal{B}_S^{\text{hw},r}$, then check that this correspondence satisfies the properties listed in Definition 1.6. The first results in this direction are contained in Tyurin [22]. There I find a natural linear map and then prove by direct calculations that this map is a Lie algebra homomorphism.

The tangent space to the moduli space $\mathcal{B}_S^{\text{hw},1}$ at a point (S, θ) is represented by pairs (ψ_1, ψ_2) with $\psi_i \in C^\infty(S, \mathbb{R})$ such that

$$\int_S \psi_i \theta^2 = 0.$$

For any two tangent vectors $v_1 = (\psi_1, \psi_2), v_2 = (\phi_1, \phi_2)$ at the point (S, θ) the symplectic form Ω is

$$\int_S [\psi_1 \phi_2 - \psi_2 \phi_1]\theta^2.$$

One can introduce some functions over the moduli space induced by smooth functions from $C^\infty(M, \mathbb{R})$, following [22]. For any $f \in C^\infty(M, \mathbb{R})$ one has

$$F_f \in C^\infty(\mathcal{B}_S^{\text{hw},1}, \mathbb{R}),$$

defined absolutely canonically. Indeed, at each point (S, θ) it is given by

$$F_f(S, \theta) = \tau \int_S f|_S \theta^2 \in \mathbb{R},$$

where τ is a real parameter. This formula gives a map

$$\mathcal{F}_\tau \colon C^\infty(M, \mathbb{R}) \to C^\infty(\mathcal{B}_S^{\text{hw},1}, \mathbb{R})$$

which is obviously linear. The main fact, established in [22], is that $\mathcal{F}_\tau$ is a Lie algebra homomorphism. The original symplectic structure

ω defines the Poisson bracket on the source space while the 'quantum symplectic structure' Ω thus constructed defines the quantum Poisson bracket on the target space. And as we will see, $\mathcal{F}_\tau$ transforms the classical bracket into the quantum bracket up to a constant depending on our real parameter τ.

Proposition 2.4 ([22]). *For any smooth functions $f, g \in C^\infty(M, \mathbb{R})$ the identity*

$$\{F_f, F_g\}_\Omega = 2\tau F_{\{f,g\}_\omega}$$

holds, where F_f, F_g are the images of f, g under $\mathcal{F}_\tau$.

We first remark that the map $\mathcal{F}_\tau$ does not preserve the standard algebraic structure on $C^\infty(M, \mathbb{R})$, defined by usual pointwise multiplication. This follows from the same property of the integral: the integral of a product $f \cdot g$ is usually not the same thing as the product of the two integrals of f and g. Thus

$$F_f \cdot F_g \neq F_{f \cdot g}.$$

At the same time, by Proposition 2.4 the image

$$\operatorname{Im} \mathcal{F}_\tau \subset C^\infty(\mathcal{B}_S^{\mathrm{hw},1}, \mathbb{R})$$

is a Lie subalgebra. Assume that the given classical mechanical system is integrable, so that (M, ω) admits a set of n algebraically independent smooth functions in involution. Since

$$\{f_i, f_j\}_\omega = 0$$

the induced functions $F_{f_1}, \ldots, F_{f_n}$ commute under the quantum Poisson bracket over the moduli space $\mathcal{B}_S^{\mathrm{hw},1}$. But the same is true for the set which consists of the functions of the following shape

$$F_{f_1^{r_1} \ldots f_n^{r_n}} \quad \text{with} \quad r_i \in \mathbb{Z}.$$

The corresponding preimages, of course, lie in the algebraic span of $\{f_i\}$. But according to our remark the last function does not belong to the algebraic span of $F_{f_1}, \ldots, F_{f_n}$. It means that for any integrable classical system the corresponding moduli space (read: quantum system) is also integrable. Now the question arises: if our given classical system was completely integrable (so $\dim M = 2n$), is the same true for the quantum system? Roughly, the space of commuting functions over $\mathcal{B}_S^{\mathrm{hw},1}$ has dimension $\mathbb{Z}^n$ while the moduli space itself has dimension $2 \cdot C^\infty(S, \mathbb{R}) - 2$ thus it seems that in general the question is not quite obvious.

Digression: integer and real parameters

Let us see how the identity of Proposition 2.4 changes when we vary integer and real parameters, contained in the picture. Recall that there are two real continuous parameters r and τ, and one integer parameter k. We start with level $k = 1$.

The first level ($k = 1$). In this case the Poisson brackets are proportional with coefficient 2τ. It is clear that this coefficient does not depend on the volume of cycles. On the other hand, $\mathcal{F}_\tau$ maps

$$f \equiv \text{const.} = c \quad \longmapsto \quad F_f \equiv \text{const.} = \tau \cdot r \cdot c.$$

Therefore if one wants to establish the situation when all numerical quantization requirements from the Dirac list are satisfied (meaning that $\tau \cdot r = 1$ and $2\tau = 1$), we need to take

$$\tau = \frac{1}{2} \quad \text{and} \quad r = 2.$$

At this step, we see that, in any case, the product $\tau \cdot r$ must equal 1 while 2τ can vary according to the question about the Planck constant.

General level. One can do the same constructions for any level. We fix any $k \in \mathbb{N}$ and construct the moduli space $\mathcal{B}^{\text{hw},r}_{S,k}$ in the same way as $\mathcal{B}^{\text{hw},r}_{S}$ starting with the prequantization data (L^k, a_k). Then one has a natural inclusion

$$\mathcal{B}^{\text{hw},r}_{S} \hookrightarrow \mathcal{B}^{\text{hw},r}_{S,k}$$

(see Gorodentsev and Tyurin [8]). The Kähler structures on the two moduli spaces are slightly different; this means that for example, the symplectic form Ω on the level 1 moduli space does not coincide with the restriction of the symplectic form Ω_k defined on that of level k: this is a crucial point that comes from the difference between the canonical Darboux-Weinstein coordinates for $\mathcal{B}^{\text{hw},r}_{S}$ and $\mathcal{B}^{\text{hw},r}_{S,k}$. Indeed, if (S, θ) is an original Bohr-Sommerfeld Lagrangian cycle for (L, a) then it clearly remains Bohr-Sommerfeld for (L^k, a_k). But the canonical complex Darboux-Weinstein coordinates for the level 1 moduli space are given by $\omega^{-1}(df) \oplus df$ where f lives on S, while for the level k moduli space they are given by $(k\omega)^{-1}(df) \oplus df = k^{-1}\omega^{-1}(df) \oplus df$. This means that one rescales one half of the first coordinate system to get the second. Locally the difference can be recognized as follows. As usual we write T^*S for the tangent bundle to S. Then this tangent bundle admits not just one symplectic structure (the canonical one) but a family of symplectic structures depending on a real parameter λ. Indeed, one can rescale by λ the classical formula giving the basic definition of the canonical 1-form (see Abraham and Marsden [1], Arnol'd and Givental [2, 8]) to obtain

$$\Gamma_\alpha^* \eta_\lambda = \lambda \alpha \quad \text{for } \alpha \in \Omega_S^1.$$

Thus for any $\lambda \in \mathbb{R}$ one gets an 'almost' canonical 1-form η_λ which is nondegenerate and gives a nondegenerate 2-form $\omega_\lambda = d\eta_\lambda$. This 'almost' canonical symplectic form looks like the canonical one; it is not hard to find an appropriate symplectomorphism

$$\Psi_\lambda \colon (T^*S, d\eta) \to (T^*S, d\eta_\lambda),$$

simply multiplying the fibre of the canonical projection by λ. The family $d\eta_\lambda$ is a possible degeneration of the canonical symplectic structure on T^*S. At the same time an interesting effect appears: due to the canonical form of the symplectic structure in the canonical coordinates, $d\eta$ and $d\eta_\lambda$ are proportional, while the same is not true for the canonical coordinates. The proportionality coefficient is just λ. Turning to the canonical Poisson brackets one sees that the corresponding skew symmetric pairings on the function space are also proportional; the ratio is λ^{-1}. Returning to the moduli spaces we infer that Ω and $\Omega_k|_{\mathcal{B}_S^{\mathrm{hw},r}}$ are proportional with coefficient k. Therefore defining induced functions on $\mathcal{B}_{S,k}^{\mathrm{hw},r}$ one constructs a similar map

$$\mathcal{F}_\tau^k \colon C^\infty(M, \mathbb{R}) \to C^\infty(\mathcal{B}_{S,k}^{\mathrm{hw},r}, \mathbb{R}),$$

given by the same formula

$$F_f(S, \theta) = \tau \int_S f|_S \theta^2 \in \mathbb{R},$$

so that

$$\{F_f, F_g\}_{\Omega_k}|_{\mathcal{B}_S^{\mathrm{hw},r}} = \frac{1}{k}\{F_f, F_g\}_\Omega.$$

This does not lead to a contradiction since

$$\{f, g\}_{k\omega} = \frac{1}{k}\{f, g\}_\omega,$$

and in the final analysis it gives

$$F_{\{f,g\}_\omega} = \frac{k}{2\tau}\{F_f, F_g\}_{\Omega_k}$$

over $\mathcal{B}^{\mathrm{hw},r}_{S,k}$. Now to satisfy the Dirac conditions we need

$$\frac{2\tau}{k} = \hbar \quad \text{and} \quad \tau \cdot r = 1.$$

Of course we could say that 2τ must equal 2 and r must equal $\frac{1}{2}$ as above. On the other hand, we can fix the ratio

$$\frac{2\tau}{k} = \text{const.}$$

which implies that

$$\tau \to \infty \quad \Longrightarrow \quad r \to 0,$$

and in the limit one gets the moduli space of unweighted Lagrangian cycles. In fact, as k tends to ∞ the moduli space of unweighted Bohr-Sommerfeld cycles covers the moduli space of all Lagrangian cycles as a dense set (in the same way that rational points cover the Jacobian $b_1(M)$-torus). At the same time, the weights tend to zero (since $r \to 0$), so in the limit one could forget the second components in the pairs (S, θ). Anyway one cannot define, say, a Poisson structure on the moduli space of Lagrangian cycles as the limit of the symplectic structures on the moduli spaces of different levels since, as we have seen, the symplectic structure on the moduli space $\mathcal{B}^{\mathrm{hw},r}_{S,k}$ degenerates as $k \to \infty$.

3 The dynamic correspondence in algebraic Lagrangian geometry

In this section we follow Tyurin [21, 23].

The previous section constructed a map from the space of smooth functions over M to the space of smooth functions over the moduli space of half-weighted Bohr-Sommerfeld cycles of fixed volume. But as we saw in Section 1, the framework of quantization is only concerned with special types of smooth functions; quantum observables are required to satisfy the property that their Hamiltonian vector fields preserve all the kinematic data of quantum phase space. Thus to continue the story we must show that all these induced smooth functions are quasisymbols over the moduli space. The following is the main result of this section.

Proposition 3.1 ([23, 24]). *Let (M,ω) be a simply-connected compact symplectic manifold with an integral symplectic class, let $[S] \in H_n(M,\mathbb{Z})$ be a middle-dimensional homology class and $\mathcal{B}^{\mathrm{hw},r}_S$ the corresponding moduli space of half-weighted Bohr-Sommerfeld Lagrangian cycles of* *fixed volume with the corresponding Kähler triple (G, I, Ω). Then the linear map $\mathcal{F}_\tau$ satisfies the following properties:*

(1) *for any $f \in C^\infty(M, \mathbb{R})$ the induced function F_f is a quasisymbol over the moduli space (Definition 1.4);*
(2) *the correspondence principle holds in the form*

$$\{F_f, F_g\}_\Omega = 2\tau F_{\{f,g\}_\omega};$$

(3) *the map*

$$\mathcal{F}_\tau \colon C^\infty(M, \mathbb{R}) \to C_q^\infty(\mathcal{B}_S^{\mathrm{hw},r}, \mathbb{R})$$

gives an irreducible representation of the Poisson algebra.

Property 2) is already known from Section 2, but the construction given below in terms of the *dynamic correspondence* gives the identity again for free. Of course, we could also check that the induced function $F_f \in C^\infty(\mathcal{B}_S^{\mathrm{hw},r}, \mathbb{R})$ is a quasisymbol using direct calculations as in Section 2. But we are working in the set-up of a classical mechanical system, so it is natural to exploit some dynamic properties of our construction.

We mentioned above that while the Lagrangian condition seems to be local and static, the Bohr-Sommerfeld condition is dynamic: local deformations of Bohr-Sommerfeld cycles correspond precisely to Hamiltonian deformations induced by Hamiltonian dynamics. Thus the dynamic property of the system defines a correspondence between Hamiltonian vector fields on the base manifold and some special vector fields on any derived object. Indeed, suppose that we construct some object in terms intrinsic to our symplectic manifold (M, ω). This must then be stable under (infinitesimal) automorphisms of (M, ω). But this means that any Hamiltonian vector field induces a vector field (infinitesimal automorphism) of this derived object.

Example 3.2 (the Souriau-Kostant method). The naturality of the method comes from the following fact: every infinitesimal symplectomorphism of (M, ω) can be lifted almost uniquely to an infinitesimal automorphism of the bundle (L, a), 'almost' meaning that it can be done up to a canonical $U(1)$-transformation. But consideration of the projectivization of $\Gamma(M, L)$ kills this ambiguity. The resulting projective space $\mathbb{P}$ is an invariant object over (M, ω). For any smooth function $f \in C^\infty(M, \mathbb{R})$, consider the corresponding infinitesimal deformation, given by the Hamiltonian vector field X_f. Then this infinitesimal deformation generates an infinitesimal automorphism of the projective space.

The point is that this induced vector field $\Theta(f) \in \text{Vect}\,\mathbb{P}$ is Hamiltonian and preserves all the structures on $\mathbb{P}$. Thus it is the Hamiltonian vector field of a *symbol* over $\mathbb{P}$, and the corresponding smooth function Q_f, normalized in an appropriate way, gives exactly the symbol, induced by the Souriau-Kostant operator $\widehat{Q}_f$.

We follow the same strategy in the study of the moduli space $\mathcal{B}_S^{\text{hw},r}$. The derived object, the moduli space $\mathcal{B}_S^{\text{hw},r}$, is preserved by Hamiltonian deformations of the base symplectic manifold. Moreover, the Kähler structure is also preserved by deformations. This gives a map

$$\Theta_{\text{DC}} \colon \text{Vect}_\omega(M) \equiv C^\infty(M,\mathbb{R})/\text{const.} \to \text{Vect}(\mathcal{B}_S^{\text{hw},r}),$$

called the *dynamic correspondence*. It can be constructed as follows: any function f on M induces a Hamiltonian vector field, whose dynamics preserves (M,ω). Moreover, on choosing prequantization data the dynamics lifts to this set-up almost uniquely (up to canonical gauge transformations). Thus this dynamics preserves all the data, hence defines a germ of automorphism of the moduli space $\mathcal{B}_S^{\text{hw},r}$. This defines a vector field on the moduli space with additional properties, reflecting the fact that it preserves the Kähler structure, since its definition was invariant. Since M is compact, every function f defines a germ of symplectomorphisms, so that every Hamiltonian vector field induces an infinitesimal transformation of the moduli space. Generalizing over the space of all Hamiltonian vector fields gives the map, which is clearly linear. The construction gives the next result:

Proposition 3.3 ([23]). *The image* $\text{Im}\,\Theta_{\text{DC}}$ *is contained in* $\text{Vect}_K(\mathcal{B}_S^{\text{hw},1})$, *the space of vector fields on the moduli space preserving the Kähler structure.*

To make the story even more specific, we compute the coordinates of any special vector field generated by a smooth function $f \in C^\infty(M,\mathbb{R})$. First this function induces the vector field X_f on M. Let $(S,\theta) \in \mathcal{B}_S^{\text{hw},r}$ be a half-weighted Bohr-Sommerfeld cycle. The vector field X_f decomposes over the support of S into inner and outer parts:

$$X_f = X_{\text{ex}} + X_{\text{in}},$$

with $X_{\text{in}} \in TS$ the tangent component. Of course, we met this decomposition in Section 2, where we denoted it by 'ver-hor'. Since we have fixed a smooth structure on $\mathcal{B}_S^{\text{hw},r}$ (and we write our formulas in the corresponding coordinate system), the vector field splits at the points of any Bohr-Sommerfeld Lagrangian submanifold (see [25]). Notice that forgetting the half-weight parts leads to some degeneration of the picture: in this case the deformation is defined only by the restriction of the source function to S, but in the half-weighted case two functions with the same restriction to S give different deformations of the pair (S, θ) if they differ in a small neighbourhood of S.

Therefore the deformation induced by X_f are expressed as follows in the canonical complex Darboux-Weinstein coordinates.

Proposition 3.4 ([23]). *The vector field induced by X_f under the dynamic correspondence, at a point (S, θ), has coordinates (ψ_1, ψ_2) given by*

$$\psi_1 = f|_S - \int_S f|_S \theta^2 \quad and \quad \psi_2 = \frac{\mathrm{Lie}_{X_{\mathrm{in}}} \theta}{\theta}.$$

Thus we see that the vector field on the moduli space induced by X_f under the dynamic correspondence is naturally expressed in terms of the canonical 'dynamic' coordinates. The dynamic properties of the given system give us the following additional condition satisfied by the vector fields in $\mathrm{Im} \, \Theta_{\mathrm{DC}}$.

Proposition 3.5 ([23]). *For any pair of smooth functions f, g the identity*

$$\Theta_{\mathrm{DC}}([X_f, X_g]) = [\Theta_{\mathrm{DC}}(X_f), \Theta_{\mathrm{DC}}(X_g)]$$

holds, where on the right-hand side one takes the standard commutator of vector fields over the moduli space.

The natural question now arises: does the above dynamic correspondence Θ_{DC} admit a lift to the level of functions? At the same time, for any smooth function $f \in C^\infty(M, \mathbb{R})$ one has two *a priori* different vector fields on the moduli space $\mathcal{B}_S^{\mathrm{hw},r}$: the Hamiltonian vector field X_{F_f} for the induced function $F_f \in C^\infty(\mathcal{B}_S^{\mathrm{hw},r}, \mathbb{R})$ and the dynamically induced $\Theta_{\mathrm{DC}}(f)$, and it is natural to compare these two vector fields. The answer is what one would expect.

Proposition 3.6 ([23]). *We have*

$$X_{F_f} = 2\tau \Theta_{\mathrm{DC}}(X_f) \quad for \ any \ smooth \ function \ f \in C^\infty(M, \mathbb{R}).$$

This key statement is proved in [23] by direct computations. The statements of Proposition 3.1 follow easily from Propositions 3.3–3.6.

(1) follows from the definition of quasisymbols and Propositions 3.3

and 3.6: for any function f the induced function F_f generates the Hamiltonian vector field proportional to a vector field in $\operatorname{Im}\Theta_{DC}$. Hence X_{F_f} preserves the whole Kähler structure over the moduli space and F_f is a quasisymbol.

(2) follows from Propositions 3.5–3.6: indeed, one continues the equality of Proposition 3.5 in both directions, substituting the equality of Proposition 3.6

$$\frac{1}{2\tau}X_{F_{\{g,f\}_\omega}} = \Theta_{DC}([X_f,X_g]) = [\Theta_{DC}(X_f),\Theta_{DC}(X_g)] = \frac{1}{4\tau^2}[X_{F_f},X_{F_g}],$$

and the last term has the standard representation as a Hamiltonian vector field. This gives the correspondence principle in the familiar form

$$\{F_f,F_g\}_\Omega = 2\tau F_{\{f,g\}_\omega}.$$

(3) is very important for us. As explained above, the irreducibility condition in our nonlinear algebro-geometric set-up consists of two items: the first saying that $\ker\mathcal{F}_\tau = 0$, and the second that there is no smooth proper submanifold in the moduli space to which every Hamiltonian vector field X_{F_f} is tangent. We begin by checking the second condition, in fact in a much stronger form. Namely we show that, for every point $(S,\theta)\in\mathcal{B}_S^{hw,r}$ and every tangent vector $v = (\psi_1,\psi_2)\in T_{(S,\theta)}\mathcal{B}_S^{hw,r}$, there exists a smooth function $f\in C^\infty(M,\mathbb{R})$ such that the Hamiltonian vector field of the induced function F_f gives this vector at this point, that is:

$$X_{F_f}(S,\theta) = v.$$

Of course, this stronger condition could be exploited in the discussion on properties of our Kähler metric over the moduli space (for example, is it a metric of constant holomorphic sectional curvature or not) but these questions come outside of the main theme of our text. One verifies this condition using Propositions 3.4 and 3.6; namely, for any pair of smooth functions $\psi_1,\psi_2\in C^\infty(S,\mathbb{R})$ one has to find a smooth function f over the whole of M such that

$$\psi_1 = f|_S - \text{const.} \quad\text{and}\quad \psi_2 = \frac{\operatorname{Lie}_{X_{in}}\theta}{\theta},$$

The first equation is easily solved by taking f to be any extension of ψ_1 over M. The second is more delicate: this question extracts from the set of possible extensions those which are appropriate and a priori one cannot say whether or not such extensions exist. We reduce the question to the following simple result.

Lemma 3.7. *Let S be any smooth compact real oriented manifold and η a volume form. Then for any real smooth function $\psi \in C^\infty(S, \mathbb{R})$ with zero integral:*

$$\int_S \psi \eta = 0,$$

there exists a vector field Y such that

$$\psi = \frac{\mathrm{Lie}_Y \, \eta}{\eta}.$$

Now we can apply this lemma in our context via the following trick. Since our S is oriented (see the first step in the construction of the moduli space $\mathcal{B}_S^{\mathrm{hw},r}$) we can consider the corresponding volume form η instead of the square θ^2. The Lie derivatives of θ and η are related by the identity

$$\frac{\mathrm{Lie}_Y \, \theta}{\theta} = \frac{1}{2} \frac{\mathrm{Lie}_Y \, \eta}{\eta} \, ;$$

thus Lemma 3.7 ensures that for every ψ_2 there exists a vector field Y on the Bohr-Sommerfeld cycle S such that the second equation is satisfied. It remains to construct an extension of ψ_1 to a neighbourhood of S in M such that the corresponding Hamiltonian vector field gives us our Y as its X_{in}. Since the considerations are local we will construct such extensions for a small neighbourhood of the zero section in T^*S. The desired function $\widetilde{f}$ has the form

$$\widetilde{f}(x, p) = \psi_1(x) + p_x(Y_x),$$

where x is the S-coordinate, p the coordinate along the fibre, identified simultaneously with the corresponding cotangent vector p_x, and Y is the vector field on S defined by ψ_2, which exists by Lemma 3.7. The standard isomorphism of a neighbourhood of S in M and the neighbourhood of the zero section in T^*S maps $\widetilde{f}$ to a function that we denote f; we claim that it possesses the desired properties.

Thus we can deform any fixed point (S, θ) in any direction along the moduli space, acting by an appropriate induced quasisymbol. On the other hand, here we want to strengthen the statement proved in [23] for homogeneous symplectic manifolds, that $\mathcal{F}_\tau$ is an inclusion. We establish this here in full generality by slightly extending the arguments of [23]. Namely, $\mathcal{F}_\tau$ could have a kernel if there were a point $x \in M$ with a neighbourhood $\mathcal{O}(x)$ disjoint from every Bohr-Sommerfeld Lagrangian cycle $S \in \mathcal{B}_S$, that is, with

$$S \cap \mathcal{O}(x) = \emptyset.$$

This means that if we take a smooth bump function concentrated in $\mathcal{O}(x)$, it restricts trivially to any Bohr-Sommerfeld cycle, and therefore belongs to the kernel. But this is not possible if $\mathcal{B}_S^{hw,r}$ is nonempty. Indeed, for any point x we can arrange a Bohr-Sommerfeld Lagrangian cycle passing through any fixed neighbourhood of x. If x is any specified point and S a Bohr-Sommerfeld cycle (which exists since we assume that $\mathcal{B}_S^{hw,r} \neq \emptyset$) then it is not hard to construct a smooth function f such that the flow generated by the Hamiltonian vector field X_f moves S to the place of x. By the dynamic property of the Bohr-Sommerfeld condition, the image of the Bohr-Sommerfeld cycle should again be Bohr-Sommerfeld; thus for any compact smooth symplectic manifold, if $\mathcal{B}_S^{hw,r}$ is nonempty then its 'points' cover the entire base manifold. This remark is extremely important if we wish to exploit a 'universal cycle' induced by the construction (and we really intend to do this in future) – this remark hints that such a cycle does exist. Now, after this fact is understood, we note that if for every smooth function $f \in C^\infty(M, \mathbb{R})$ there exists a Bohr-Sommerfeld Lagrangian cycle S for which the restriction $f|_S$ is nontrivial, then we can choose a half-weight θ over S such that

$$\int_S f|_S \theta^2 \neq 0.$$

It is not hard to see that we can make this choice (see [24]).

Thus any nonempty moduli space $\mathcal{B}_S^{hw,r}$ solves the problem of algebro-geometric quantization of any given integral compact symplectic manifold.

Digression: induced dynamics

Any given symplectic manifold (M, ω) can be viewed as the phase space of a classical mechanical system. In classical mechanics, we usually study the motion of points in M: we fix a point and then switch on the dynamics of the system, generated by an appropriate Hamiltonian. This point then moves in M, and its trajectory gives us the integral trajectory – the solution in the classical set-up. At the same time we could consider not just a point, but *any* submanifold of M, and study the dynamics of this submanifold for an appropriate Hamiltonian. But this dynamics is quite hard to describe in the general case. We need, first, a suitable description of the moduli space of possible locations of the submanifold and, second, a reasonable equation of motion for this type of submanifolds which can be solved.

The induced dynamics of half-weighted Bohr-Sommerfeld Lagrangian cycles gives us a (generalized) quantum system closely related to the given classical system. Thus the classical mechanical system hides some quantum induced system whose dynamics is generated by its classical dynamics. This picture seems to be in complete harmony with the ideas from the Copenhagen programme mentioned in the introduction.

4 Reduction of ALG(a)-quantization

The new method of quantization, based on algebraic Lagrangian geometry, is called ALG(a)-quantization. Although new, it is quite compatible with the known methods of geometric quantization. In this section we discuss how ALG(a)-quantization can be reduced in the cases when our given symplectic base manifold is equipped with an additional structure, a (real or complex) polarization.

Real polarization

Assume now that our symplectic manifold (M, ω) admits an appropriate real polarization. This means that (M, ω) can be given a Lagrangian distribution, that is, a field of Lagrangian subspaces in the complexified tangent bundle which is integrable. The complex and the real case differ by the nature of these Lagrangian subspaces: in the real case they are all real, while in the complex case they are pure complex. In Section 1 we recalled the methods of quantization applicable in these cases. We understand a real polarization as the case when M has a set of smooth functions $f_1, \ldots, f_n$ such that

$$\{f_i, f_j\}_\omega = 0 \quad \text{for all } i, j,$$

defining a Lagrangian fibration

$$\pi \colon M \to \Delta,$$

where $\Delta \subset \mathbb{R}^n$ is a convex polytope. For each internal point

$$(t_1, \ldots, t_n) \in \Delta \setminus \partial\Delta$$

the corresponding fibre

$$\pi^{-1}(t_1, \ldots, t_n) = f_1^{-1}(t_1) \cap \cdots \cap f_n^{-1}(t_n)$$

is a smooth Lagrangian cycle. The degenerations at the faces of Δ are regular, so the internal part of each $n - k$-face corresponds to $n - k$-dimensional isotropic submanifolds. In this case the quantization scheme usually distinguishes some special fibres of π, namely the Bohr-Sommerfeld fibres (see Śniatycki [16]).

Now we apply the ALG(a)-programme to our completely integrable system, taking $[S]$ as the homology class of the fibre. Then one gets the corresponding moduli space $\mathcal{B}_S^{\mathrm{hw},1}$, where

$$[S] = [\pi^{-1}(\mathrm{pt.})] \in H_n(M, \mathbb{Z});$$

we take volume 1 just for simplicity. The set $(f_1, \ldots, f_n)$ defines quasisymbols $F_{f_1}, \ldots, F_{f_n}$ which are again in involution (moreover, we can find infinitely many functions in involution by taking all finite monomials in $f_1, \ldots, f_n$ and mapping these by $\mathcal{F}_\tau$). We write $\mathrm{Crit}(F_{f_i})$ for the set of critical points of F_{f_i}. Consider the following intersection

$$P = \mathrm{Crit}(F_{f_1}) \cap \cdots \cap \mathrm{Crit}(F_{f_n}) \subset \mathcal{B}_S^{\mathrm{hw},1},$$

that is, the mutual critical set. We have the following result.

Proposition 4.1 ([23]). *P is a 2-to-1 cover of the set of Bohr-Sommerfeld fibres $\{S_i\}$.*

Thus in general we can recover the well-known method of Śniatycki [16]: one just takes the mutual critical set for the distinguished functions which preserve the given real polarization; their support then corresponds to a basis of $\mathcal{H}$. Moreover, the proposition is true in the general noncompact case (but we do not consider it here) so one could try to exploit this correspondence in some other contexts.

At the same time, we can add the following remark to the standard set-up of geometric quantization in the case of real polarization.

Proposition 4.2 ([23]). *Let (M, ω) be a symplectic manifold with an integral symplectic form, and admitting a Lagrangian fibration*

$$\pi \colon M \to \Delta$$

with compact fibres. Then the set of smooth Bohr-Sommerfeld fibres is discrete. Moreover if M is compact it follows that the set is finite.

In the general case one uses the following geometric argument: if S_0 is a Bohr-Sommerfeld fibre:

$$S_0 = \pi^{-1}(p_0), \quad \text{for some} \quad p_0 \in \Delta,$$

then there exists a neighbourhood $\mathcal{O}(p_0)$ of the point in Δ such that $\pi^{-1}(\mathcal{O}(p_0)$ is a Darboux-Weinstein neighbourhood of S_0 in M. Thus if we suppose that there is another Bohr-Sommerfeld fibre S projecting to $p \subset \mathcal{O}(p_0) \subset \Delta$, there should exist a smooth function $\psi \in C^\infty(S_0, \mathbb{R})$ such that S coincides with the graph of $d\psi$ in this Darboux-Weinstein neighbourhood. Since S_0 and S have zero intersection (they are two different fibres) the differential $d\psi$ must be everywhere nonvanishing. But any smooth function on a compact set has at least two extremum points, its minimum and maximum. This means that $d\psi$ has to vanish somewhere, which leads to a contradiction. Therefore if our S_0 is Bohr-Sommerfeld, there exists a neighbourhood of $\pi(S_0) = p_0$ in Δ such that $p_0 \in \mathcal{O}(p_0)$ is the unique 'Bohr-Sommerfeld point' in this neighbourhood. Thus, globally, every Bohr-Sommerfeld fibre of π is separated by such a neighbourhood and hence the set of Bohr-Sommerfeld fibres is discrete.

Remark 4.3. It is very natural and reasonable to continue here the observation given at the end of Section 2, where we discussed the case of completely integrable systems: in this case one could construct, starting from the given set of first integrals $\{f_1, \ldots, f_n\}$, an infinite set of commuting quantum observables on the moduli space $\mathcal{B}_S^{\mathrm{hw},r}$. Indeed, one just takes the powers

$$\{F_{f_j^k}\}, \quad \text{for } j = 1, \ldots, n, \text{ and } k > 0$$

and the quantum Poisson bracket of every pair from this set vanishes. However, it is clear that the mutual critical set P is the same for every degree k and any combination of the first integrals (while the corresponding quantum observables are no longer algebraically dependent). In fact, the conditions

$$f_j|_S = \text{const.} \quad \text{and} \quad f_j^k|_S = \text{const.}$$

are completely equivalent (since our functions are real and smooth). Although the quasisymbols of type $\{F_{f_j^k}\}$ are algebraically independent, their critical values *are* algebraically dependent in mutual critical points. Indeed, every first integral f_j gives the following critical values (via the powers of f_j):

$$c = f_j|_S, c^2, \ldots, c^k, \ldots,$$

and it is clear that this set is algebraically dependent. Hence we cannot derive additional geometric information for the completely integrable systems using our method (at least in the present discussion).

Complex polarization

This is exactly the case of algebraic geometry. A complex polarization on a symplectic manifold (M, ω) is the choice of an integrable complex structure I compatible with ω, making M into an algebraic manifold; we suppose that such a structure exists. This implies that (M_I, ω) is a Kähler manifold and the integrability condition for ω ensures that the Kähler form is of Hodge type, hence (M_I, ω) is an algebraic variety. While any symplectic manifold admits an infinite set of almost complex structures, the possibility of choosing an integrable complex structure considerably restricts the horizon of the examples.

We discussed the known quantization methods in Section 1; as mentioned there, they are all based on some reductions of the basic Souriau-Kostant method. Here we use both reductions: the Berezin-Rawnsley method is more appropriate for dynamic coherence, while the Berezin-Toeplitz method is described by explicit formulas. In either case, the corresponding Hilbert space is the same – the space of holomorphic sections of the prequantization line bundle with respect to the prequantization connection. We projectivize the space following the strategy of the geometric formulation of quantum mechanics. The first step is to relate the quantum phase space of the known method with ALG(a)-quantization. The construction is due to Gorodentsev and Tyurin [8, 19]. The desired relation is given by the so-called BPU map ('BPU' stands for Borthwick, Paul and Uribe [5]). It fibres the moduli space $\mathcal{B}_S^{\mathrm{hw},r}$ over the projective space:

$$\mathrm{BPU} \colon \mathcal{B}_S^{\mathrm{hw},r} \to \mathbb{P}H^0(M_I, L).$$

This BPU map defines a reduction of the quantum phase space of ALG(a)-quantization to one of the known methods.

Suppose further that a smooth function f is a quantizable observable for the given complex polarization. Then the Hamiltonian vector field X_f preserves the complex structure I and its image on $\mathbb{P}(\Gamma(M, L))$ under the dynamic correpondence preserves the finite-dimensional piece $\mathbb{P}H^0(M_I, L)$. Moreover, the field $\Theta_{\mathrm{DC}}^p(X_f) \in \mathrm{Vect}\,\mathbb{P}(H^0)$ preserves the whole Kähler structure and corresponds to some smooth function Q_f (see Section 1; we stress again that this holds because f is quantizable). On the other hand, the Hamiltonian vector field defines infinitesimal transformations on both side of the BPU map. For each quantizable function, the corresponding dynamic actions on the source and target space must be compatible. Thus for any quantizable function (in the sense of the Rawnsley-Berezin method) one has:

1) a pair of quasisymbols F_f and Q_f on the source and the target spaces respectively;

2) a pair of vector fields $\Theta_{\mathrm{DC}}(f)$ and $\Theta_{\mathrm{DC}}^p(f)$ on the source and the target spaces that correspond under the dynamic correspondence.

Taking into account the dynamic arguments one finds that the differential of the BPU map transforms our special vector field $\Theta_{\mathrm{DC}}(f)$ into the special vector field $\Theta_{\mathrm{DC}}^p(f)$. For this situation one has this result.

Proposition 4.4 ([23]). *For any quantizable function f the Hamiltonian vector fields of quasisymbols F_f and Q_f are related by*

$$d\mathrm{BPU}(X_{F_f}) = c \cdot X_{Q_f},$$

where c is a real constant.

Thus one reduces ALG(a)-quantization to a well-known method in the complex polarization case. Proposition 4.4 gives a method for finding the eigenstates of the quantum observable Q_f having the eigenstates of F_f. The relation is very similar to the answer in the preceding real case.

Corollary 4.5 ([23]). *The BPU map projects the set of eigenstates of the quantum observable F_f to the set of eigenstates of Q_f.*

Digression: the quantum Morse inequality

Each case gives us some important information about the existence of eigenstates for some distinguished quantum observables. But the problem in the general case is hard: we must establish that a quasisymbol $F_f \in C^\infty(\mathcal{B}_S^{\mathrm{hw},r}, \mathbb{R})$ has eigenstates $\{(S_i, \theta_i)\}$ and that there enough of these to perform the measurement process described in Section 1. There is no problem with the half-weight part, so we can reformulate the problem of eigenstates in the usual symplectic set-up. Let (M, ω) be an integral compact simply-connected symplectic manifold and $f \in C^\infty(M, \mathbb{R})$ some sufficiently generic smooth function. Can we estimate the number of Bohr-Sommerfeld Lagrangian submanifolds of fixed topological type that are stable with respect to the Hamiltonian vector field X_f? It was shown that for a generic smooth function the set is discrete (see [23]); on the other hand in the simplest case we have an appropriate bound. Indeed, consider the *toy example* of Gorodentsev and Tyurin [8] – the sphere S^2 with the standard integral symplectic structure ω (in this case, by a theorem of Moser, it is the unique symplectic invariant, the symplectic volume, see Abraham and Marsden [1]). In this case any smooth loop $\gamma \subset S^2$ is Lagrangian, but a loop γ is Bohr-Sommerfeld for $k = 1$ if and only if it divides the surface of the sphere into two pieces of equal area with respect to ω. For level $k > 1$, the Bohr-Sommerfeld loops are the level curves of f that divide the area of S^2 in the ratios $i : k + 1 - i$, for $i \in [1, k]$. Thus for a generic function $f \in C^\infty(S^2, \mathbb{R})$ the number of them is

$$\#\left\{\gamma \in \mathcal{B}_S^k \mid f|_\gamma = \text{const.}\right\} = k.$$

Of course, there may be functions for which either Bohr-Sommerfeld loops do not exist at all, or the set of the Bohr-Sommerfeld loops is bigger than in the general case, just as in the classical case when the set of Morse functions does not exhaust the space of all smooth functions.

Thus we see that increasing the level we can reach the situation when we have enough eigenstates over S^2. Is there any bound on this number in the general case? The problem can be understood as a 'quantum Morse inequality' although in this case the index of a critical point cannot be defined; thus in some sense the story is simplified, which is not unexpected due to the standard slogan 'quantization removes degenerations'. On the other hand, this problem is a generalization of a classical problem of Poincaré from classical mechanics.

ALG(a)-quantization and its quasiclassical limit

To finish our list of reductions, we would like to mention a different type of reduction that we understand as a quasiclassical limit of ALG(a)-quantization.

Following Berezin, we view the level k as inversely proportional to the Planck constant. This kind of dependence was mentioned in Section 2 when we introduced the parameter τ in the definition of $\mathcal{F}_\tau$; now we take τ proportional to k. This makes the formulation of the Dirac quantization principle more familiar (see Proposition 2.4). At the same time, we said that in the limit the volume of half-weighted cycles goes to zero. So the upshot is the limit space $\mathcal{B}_{\lim}$, which is dense in the space of all Lagrangian submanifolds of the specified topological type. Indeed, attaching the character of the restriction to a Lagrangian submanifold of the prequantization data (L, a) defines a map

$$\chi \colon \mathcal{L}_S \to J_S,$$

where $\mathcal{L}_S$ is the space of all Lagrangian submanifolds of specified topological type and J_S the Jacobian torus of S. But when we raise the level $k > 1$, the $\chi^{-1}(\theta_i)$ become Bohr-Sommerfeld cycles for the points of order k, which are classically denoted by θ_i. As $k \to \infty$, the points of order k cover J_S densely, hence at the same time the space of Bohr-Sommerfeld submanifolds becomes dense in $\mathcal{L}_S$.

Now what happens to the quantum observables F_f in this process? And how about the limiting Poisson bracket on the moduli space $\mathcal{L}_S$? Answers to both questions come by consideration of the following objects over the moduli space of Lagrangian cycles. Namely a smooth function $f \in C^\infty(M, \mathbb{R})$ generates a special object on $\mathcal{L}_S$ having a dual nature. On one hand we have the vector field Y_f defined by restricting f to Lagrangian cycles. Recall that the restriction of f to $S \in \mathcal{L}_S$ gives the corresponding Hamiltonian (isodrastic) deformation of S, given in a Darboux-Weinstein neighbourhood by $d(f|_S)$. Thus Y_f simply equals this tangent vector at the point S. On the other hand, Y_f is not just a vector field: it also takes well-defined numerical values at points where Y_f vanishes as a vector field. Indeed, Y_f vanishes at S as a vector field if and only if it restricts to a constant on S; but this constant then defines a value. Thus the induced object Y_f is described by a pair

$$Y_f = (Y_f^0, Y_f^1),$$

where Y_f^0 is a real function (rather singular, of course) and Y_f^1 a vector field (absolutely smooth, of course). We write $C^q(\mathcal{L}_S)$ for the set of all such objects arising from smooth functions on M. Then one has the following result.

Proposition 4.6. _The set $C^q(\mathcal{L}_S)$ is a Lie algebra._

To see this fact, note first that the correspondence $f \mapsto Y_f$ is obviously linear. The bracket $[\cdot, \cdot]$ is given tautologically by the formula

$$[Y_f, Y_g] = Y_{\{f,g\}_\omega}.$$

The Jacobi identity is satisfied by definition.

On the other hand, we should emphasize the following.

Proposition 4.7(1) _For every f the corresponding object Y_f is the natural result of the limiting procedure, applied to the quasisymbol F_f;_

(2) _the Lie bracket defined above is the natural result of the limiting procedure, applied to the quantum Poisson bracket $\{\,\cdot\,,\,\cdot\,\}_\Omega$._

Further, we see that the system based on $\mathcal{L}_S$ carries dynamical properties arising from the classical dynamics of the given classical mechanical system. Indeed, a Hamiltonian $H \in C^\infty(M, \mathbb{R})$ generates dynamics on $\mathcal{L}_S$, preserving the Lie bracket on the space of objects over the moduli space. Therefore we understand the process as an appropriate quasiclassical limit of ALG(a)-quantization method: during the limiting procedure we lose the measurement aspects but we keep the dynamical properties compatible with the dynamics of the given system (more than compatible, we would say).

To conclude, we have just presented some elements of a new method of quantization, skipping many additional questions and details to be clarified and established in future. We have set up the problem with certain (possibly accidental) features in mind, but we are confident that the study of algebraic Lagrangian geometry introduced by Gorodentsev and Tyurin [8, 19] will lead to new and interesting results.

Bibliography

[1] R. Abraham and J. E. Marsden. *Foundations of mechanics.* Benjamin/Cummings, Reading, Mass., 1978. Second edition.

[2] V. I. Arnol'd and A. B. Givental'. Symplectic geometry. In *Current problems in mathematics. Fundamental directions, Vol. 4*, Itogi Nauki i Tekhniki, pages 5–139, 291. Akad. Nauk SSSR Vsesoyuz. Inst. Nauchn. i Tekhn. Inform., Moscow, 1985.

[3] A. Ashtekar and T. Schilling. Geometrical formulation of quantum mechanics. arXiv: `gr-qc/9706069`.

[4] M. Bordemann, E. Meinrenken, and M. Schlichenmaier. Toeplitz quantization of Kähler manifolds and $gl(N)$, $N \to \infty$ limits. *Comm. Math. Phys.*, 165:281–296, 1994.

[5] D. Borthwick, T. Paul, and A. Uribe. Legendrian distributions with applications to relative Poincaré series. *Invent. Math.*, 122:359–402, 1995.

[6] S. K. Donaldson. Symplectic submanifolds and almost complex geometry. *J. Diff. Geom.*, 44(4):666–705, 1990.

[7] A. L. Gorodentsev and A. N. Tyurin. ALAG, 2000. Preprint MPI (Bonn) no. 7.

[8] A. L. Gorodentsev and A. N. Tyurin. Abelian Lagrangian algebraic geometry. *Izvestiya: Math.*, 65: 3:437–467, 2001.

[9] P. Griffiths and J. Harris. *Principles of algebraic geometry.* Wiley-Interscience [John Wiley & Sons], New York, 1978. Pure and Applied Mathematics.

[10] N. E. Hurt. *Geometric quantization in action*, volume 8 of *Mathematics and Its Applications (East European Series)*. D. Reidel Publishing Co., Dordrecht, 1983.

[11] A. Kapustin and D. Orlov. Lectures on Mirror Symmetry, derived categories and D-branes. arXiv: `math.AG/0308173`.

[12] B. Kostant. Quantization and unitary representations. I. Prequantization. In *Lectures in modern analysis and applications, III*, pages 87–208. Lecture Notes in Math., Vol. 170. Springer, Berlin, 1970.

[13] L. D. Landau and E. M. Lifschitz. *Lehrbuch der theoretischen Physik. Band III: Quantenmechanik.* Akademie-Verlag, Berlin, 1985.

[14] J. Rawnsley, M. Cahen, and S. Gutt. Quantization of Kähler manifolds. I. Geometric interpretation of Berezin's quantization. *J. Geom. Phys.*, 7(1):45–62, 1990.

[15] T. Schilling. *The geometry of quantum mechanics.* PhD thesis, Penn. State Univ., 1996.

[16] J. Śniatycki. *Geometric quantization and quantum mechanics*, volume 30 of *Applied Mathematical Sciences*. Springer-Verlag, New York, 1980.

[17] J.-M. Souriau. *Structure des systèmes dynamiques.* Maîtrises de mathématiques. Dunod, Paris, 1970.

[18] G. M. Tuynman. Quantization: towards a comparison between methods. *J. Math. Phys.*, 28:2829–2840, 1987.

[19] A. N. Tyurin. Complexification of Bohr-Sommerfeld conditions, 1999. Preprint, Inst. of Math. Oslo Univ., no. 15.

[20] A. N. Tyurin. Special Lagrangian geometry as a slight deformation of algebraic geometry. *Izvestiya: Math.*, 64 : 2:363–437, 2000.

[21] N. A. Tyurin. ALAG-quantization, 2001. Preprint KIAS (Seoul) M010002; arXiv: SG/0106004.

[22] N. A. Tyurin. The correspondence principle in Abelian Lagrangian geometry. *Izvestiya: Math.*, 65: 4:823–834, 2001.

[23] N. A. Tyurin. Dynamical correspondence in algebraic Lagrangian geometry. *Izvestiya: Math.*, 66: 3:611–629, 2002.

[24] N. A. Tyurin. Irreducibility of the ALG(a)-quantization. *Proc. Steklov Inst. Math.*, 241(2):249–255, 2003.

[25] N. A. Tyurin. Letter to editors. *Izv. Math.*, 68(3):219–220, 2004.

[26] A. Weinstein. Lagrangian submanifolds and Hamiltonian systems. *Ann. of Math. (2)*, 98:377–410, 1973.

[27] N. Woodhouse. *Geometric quantization.* The Clarendon Press Oxford University Press, New York, 1980. Oxford Mathematical Monographs.

Bogolyubov Laboratory of Theoretical Physics
Joint Institute for Nuclear Research
141980, Dubna, Moscow Region
Russia
ntyurintheor.jinr.ru
jtyurinmpim-bonn.mpg.de

Printed in the United States
by Baker & Taylor Publisher Services

Printed in the United States
by Baker & Taylor Publisher Services